Selvadurai: Partial Differential Equations in Mechanics 1

Springer
Berlin
Heidelberg
New York
Barcelona
Hong Kong
London
Milan
Paris
Singapore
Tokyo

A.P. S. Selvadurai

Partial Differential Equations in Mechanics 1

Fundamentals, Laplace's Equation, Diffusion Equation, Wave Equation

With 162 Figures

Springer

Professor Dr. A.P. S. Selvadurai
McGill University
Department of Civil Engineering and Applied Mechanics
Macdonald Engineering Building
817 Sherbrooke Street West
Montreal, QC H3A 2K6
Canada
e-mail: apss@civil.lan.mcgill.ca
and
Humboldt-Forschungspreisträger
Institut A für Mechanik
Universität Stuttgart
Pfaffenwaldring 9
70550 Stuttgart, Germany

Library of Congress Cataloging-in-Publication Data. Selvadurai, A.P.S. Partial differential equations in mechanics/A.P.S. Selvadurai. p. cm. Includes bibliographical references and indexes. Contents: 1. Fundamentals, Laplace's equation, diffusion equation, wave equation – 2. The biharmonic equation, Poisson's equation. ISBN 3540672834 (v.1: acid-free paper) – ISBN 3540672842 (v.2: acid-free paper) 1. Mechanics, Analytic. 2. Differential equations, Partial. I. Title. QA 805.S45 2000 531'.01'515353–dc21 00-044024

ISBN 3-540-67283-4 Springer-Verlag Berlin Heidelberg New York

Springer-Verlag Berlin Heidelberg New York
a member of BertelsmannSpringer Science+Business Media GmbH

Printed in Germany

Typesetting: Data conversion by LE-TeX, Leipzig
Cover design: de'blik, Berlin

Printed on acid-free paper
SPIN: 10764494 62/3141/ba – 5 4 3 2 1 0

To the memory of my parents

K.S. Selvadurai and W.M.A. Selvadurai

and to

Sally, Emily, Paul, Mark

and

Elizabeth

Preface

"For he who knows not mathematics cannot know any other sciences; what is more, he cannot discover his own ignorance or find its proper remedies." [Opus Majus] Roger Bacon (1214-1294)

The material presented in these monographs is the outcome of the author's long-standing interest in the analytical modelling of problems in mechanics by appeal to the theory of partial differential equations. The impetus for writing these volumes was the opportunity to teach the subject matter to both undergraduate and graduate students in engineering at several universities. The approach is distinctly different to that which would adopted should such a course be given to students in pure mathematics; in this sense, the teaching of partial differential equations within an engineering curriculum should be viewed in the broader perspective of *"The Modelling of Problems in Engineering"* . An engineering student should be given the opportunity to appreciate how the various combination of balance laws, conservation equations, kinematic constraints, constitutive responses, thermodynamic restrictions, etc., culminates in the development of a partial differential equation, or sets of partial differential equations, with potential for applications to engineering problems. This ability to distill all the diverse information about a physical or mechanical process into partial differential equations is a particular attraction of the subject area. A second aspect of the teaching of partial differential equations to engineering students must cover topics that should enable them to pose an engineering problem as a correct mathematical statement. This process can include the mathematical structuring of the physical boundary conditions, regularity conditions, initial conditions and uniqueness theorems to generate a well-posed problem in partial differential equations. Thirdly, the presentation should include an introduction to the solution schemes that will highlight the basic structure of the solutions associated with the different classes of partial differential equations, by appeal to suitable idealizations of engineering problems. These volumes are also intended to illustrate the extensive range of applicability of the basic linear partial differential equations in a *multidisciplinary sense*, with special emphasis on applications to problems in mechanics encountered in civil engineering, mechanical engineering, theoretical and applied mechanics, chemical engineering, geological engineering, earth sciences, etc., covering topics such as fluid flow, diffusion and mass transport in porous media, pressure transients and moisture diffusion in porous geomaterials, heat conduction in solids, waves in elastic solids, fluids and membranes, elasto-mechanics of solids and structural elements and mechanics of viscous fluids.

These companion volumes contain a total of nine chapters, which introduce the basic concepts of partial differential equations with the central theme of modelling and applications in mechanics. The division of the presentation into two companion volumes with chapter continuity achieves two purposes. The material contained in Volume I, chapters 1 to 7, can form the subject matter of a senior level undergraduate course or a graduate level course in applications of partial differential equations in mechanics for engineering students. The material contained in Volume II, chapters 8 and 9, is ideally suited for a graduate level course devoted to applications of partial differential equations in mechanics of solids. The introductory chapter gives a review of the mathematical preliminaries, including vector calculus, Fourier series and integral transforms. This chapter is not intended to provide an exhaustive coverage of these topics and the interested reader can further review this material by consulting the bibliography cited at the end of these volumes. The presentation in chapter 1 is kept to a reasonable length by introducing as brief a derivation of the salient results and procedures as possible. Chapter 2 introduces partial differential equations and definitions of order, linearity, homogeneity, and well posedness. Chapter 3 provides a brief account of first-order partial differential equations with applications that involve characteristic equations. Applications of integral transform techniques to the solution of elementary problems involving transport in porous media are also discussed. Chapter 4 deals with the classification of partial differential equations of the second-order with procedures for their reduction to canonical forms. The application of Laplace's equation to problems of steady state heat conduction, ideal fluid flow, flow in porous media and applications to the study of deflections of stretched membranes are detailed in chapter 5. Chapter 6 examines the diffusion equation in relation to transient heat conduction, mass transport in porous media and pressure transients in porous media. Chapter 7 deals with the wave equation and its application to wave propagation in infinite and finite strings, vibrations of membranes and elementary one-dimensional vibration problems in solid mechanics. This chapter also examines the application of the wave equation to the study of shallow water waves. Chapter 8 presents a very complete exposition of the application of the biharmonic equation to problems in mechanics. A perusal of many existing texts on partial differential equations reveals the conspicuous absence of any detailed treatment of the biharmonic equation. Yet, the biharmonic equation is one of the most important partial differential equations in applied mechanics, with applications in the theory of elasticity, mechanics of elastic plates and the theory of slow flows of viscous fluids, all subject areas of fundamental importance to the engineering sciences. Chapter 9 deals with Poisson's equation. Many expositions of Poisson's equation present themselves as appendages to Laplace's equation. The objective of this chapter is to demonstrate that Poisson's equation has significance in its own right and has extensive applications to engineering problems dealing with steady state heat conduction in heat generating media, groundwater flow with recharge or depletion, mechanics of stretched, loaded membranes and in the study of the theory of torsion of prismatic elastic bodies.

Each chapter contains a detailed discussion of the application of the relevant partial differential equation, its derivation in a generalized fashion and the formulation of consistent boundary and/or initial conditions required for their well posedness. The proof of the relevant uniqueness theorems, maximum principles and other topics of general interest to identifying the qualitative aspects of the specific partial differential equations are also discussed. Worked examples within the text and problem sets at the end of each chapter highlight engineering applications of the theories and relevant analytical developments. The volumes are reasonably self-contained, in the sense that all necessary material, including developments of the governing partial differential equations, proofs of general theorems and applications, are presented in their entirety without recourse to references within the text. There is, of course, a wealth of information available in other texts and treatises devoted to the subject of partial differential equations and the reader, and students in particular, should at least aware of these developments. To this end, these volumes contain an extensive bibliography divided into topics covering general engineering mathematics, ordinary differential equations, partial differential equations, Fourier series, integral transforms and special functions, boundary value problems and mathematical methods. The bibliography also contains titles of interest to the separate chapters, including first-order partial differential equations, Laplace's equation, the diffusion equation, the wave equation, the biharmonic equation and Poisson's equation. Although not required for the subject matter covered in these volumes, the bibliography also contains titles related to non-linear partial differential equations, numerical methods for the solution of partial differential equations, applications of computer based symbolic manipulation methods to mathematics and references to historical material in both mechanics and mathematics. Symbolic computer methods are gaining popularity in many engineering curricula; they should be considered as a useful complement to carrying out mathematical operations in a time effective manner, particularly as computer laboratory exercises of examples given in these volumes.

These monographs, by design, emphasize analytical procedures for the solution of the various partial differential equations. This should not be construed as an opportunity to de-emphasize numerical and computational techniques. On the contrary, any realistic engineering application of even the simplest of theories, such as those described by the linear partial differential equations treated in these monographs will invariably require access to sophisticated numerical schemes. These schemes can include finite difference, finite element and boundary integral equation techniques that are gaining considerable popularity in engineering curricula, both at the undergraduate and graduate levels. The teaching of these numerical techniques can only benefit by instilling in undergraduate engineering students the confidence in both the mathematical aspects of partial differential equations and their applications potential. Indeed, in advanced finite element formulations based on the Galerkin technique, the governing partial differential equations are a prerequisite; similarly, knowledge of the Green's function applicable to a particular partial differential equation is a requirement for the application of boundary integral equation techniques. The subject of partial differential equations has a long and rich tradition in mathematics and mechanics, and, as the bibliography demonstrates, has at its disposal an extensive collection of texts and treatises devoted to the subject. The bibliography is certainly not meant to be all encompassing and up to date. The texts cited are considered sufficient as supplementary reading material. The present volumes, however, differ from many traditional presentations in that the subject matter is introduced within the context of mathematical rigour, modelling in mechanics and the applications of elementary solutions to problems in engineering mechanics.

The need for engineers to understand fundamental concepts associated with partial differential equations and techniques available for their solution becomes apparent when mathematical modelling and analysis of engineering problems are set in their modern context. Engineers are constantly exposed to new engineering software tools, usually involving numerical methods that enable them to carry out engineering computations for practical problems with speed and accuracy. In a majority of these cases, the numerical schemes are designed to solve partial differential equations with complicated coupled phenomena, which usually have a non-linear character. The validity and success of the numerical procedures in such software tools cannot be confirmed in a universal sense. The opportunity, however, exists for engineers to conduct calibrations of such numerical schemes by recourse to solutions developed for classes of linearized problems, albeit simplified. The modelling of an engineering problem as a mathematical statement can be regarded as the modern equivalent of a design exercise. These companion volumes highlight the point that even elementary partial differential equations are a powerful tool for posing and solving practical problems in engineering. The mathematical rigour of the presentations is balanced by an extensive variety of applications in mechanics. The practical nature of the problems examined emphasizes mathematical modelling as an important aspect of the training of engineering students. In this sense, the subject of partial differential equations has much to offer to enhance the quality of the broader mathematical education of students in engineering and to effectively integrate mathematics into the mainstream of modelling of engineering problems.

Acknowledgements

It is a pleasure for me to record my appreciation to a number of individuals who have helped me either directly or indirectly, in achieving the task of initiating and completing these volumes. I am deeply grateful to Alexander von Humboldt Stiftung for the award of a Preistrager Fellowship, which enabled me to visit the University of Stuttgart during the completion of this work. I would like to thank Professor Lothar Gaul, Director of the Institut A für Mechanik, Universität Stuttgart, his colleagues and staff for their kind hospitality and for the productive intellectual environment that I enjoyed at the Institute. The support of Professor Pieter Vermeer of the Institut für Geotechnik, Universität Stuttgart is gratefully acknowledged. A significant part of the latter chapters and revisions to the manuscript were done during my stay at the University of Canterbury, Christchurch, New Zealand. I am grateful to the University for the award of an Erskine Fellowship and to Professor Rob Davis, of the Department of Civil Engineering, for making the visit possible. His critical comments on many of the chapters were invaluable. I record my thanks to Professor Marc Boulon and his colleagues at the Laboratoire 3S, Université Joseph Fourier, Grenoble, France, for the valuable interactions and for the hospitality during my brief sojourn in Grenoble.

The debt I owe to my *gurus* Professor A.J.M. Spencer FRS, Emeritus Professor of Theoretical Mechanics, University of Nottingham, and Professor R.E. Gibson, Emeritus Professor, University of London, is particularly substantial. Quite apart from their mentoring, support and encouragement, their contributions to continuum mechanics and theoretical geomechanics continue to be a source of inspiration to me personally, and to many other researchers who enjoy practicing the art of applying mathematics to the modelling and solution of problems in engineering. These volumes are in some small way a tribute to their substantial scholarly achievements. I would like to thank John Adjeleian, Professor Emeritus at Carleton University, Ottawa and Branko Ladanyi FRSC, Professor Emeritus at École Polytechnique, Montreal, for the encouragement and support that I have received over the many years. It is also a pleasure to record the support of my friends in the geomechanics fraternity, who appreciate my continuing preoccupation with mathematical modelling, but most importantly, recognize the valuable role mathematics and modelling plays in the education of engineers. The opportunity to teach the material contained in the first seven chapters as parts of a required course to undergraduate engineering students and an elective course to graduate students in civil engineering at McGill and parts of chapter eight as a graduate course at Carleton is appreciated. The presentation of the material in these volumes has greatly benefitted from this teaching activity.

I would like to thank Dr. Dietrich Merkle, Engineering Editor, at Springer-Verlag, Heidelberg and his staff for their very positive support for the project and for the assistance they have provided in allowing me to present the manuscript in a form that would be of benefit to the readership. Although I had the pleasure of "writing" the manuscript the arduous task of preparing several versions of it into LaTeX forms was performed by Ms. Livia Nardini. Mr. Nick Vannelli converted my "sketches" to Corel Draw equivalents. I am also grateful to a number of technical assistants, notably, Marwan Al-Habbal and Todd Hirtle who were involved in preparing the final manuscripts to conform to Springer guidelines. As former students in the PDEs course, their familiarity with the course material was a great asset in this exercise. Every attempt has been made to check the accuracy of the material contained in these volumes. Some errors are bound to exist despite my best efforts. I would therefore be most grateful to the readers if they would be kind enough to bring these to my attention.

Last but not least, my appreciation goes to my wife Sally, not only for her editorial assistance, but also for her willingness to attend to the demands of a family when I was engrossed, absent or preoccupied with the task of completing these volumes against self-imposed deadlines. It is fair to say that all the family appreciates the completion of the project.

Montréal, April 2000 *A.P.S. Selvadurai*

Contents

Chapter 1

Mathematical preliminaries

The objective of this chapter is to summarize certain results related to vector algebra, vector calculus, integral transforms, etc., which will be useful in aspects related to material covered in these volumes. The results are given in summary form only and details of the derivations and proofs etc., are generally covered in texts devoted to either advanced calculus or engineering mathematics or mathematical methods.

1.1 Components of a vector

If $\mathbf{i}$, $\mathbf{j}$, $\mathbf{k}$ are respectively the unit vectors in the directions of the positive x, y, z axes, then, the vector $\mathbf{A}$ can be written as

$$\mathbf{A} = A_1\mathbf{i} \ + \ A_2\mathbf{j} \ + \ A_3\mathbf{k} \tag{1.1}$$

where $A_1\mathbf{i}$, $A_2\mathbf{j}$, $A_3\mathbf{k}$ are component vectors of $\mathbf{A}$ in the $\mathbf{i}, \mathbf{j}, \mathbf{k}$ directions respectively, and A_1, A_2, A_3 are the components of $\mathbf{A}$.

1.2 Dot or scalar product

We define the dot or scalar product of vectors $\mathbf{A}$ and $\mathbf{B}$ by

$$\mathbf{A} \cdot \mathbf{B} = |\mathbf{A}||\mathbf{B}| \ \cos\theta \tag{1.2}$$

where $|\mathbf{A}| = \left[A_1^2 + A_2^2 + A_3^2\right]^{1/2}$; $|\mathbf{B}| = \left[B_1^2 + B_2^2 + B_3^2\right]^{1/2}$ where θ is the angle between $\mathbf{A}$ and $\mathbf{B}$. Also $\mathbf{A} \cdot \mathbf{B} = \mathbf{B} \cdot \mathbf{A}$ (commutative law)

$$\mathbf{A} \cdot (\mathbf{B} + \mathbf{C}) = \mathbf{A} \cdot \mathbf{B} + \mathbf{A} \cdot \mathbf{C} \quad \text{(distributive law)} \tag{1.3}$$

$$\mathbf{A} \cdot \mathbf{B} = A_1 B_1 + A_2 B_2 + A_3 B_3 \tag{1.4}$$

1.3 Cross or vector product

We define the cross or vector product of vectors **A** and **B** as

$$\mathbf{A} \times \mathbf{B} = |\mathbf{A}||\mathbf{B}| \sin\theta \, \mathbf{u} \quad ; \quad 0 \le \theta \le \pi \tag{1.5}$$

where θ is the angle between **A** and **B** and **u** is a unit vector perpendicular to the plane of **A** and **B** such that **A**, **B** and **u** form a right handed system (Fig.1.1).

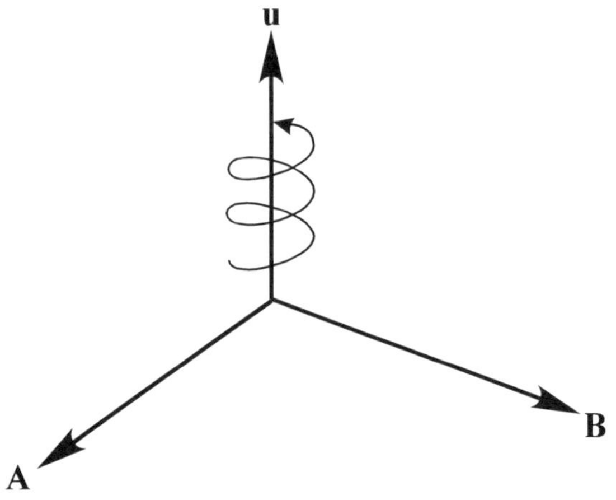

Figure 1.1: Graphical representation of vector product

The cross or vector product $\mathbf{A} \times \mathbf{B}$ can be determined from the *mnemonic operation*

$$\mathbf{A} \times \mathbf{B} = \begin{vmatrix} \mathbf{i} & \mathbf{j} & \mathbf{k} \\ A_1 & A_2 & A_3 \\ B_1 & B_2 & B_3 \end{vmatrix} \tag{1.6}$$

Also

$$
\begin{aligned}
&\text{(i)} && \mathbf{A} \times \mathbf{B} = -\mathbf{B} \times \mathbf{A} \\
&\text{(ii)} && \mathbf{A} \times (\mathbf{B} + \mathbf{C}) = \mathbf{A} \times \mathbf{B} + \mathbf{A} \times \mathbf{C} \\
&\text{(iii)} && \mathbf{A} \times (\mathbf{B} \times \mathbf{C}) = \mathbf{B}\,(\mathbf{A} \cdot \mathbf{C}) - \mathbf{C}\,(\mathbf{A} \cdot \mathbf{B}) \\
&\text{(iv)} && (\mathbf{A} \times \mathbf{B}) \times \mathbf{C} = \mathbf{B}\,(\mathbf{A} \cdot \mathbf{C}) - \mathbf{A}\,(\mathbf{B} \cdot \mathbf{C}) \\
&\text{(v)} && (\mathbf{A} \times \mathbf{B}) \cdot (\mathbf{C} \times \mathbf{D}) = (\mathbf{A} \cdot \mathbf{C})\,(\mathbf{B} \cdot \mathbf{D}) - (\mathbf{A} \cdot \mathbf{D})\,(\mathbf{B} \cdot \mathbf{C}) \qquad (1.7) \\
&\text{(vi)} && \mathbf{A} \times \mathbf{A} = \mathbf{0} \\
&\text{(vii)} && \mathbf{A} \cdot (\mathbf{A} \times \mathbf{B}) = 0 \\
&\text{(viii)} && \mathbf{B} \cdot (\mathbf{A} \times \mathbf{B}) = 0 \\
&\text{(ix)} && \mathbf{A} \times \mathbf{B} \cdot \mathbf{C} = \mathbf{A} \cdot \mathbf{B} \times \mathbf{C}
\end{aligned}
$$

1.4 Derivative of a vector

The derivative of a vector function $\mathbf{A}(t) = A_1(t)\,\mathbf{i} + A_2(t)\,\mathbf{j} + A_3(t)\,\mathbf{k}$ of the scalar variable t is given by

$$
\frac{d\mathbf{A}}{dt} = \operatorname*{Lim}_{\Delta t \to 0} \frac{\mathbf{A}(t + \Delta t) - \mathbf{A}(t)}{\Delta t} = \frac{dA_1}{dt}\mathbf{i} + \frac{dA_2}{dt}\mathbf{j} + \frac{dA_3}{dt}\mathbf{k} \qquad (1.8)
$$

If $\mathbf{A} = \mathbf{A}(u, v, w)$ we can define the partial derivatives of $\mathbf{A}$ in the same way.

1.5 Results involving derivatives

Chain-rule for derivatives give the following

$$
\frac{d}{dt}(\mathbf{A} \cdot \mathbf{B}) = \mathbf{A} \cdot \frac{d\mathbf{B}}{dt} + \frac{d\mathbf{A}}{dt} \cdot \mathbf{B} \qquad (1.9)
$$

$$
\frac{d}{dt}(\mathbf{A} \times \mathbf{B}) = \mathbf{A} \times \frac{d\mathbf{B}}{dt} + \frac{d\mathbf{A}}{dt} \times \mathbf{B} \qquad (1.10)
$$

Combining (1.9) and (1.10) we obtain

$$\frac{d}{dt}\{\mathbf{A}\cdot(\mathbf{B}\times\mathbf{C})\} = \frac{d\mathbf{A}}{dt}\cdot(\mathbf{B}\times\mathbf{C}) + \mathbf{A}\cdot\left(\frac{d\mathbf{B}}{dt}\times\mathbf{C}\right) + \mathbf{A}\cdot\left(\mathbf{B}\times\frac{d\mathbf{C}}{dt}\right) \tag{1.11}$$

1.6 Partial derivatives of vectors

Let $\mathbf{r}(x, y, z)$ represent the position vector of an arbitrary point referred to the rectangular Cartesian (named after the French philosopher and mathematician Rene du Perron Descartes (1596- 1650)) coordinate system x, y, z. Let us suppose that there is a *scalar function* g which assigns a value to each point in space. We have

$$g = g(x, y, z) \quad \text{or} \quad g = g(\mathbf{r}) \tag{1.12}$$

If there is a *vector function* which assigns a vector to each point in space then we can write

$$\mathbf{A} = \mathbf{A}(x, y, z) \quad \text{or} \quad \mathbf{A} = \mathbf{A}(\mathbf{r}) \tag{1.13}$$

The values of $\mathbf{A}$ associated with all points in space constitutes a vector field. In science and engineering we frequently encounter scalar and vector fields which vary with time, i.e.

$$g = g(\mathbf{r}, t) \tag{1.14}$$

$$\mathbf{A} = \mathbf{A}(\mathbf{r}, t) \tag{1.15}$$

e.g. Scalar Fields: *Temperature, Fluid Pressure*
Pollutant Concentration
Vector Fields: *Magnetic Fields, Gravity Fields*
Fluid Velocity, Displacements, Tractions.

1.6.1 The gradient of a scalar field

We consider the scalar function $f(x, y, z)$ and assume that the partial derivatives of the function are continuous throughout some region of space. In such a region

$$df = \frac{\partial f}{\partial x}dx + \frac{\partial f}{\partial y}dy + \frac{\partial f}{\partial z}dz \tag{1.16}$$

Equation (1.16) can be regarded as the scalar product of two vectors

$$df = \left(\frac{\partial f}{\partial x}, \frac{\partial f}{\partial y}, \frac{\partial f}{\partial z}\right) \cdot (dx,\ dy,\ dz) \tag{1.17}$$

or

$$df = \left(\mathbf{i}\frac{\partial f}{\partial x} + \mathbf{j}\frac{\partial f}{\partial y} + \mathbf{k}\frac{\partial f}{\partial z}\right) \cdot (\mathbf{i}dx + \mathbf{j}dy + \mathbf{k}dz) \tag{1.18}$$

The operator

$$\nabla = \left(\mathbf{i}\frac{\partial}{\partial x} + \mathbf{j}\frac{\partial}{\partial y} + \mathbf{k}\frac{\partial}{\partial z}\right) \tag{1.19}$$

is called the "*del*" or "*gradient*" operator and

$$\nabla f = \text{grad } f = \left(\mathbf{i}\frac{\partial}{\partial x} + \mathbf{j}\frac{\partial}{\partial y} + \mathbf{k}\frac{\partial}{\partial z}\right) f \tag{1.20}$$

is defined as the gradient of f. Hence, since $d\mathbf{r} = \mathbf{i}dx + \mathbf{j}dy + \mathbf{k}dz$, we have

$$df = \nabla f \cdot d\mathbf{r} \tag{1.21}$$

The ∇ operator can also be given a geometric interpretation (Fig.1.2).

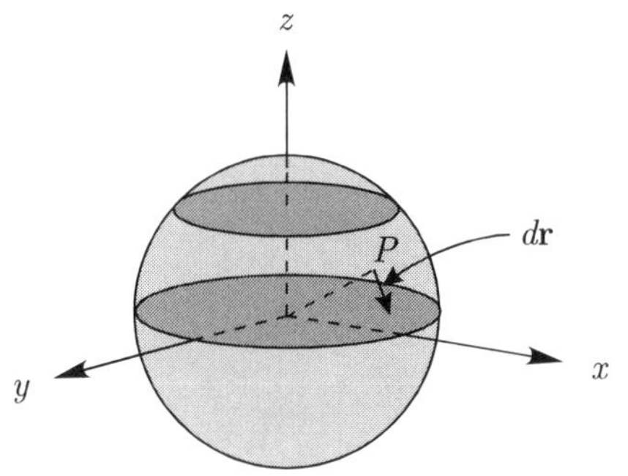

Figure 1.2: Variation of function $f\,(x, y, z) = 0$ on surface of sphere

1. The equation $f(x, y, z) = 0$ is the equation of a surface in space.

2. For example, $x^2 + y^2 + z^2 - a^2 = 0$, is the equation of a sphere of radius a centered at the origin. For all points on the surface $f(x, y, z) = 0$ and $df = 0$. Hence

$$\nabla f \cdot d\mathbf{r} = 0 \tag{1.22}$$

3. We now choose a point P on the surface and let $d\mathbf{r}$ represent any infinitesimal displacement vector which begins at P and is tangent to $f(x, y, z) = 0$ at P, and $d\mathbf{r}$ is tangential to the surface.
4. The result (1.22) states that ∇f must be perpendicular to vector $d\mathbf{r}$.

5. The orientation of $d\mathbf{r}$ is arbitrary and only subject to the constraint that it be tangential to the surface at P.
6. Hence we can conclude that ∇f is normal to the surface $f = 0$ at any point on the surface.

1.6.2 The divergence of a vector field

The divergence of a vector field is defined as the scalar product of the ∇ operator with the field. Hence if

$$\mathbf{A} = A_x\mathbf{i} + A_y\mathbf{j} + A_z\mathbf{k}$$

then

$$\nabla \cdot \mathbf{A} = \text{div } \mathbf{A} = \frac{\partial A_x}{\partial x} + \frac{\partial A_y}{\partial y} + \frac{\partial A_z}{\partial z} \tag{1.23}$$

1.6.3 The Laplacian of a scalar or vector field

The Laplacian of a scalar function f is defined by

$$\nabla^2 f = \text{div grad } f = \frac{\partial^2 f}{\partial x^2} + \frac{\partial^2 f}{\partial y^2} + \frac{\partial^2 f}{\partial z^2} \tag{1.24}$$

The Laplacian of a vector field $\mathbf{A}$ is defined by

$$\nabla^2 \mathbf{A} = \mathbf{i}\nabla^2 A_x + \mathbf{j}\nabla^2 A_y + \mathbf{k}\nabla^2 A_z \tag{1.25}$$

Also ∇^2 is referred to as Laplace's operator(named after Pierre Simon Marquis de Laplace (1749-1827)).

1.6.4 The curl of a vector field

If $\mathbf{A} = A_x\mathbf{i} + A_y\mathbf{j} + A_z\mathbf{k}$, then we define the curl of $\mathbf{A}$ by

$$\nabla \times \mathbf{A} = \text{curl } \mathbf{A} = \det \begin{vmatrix} \mathbf{i} & \mathbf{j} & \mathbf{k} \\ \dfrac{\partial}{\partial x} & \dfrac{\partial}{\partial y} & \dfrac{\partial}{\partial z} \\ A_x & A_y & A_z \end{vmatrix} \tag{1.26}$$

1.6.5 Other formulae involving ∇

$$\begin{aligned}
&\nabla \cdot (\mathbf{A} + \mathbf{B}) = \nabla \cdot \mathbf{A} + \nabla \cdot \mathbf{B} \\
&\nabla \times (\mathbf{A} + \mathbf{B}) = \nabla \times \mathbf{A} + \nabla \times \mathbf{B} \\
&\nabla \cdot (U\mathbf{A}) = (\nabla U) \cdot \mathbf{A} + U\,(\nabla \cdot \mathbf{A}) \\
&\nabla \times (U\mathbf{A}) = (\nabla U) \times \mathbf{A} + U\,(\nabla \times \mathbf{A}) \\
&\nabla \cdot (\mathbf{A} \times \mathbf{B}) = \mathbf{B} \cdot (\nabla \times \mathbf{A}) - \mathbf{A} \cdot (\nabla \times \mathbf{B}) \\
&\nabla \times (\mathbf{A} \times \mathbf{B}) = (\mathbf{B} \cdot \nabla)\,\mathbf{A} - \mathbf{B}\,(\nabla \cdot \mathbf{A}) - (\mathbf{A} \cdot \nabla)\,\mathbf{B} + \mathbf{A}\,(\nabla \cdot \mathbf{B}) \\
&\nabla\,(\mathbf{A} \cdot \mathbf{B}) = (\mathbf{B} \cdot \nabla)\,\mathbf{A} + (\mathbf{A} \cdot \nabla)\,\mathbf{B} + \mathbf{B} \times (\nabla \times \mathbf{A}) + \mathbf{A} \times (\nabla \times \mathbf{B})
\end{aligned} \tag{1.27}$$

1.7 Divergence of a vector field: an application

Let us consider an example in fluid flow. Denote by $\mathbf{q} = \mathbf{i}u + \mathbf{j}v + \mathbf{k}w$, the unsteady velocity field of a compressible fluid. Let us consider the rate of mass flow out of an elemental cell $dx\ dy\ dz\ = dV$, fixed with respect to the x, y, z axes (Fig.1.3).

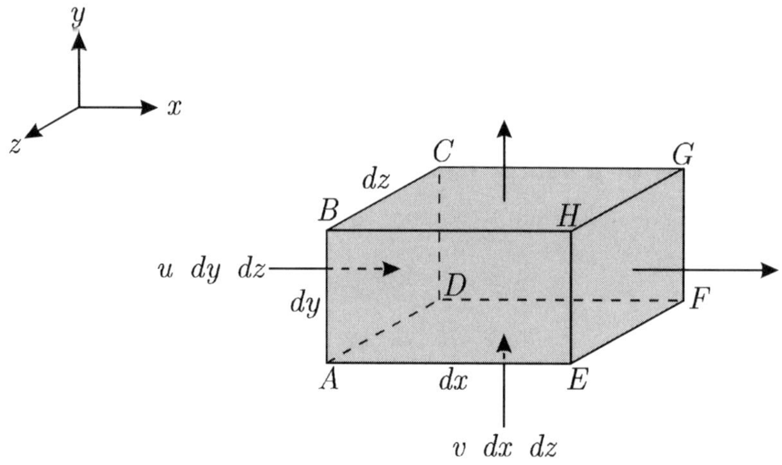

Figure 1.3: Fluid flow through an element

The mass that flows through the face $ABCD$ during time interval dt is $\rho u\ dy\ dz\ dt$ where ρ is the mass density. The mass that flows out of the face $EFGH$ during time dt is

$$\left[\rho u + \frac{\partial}{\partial x}(\rho u)\,dx\right] dy\ dz\ dt \tag{a}$$

Similar expressions can be obtained for mass flows *in* and *out* of the other faces. The net mass that passes through the cell dV during dt is

$$\left[\frac{\partial}{\partial x}(\rho u) + \frac{\partial}{\partial y}(\rho v) + \frac{\partial}{\partial z}(\rho w)\right] dV\ dt \tag{b}$$

which can be written as

$$\nabla \cdot (\rho \mathbf{q})\, dV\ dt \tag{c}$$

or

$$\text{div}\ (\rho \mathbf{q})\, dV\ dt \tag{d}$$

The product $\rho\mathbf{q}$ is called the *current density*. Now the mass in the cell at time t is ρdV. Hence the increase in mass during dt is

$$\frac{\partial \rho}{\partial t} dV\ dt$$

Since (e) must be the negative of (d) we obtain

$$\frac{\partial \rho}{\partial t} + \text{div}\ (\rho \mathbf{q}) = 0 \tag{1.28}$$

Equation (1.28) is known as the *Eulerian continuity equation for fluids* (named after the Swiss mathematician Leonhard Euler (1707-1783)). (i) For steady flow, the term $\partial \rho / \partial t$ disappears. (ii) For an incompressible fluid $\rho =$ constant. Hence the continuity equation for an incompressible fluid

$$\nabla \cdot (\mathbf{q}) = \text{div}\ (\mathbf{q}) = 0 \quad \text{or} \quad \frac{\partial u}{\partial x} + \frac{\partial v}{\partial y} + \frac{\partial w}{\partial z} = 0 \tag{1.29}$$

1.8 Divergence or Green's theorem

Consider any continuous and differentiable vector field $\mathbf{A}(x, y, z)$. We can, for the purposes of illustration, identify $\mathbf{A}$ as the current density in a fluid. From the results presented in section 1.7, $(\text{div}\ \mathbf{A})(dV)$ is the net rate at which fluid flows out of the element where dV is the volume of the element, which is fixed in space (Fig.1.4)
Therefore if V is a given fixed region that is bounded by surface S, then the net rate at which fluid passes out of V is given by

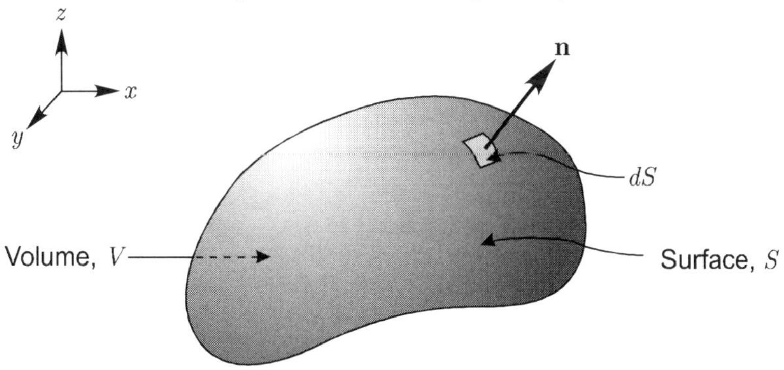

Figure 1.4: Control volume for application of Green's theorem.

$$\iiint_V \text{div } \mathbf{A}\,(dV)$$

If mass is to be conserved, then this must also be the rate at which fluid passes through the surface S. If dS is an element of area of this surface with outward directed unit normal $\mathbf{n}$, the rate of fluid flow through dS is $\mathbf{A}\cdot\mathbf{n}\,dS$. Therefore

$$\iiint_V \text{div } \mathbf{A}\,dV = \iint_S \mathbf{A}\cdot\mathbf{n}\,dS \tag{1.30}$$

This result is known as the *divergence theorem* or *Green's theorem* (named after George Green (1793-1841)). The result is purely mathematical and the reference to the fluid flow problem is to make the derivation relate to an application. Most importantly, this theorem transforms a volume integral into a surface integral and *vice versa.*

The result (1.30) can be recast in a number of forms. For example, let us assume that (L, M, N) are the components of $\mathbf{A}$. The result (1.30) can be expressed in the scalar form

$$\begin{aligned}\iiint_V \left(\frac{\partial L}{\partial x} + \frac{\partial M}{\partial y} + \frac{\partial N}{\partial z}\right) dx\,dy\,dz &= \iint_S (Ln_1 + Mn_2 + Nn_3)\,dS \\ &= \iint_S A_n dS\end{aligned} \tag{1.31}$$

where A_n denotes the projection of $\mathbf{A}$ in the direction of $\mathbf{n}$ and (n_1, n_2, n_3) are the direction cosines of the unit vector $\mathbf{n}$. There are no restriction on the functions (L, M, N) except that they must be continuous and differentiable. The surface can be composed of a finite number of smooth parts that are joined at the edges. If the vector $\mathbf{n}$ is directed inward, the sign of the right hand side of (1.31) should be reversed.

Now consider the case when

$$L = P\,(x, y, z)\;\;Q\,(x, y, z)\quad ;\quad M = N = 0 \tag{1.32}$$

Using (1.32) in (1.31) we have

$$\begin{aligned}\iiint_V P\,\frac{\partial Q}{\partial x}\,dx\,dy\,dz &= -\iiint_V Q\,\frac{\partial P}{\partial x}\,dx\,dy\,dz \\ &\quad + \iint_S PQ\;n_1 dS\end{aligned} \tag{1.33}$$

We can obtain similar expressions by setting in turn

$$M = P(x,y,z)\ Q(x,y,z) \quad ; \quad L = N = 0 \tag{1.34}$$

etc. These equations are similar, in form, to the formula for integration by parts of a single integral.
If we set $M = N = 0$, (1.31) gives

$$\int\int\int_V \frac{\partial L}{\partial x}\, dx\, dy\, dz = \int\int_S L\, n_1\, dS \tag{1.35}$$

Similar results can be obtained by setting $L = N = 0$ and $L = M = 0$ in (1.31). This reduction is referred to as *Gauss' theorem* (named after Carl Friedrich Gauss (1777-1855)). In general terms, *Gauss' theorem* can be written in the form

$$\int\int\int_V \frac{\partial F_i}{\partial x_i} dV = \int\int_S F_i n_i dS \tag{1.36}$$

where

$$F_i = F_i(x_1, x_2, x_3)$$
$$n_i = (n_1, n_2, n_3)$$
$$x_1 = x \quad ; \quad x_2 = y \quad ; \quad x_3 = z$$

and *the summation convention of Albert Einstein (1874-1955)* is invoked for repeated indices;

e.g. equation (1.36) is the compact form for

$$\int\int\int_V \left(\frac{\partial F_1}{\partial x_1} + \frac{\partial F_2}{\partial x_2} + \frac{\partial F_3}{\partial x_3}\right) dV$$
$$= \int\int_S (F_1 n_1 + F_2 n_2 + F_3 n_3)\, dS \tag{1.37}$$

A further useful result can be obtained by assuming that the vector $\mathbf{A}$ can be represented as a product of a scalar function φ and a vector $\mathbf{A}^*$, i.e.

$$\mathbf{A} = \varphi \mathbf{A}^* \tag{1.38}$$

where

$$\mathbf{A}^* = \left(A_x^*\mathbf{i} + A_y^*\mathbf{j} + A_z^*\mathbf{k}\right) \tag{1.39}$$

Hence

$$\begin{aligned} \mathrm{div}\mathbf{A} &= \varphi\ \mathrm{div}\mathbf{A}^* + \frac{\partial\varphi}{\partial x}A_x^* + \frac{\partial\varphi}{\partial y}A_y^* + \frac{\partial\varphi}{\partial z}A_z^* \\ &= \varphi\ \mathrm{div}\mathbf{A}^* + (\mathrm{grad}\ \varphi)\cdot\mathbf{A}^* \end{aligned} \tag{1.40}$$

Considering (1.31) we obtain

$$\int\!\!\int_S \varphi A_n^* dS = \int\!\!\int\!\!\int_V \left[\varphi\ \mathrm{div}\mathbf{A}^* + (\mathrm{grad}\ \varphi)\cdot\mathbf{A}^*\right] dV \tag{1.41}$$

Let us now assume that the vector $\mathbf{A}^*$ can be represented as the gradient of a second scalar function ψ i.e.

$$\mathbf{A}^* = \mathrm{grad}\ \psi \tag{1.42}$$

Then, by the definition of the scalar product of two vectors,

$$A_n^* = \frac{\partial\psi}{\partial n} = \mathbf{n}\cdot\nabla\psi = \mathbf{n}\cdot\mathrm{grad}\psi \tag{1.43}$$

and

$$\mathrm{div}\mathbf{A}^* = \mathrm{div\ grad}\ \psi = \nabla\cdot(\nabla\psi) = \nabla^2\psi \tag{1.44}$$

Using (1.43) and (1.44) in (1.41) we obtain, *Green's first identity*:

$$\int\!\!\int_S \varphi\frac{\partial\psi}{\partial n}dS = \int\!\!\int\!\!\int_V \left[\varphi\nabla^2\psi + (\mathrm{grad}\ \varphi)\cdot(\mathrm{grad}\ \psi)\right] dV \tag{1.45}$$

The result (1.45) is valid for any two scalar functions φ and ψ that are *finite*, *continuous* and *twice differentiable* in V. Interchanging φ and ψ in (1.45)

we have

$$\int\int_S \psi \frac{\partial \varphi}{\partial n} dS = \int\int\int_V \left[\psi \nabla^2 \varphi + (\text{grad}\psi) \cdot (\text{grad } \varphi)\right] dV \tag{1.46}$$

Subtracting (1.46) from (1.45) we obtain *Green's second identity*:

$$\int\int_S \left[\varphi \frac{\partial \psi}{\partial n} - \psi \frac{\partial \varphi}{\partial n}\right] dS = \int\int\int_V \left[\varphi \nabla^2 \psi - \psi \nabla^2 \varphi\right] dV \tag{1.47}$$

The equations (1.45), (1.46) and (1.47) are all referred to as *Green's identities*.
(We note here the basic definitions: **Theorem**: a demonstrable or established but not self evident principle; **Lemma**: a preliminary proposition **Identity**: An equation true for all values involved)

1.9 Green's theorem in two dimensions

Considering the two-dimensional analogue of (1.31) we have

$$\int\int_S (\nabla \times \mathbf{A}) \cdot \mathbf{n} \, dS = \oint_C \mathbf{A} \cdot \mathbf{t} \, dC, \tag{1.48}$$

where the components of $\mathbf{A}$ are any continuous and differentiable functions of x and y; S denotes a region in the (x, y) plane and C is the curve that bounds the region S.

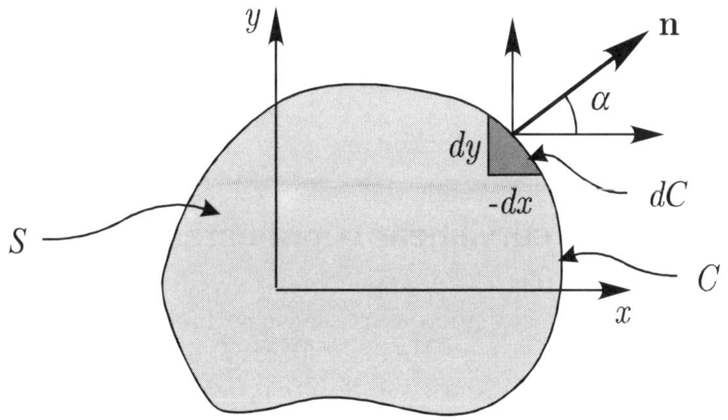

Figure 1.5: Green's theorem in two dimensions

Referring to Figure 1.5, we note that the unit normal vector $\mathbf{n} = (n_1, n_2)$ is directed outward. The element of arc length is denoted by dC. The circle on the integral sign (equation (1.48)) shows that the integration extends completely around the curve C in the counterclockwise sense. We have

$$n_1 = \cos\alpha \quad ; \quad n_2 = \sin\alpha \tag{1.49}$$

and

$$n_1 \, dC = dy \quad ; \quad n_2 \, dC = -dx \tag{1.50}$$

Hence, (1.48) can be written as

$$\int\int_S \left[\frac{\partial L}{\partial x} + \frac{\partial M}{\partial y}\right] dxdy = \oint_C [Ldy - M \, dx] \tag{1.51}$$

This result is referred to as *Green's theorem of the plane.* We can perform operations similar to those indicated previously and obtain the two dimensional equivalent of *Green's identities*: i.e.

$$\int\int_S [\varphi\nabla^2\psi + (\text{grad } \varphi)\cdot(\text{grad } \psi)] \, dS = \oint_C \varphi\frac{\partial\psi}{\partial n} dC \tag{1.52}$$

and

$$\int\int_S [\varphi\nabla^2\psi - \psi\nabla^2\varphi] \, dS = \oint_C \left[\varphi\frac{\partial\psi}{\partial n} - \psi\frac{\partial\varphi}{\partial n}\right] dC \tag{1.53}$$

where φ and ψ are functions of x and y and twice differentiable in S.

1.10 Orthogonal curvilinear coordinates

So far we have presented the results applicable to vector calculus only with reference to a system of rectangular Cartesian coordinates (x, y, z). When dealing with problems which have circular, spherical and other curvilinear boundaries it is convenient to recast the results presented previously in the relevant orthogonal curvilinear coordinate system. The main requirement on these developments is that the coordinate lines of the curvilinear coordinate system (Fig.1.6) must intersect at right angles (i.e. the coordinate system must be orthogonal).

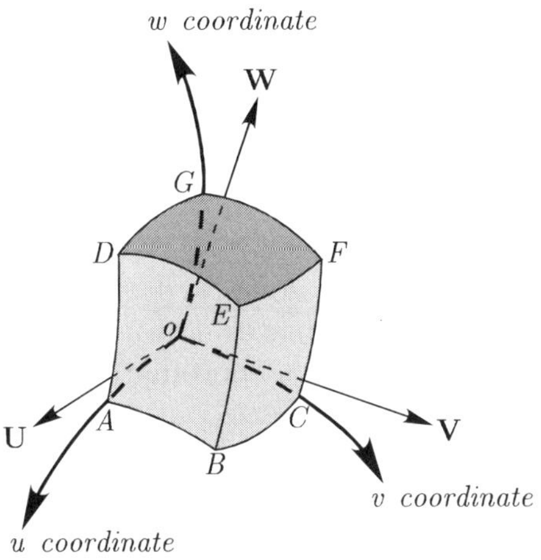

Figure 1.6: Orthogonal curvilinear coordinates

Consider three independent scalar functions (u, v, w) defined in terms of the variables (x, y, z) as follows

$$u = U(x, y, z) \quad ; \quad v = V(x, y, z) \quad ; \quad w = W(x, y, z) \tag{1.54}$$

Since u, v, w are independent functions it implies that (1.54) yield unique solutions for (x, y, z), i.e.

$$x = X(u, v, w) \quad ; \quad y = Y(u, v, w) \quad ; \quad z = Z(u, v, w) \tag{1.55}$$

For example, if we choose (x, y, z) to represent the rectangular Cartesian coordinate system and (u, v, w) to represent the cylindrical coordinate system (r, θ, z) then

$$x = u \cos v \quad ; \quad y = u \sin v \quad ; \quad z = w \tag{1.56}$$

or, in more familiar notation,

$$x = r \cos \theta \quad ; \quad y = r \sin \theta \quad ; \quad z = z \tag{1.57}$$

When (u, v, w) are assigned constant values, (1.54) gives

$$u_0 = U_0(x, y, z) \quad ; \quad v_0 = V_0(x, y, z) \quad , \quad w_0 = W_0(x, y, z) \tag{1.58}$$

The intersection of two of the surfaces determines a curve in space (e.g. intersection of $u_0 = U_0$ and $v_0 = V_0$ determines the w curvilinear coordinate line.

Let $(\mathbf{i}, \mathbf{j}, \mathbf{k})$ be, respectively, unit vectors along (x, y, z) axes. Let (u, v, w) be a system of orthogonal curvilinear coordinates. Let us assign v and w to be constant. Then the tangent vector to the u coordinate line is

$$\mathbf{U} = \frac{\partial x}{\partial u}\mathbf{i} + \frac{\partial y}{\partial u}\mathbf{j} + \frac{\partial z}{\partial u}\mathbf{k} \tag{1.59}$$

Similarly we have

$$\mathbf{V} = \frac{\partial x}{\partial v}\mathbf{i} + \frac{\partial y}{\partial v}\mathbf{j} + \frac{\partial z}{\partial v}\mathbf{k} \tag{1.60}$$

$$\mathbf{W} = \frac{\partial x}{\partial w}\mathbf{i} + \frac{\partial y}{\partial w}\mathbf{j} + \frac{\partial z}{\partial w}\mathbf{k} \tag{1.61}$$

Since the vectors arc mutually perpendicular

$$\mathbf{U} \cdot \mathbf{V} = \mathbf{V} \cdot \mathbf{W} = \mathbf{W} \cdot \mathbf{U} = 0 \tag{1.62}$$

If we denote the magnitudes of the lengths of the vectors $(\mathbf{U}, \mathbf{V}, \mathbf{W})$ by (h_1, h_2, h_3), then from the definition of scalar products

$$h_1^2 = \mathbf{U} \cdot \mathbf{U} \quad ; \quad h_2^2 = \mathbf{V} \cdot \mathbf{V} \quad ; \quad h_3^2 = \mathbf{W} \cdot \mathbf{W} \tag{1.63}$$

e.g.

$$\begin{aligned}
h_1^2 &= \left(\frac{\partial x}{\partial u}\right)^2 + \left(\frac{\partial y}{\partial u}\right)^2 + \left(\frac{\partial z}{\partial u}\right)^2 \\
h_2^2 &= \left(\frac{\partial x}{\partial v}\right)^2 + \left(\frac{\partial y}{\partial v}\right)^2 + \left(\frac{\partial z}{\partial v}\right)^2 \\
h_3^2 &= \left(\frac{\partial x}{\partial w}\right)^2 + \left(\frac{\partial y}{\partial w}\right)^2 + \left(\frac{\partial z}{\partial w}\right)^2
\end{aligned} \tag{1.64}$$

Consider a differential line element AB of length ds with coordinates $A = A(x, y, z)$ and $B = B(x + dx, y + dy, z + dz)$. We have

$$(ds)^2 = (dx)^2 + (dy)^2 + (dz)^2 \tag{1.65}$$

Considering (1.55) we have

$$\begin{aligned}
dx &= \frac{\partial x}{\partial u}(du) + \frac{\partial x}{\partial v}(dv) + \frac{\partial x}{\partial w}(dw) \\
dy &= \frac{\partial y}{\partial u}(du) + \frac{\partial y}{\partial v}(dv) + \frac{\partial y}{\partial w}(dw) \\
dz &= \frac{\partial z}{\partial u}(du) + \frac{\partial z}{\partial v}(dv) + \frac{\partial z}{\partial w}(dw)
\end{aligned} \tag{1.66}$$

From (1.64), (1.65) and (1.66) we obtain.

$$ds^2 = h_1^2 (du)^2 + h_2^2 (dv)^2 + h_3^2 (dw)^2 \tag{1.67}$$

The result (1.67) expresses the length of the differential line element ds in terms of the orthogonal curvilinear coordinates (u, v, w). The coefficients (h_1, h_2, h_3) are called *Lamé coefficients*. The *Lamé coefficients* are equal in magnitude to the lengths of the vectors $(\mathbf{U}, \mathbf{V}, \mathbf{W})$ which are tangential to the (u, v, w) coordinate lines.

1.11 Gradient and Laplacian in orthogonal curvilinear coordinates

We refer to the infinitesimal parallelepiped (see Figure 1.6) whose diagonal OE is the line element ds. The faces of the parallelepiped coincide with the planes $u =$ const; $v =$ const. and $w =$ const. The gradient of u, (∇u), has the direction normal to the surface $u =$ const. Hence (∇u) is the direction of $\mathbf{U}$ or the unit vector $\mathbf{U}/\mid \mathbf{U} \mid= \mathbf{U}/h_1$. The magnitude of ∇u is equal to the derivative of u in this direction. Considering v and w to be constant, we obtain, from (1.67), the magnitude of ∇u as

$$\frac{du}{ds} = \frac{1}{h_1} \tag{1.68}$$

Hence the gradient vector of u

$$\nabla u = \frac{1}{h_1^2}\mathbf{U} \tag{1.69}$$

Similarly, the gradients of v and w are

$$\nabla v = \frac{1}{h_2^2}\mathbf{V} \quad ; \quad \nabla w = \frac{1}{h_3^2}\mathbf{W} \tag{1.70}$$

If we denote the unit vectors in the directions $(\mathbf{U}, \mathbf{V}, \mathbf{W})$ by $(\mathbf{u}, \mathbf{v}, \mathbf{w})$

$$\mathbf{u} = \frac{\mathbf{U}}{h_1} \quad ; \quad \mathbf{v} = \frac{\mathbf{V}}{h_2} \quad ; \quad \mathbf{w} = \frac{\mathbf{W}}{h_3} \tag{1.71}$$

If $f = f(u, v, w)$ is any scalar function, then

$$\begin{aligned}
\frac{\partial f}{\partial x} &= \frac{\partial f}{\partial u}\frac{\partial u}{\partial x} + \frac{\partial f}{\partial v}\frac{\partial v}{\partial x} + \frac{\partial f}{\partial w}\frac{\partial w}{\partial x} \\
\frac{\partial f}{\partial y} &= \frac{\partial f}{\partial u}\frac{\partial u}{\partial y} + \frac{\partial f}{\partial v}\frac{\partial v}{\partial y} + \frac{\partial f}{\partial w}\frac{\partial w}{\partial y} \\
\frac{\partial f}{\partial z} &= \frac{\partial f}{\partial u}\frac{\partial u}{\partial z} + \frac{\partial f}{\partial v}\frac{\partial v}{\partial z} + \frac{\partial f}{\partial w}\frac{\partial w}{\partial z}
\end{aligned} \tag{1.72}$$

Hence, the gradient of f is

$$\nabla f = \frac{\partial f}{\partial u}\nabla u + \frac{\partial f}{\partial v}\nabla v + \frac{\partial f}{\partial w}\nabla w \tag{1.73}$$

Using (1.69), (1.70) and (1.71) in (1.73) we have

$$\nabla f = \frac{1}{h_1}\mathbf{u}\frac{\partial f}{\partial u} + \frac{1}{h_2}\mathbf{v}\frac{\partial f}{\partial v} + \frac{1}{h_3}\mathbf{w}\frac{\partial f}{\partial w} \tag{1.74}$$

Therefore the ∇ operator in orthogonal curvilinear coordinates is

$$\nabla = \frac{1}{h_1}\mathbf{u}\frac{\partial}{\partial u} + \frac{1}{h_2}\mathbf{v}\frac{\partial}{\partial v} + \frac{1}{h_3}\mathbf{w}\frac{\partial}{\partial w} \tag{1.75}$$

In order to derive an expression for the Laplacian ∇^2 we derive the expression for the divergence of a vector field $\mathbf{F} = (F_1, F_2, F_3)$ in curvilinear coordinates. If we restrict attention to the elemental volume shown in Figure 1.6, the edge lengths are $h_1 du$, $h_2 dv$ and $h_3 dw$ and the volume is $h_1 h_2 h_3 \, du \, dv \, dw$. Using Green's theorem in the form given by (1.30) we have

$$\int\int\int_V (\nabla \cdot \mathbf{F}) \, dV = \int\int_S \mathbf{F} \cdot \mathbf{n} \, dS \tag{1.76}$$

Now considering each face of the elemental volume, the contribution of the surface $OCFG$ taken in the direction of the outward normal is $-F_1 \, h_2 \, dv \, h_3 \, dw$. The contribution of the surface $ABED$ is

$$F_1 h_2 h_3 dv \, dw + \frac{\partial}{\partial u} (F_1 h_2 h_3) \, du \, dv \, dw$$

Therefore, the net contribution of the surfaces perpendicular to the u-coordinate line is

$$\frac{\partial}{\partial u} (F_1 \, h_2 \, h_3) \; du \, dv \, dw$$

Similar results can be obtained by considering the net contributions from other surfaces of the elemental volume; e.g.

$$\frac{\partial}{\partial v} (F_2 \, h_1 \, h_3) \; du \, dv \, dw, \qquad \frac{\partial}{\partial w} (F_3 \, h_1 \, h_2) \; du \, dv \, dw$$

Since the elemental volume is infinitesimal

$$\int\int\int_{V \to 0} (\nabla \cdot \mathbf{F}) \, dV \;\; \to \;\; (\nabla \cdot \mathbf{F}) h_1 \; h_2 \; h_3 \; du \; dv \; dw \tag{1.77}$$

Hence from (1.76)

$$\begin{aligned} \nabla \cdot \mathbf{F} &= \text{div } \mathbf{F} \\ &= \frac{1}{h_1 h_2 h_3} \left[\frac{\partial}{\partial u} (F_1 \, h_2 \, h_3) + \frac{\partial}{\partial v} (F_2 \, h_1 \, h_3) \right. \\ &\left. + \frac{\partial}{\partial w} (F_3 \, h_1 \, h_2) \right] \end{aligned} \tag{1.78}$$

If we now denote $\mathbf{F}$ as the gradient of a scalar function f such that

$$\mathbf{F} = \nabla f = \text{grad } f \tag{1.79}$$

then, from (1.75)

$$F_1 = \frac{1}{h_1}\frac{\partial f}{\partial u} \quad ; \quad F_2 = \frac{1}{h_2}\frac{\partial f}{\partial v} \quad ; \quad F_3 = \frac{1}{h_3}\frac{\partial f}{\partial w} \tag{1.80}$$

Consequently we obtain from (1.78) and (1.80)

$$\nabla^2 f = \nabla \cdot \nabla f = \frac{1}{h_1 h_2 h_3}\left[\frac{\partial}{\partial u}\left(\frac{h_2 h_3}{h_1}\frac{\partial f}{\partial u}\right) + \frac{\partial}{\partial v}\left(\frac{h_1 h_3}{h_2}\frac{\partial f}{\partial v}\right) + \frac{\partial}{\partial w}\left(\frac{h_1 h_2}{h_3}\frac{\partial f}{\partial w}\right)\right] \tag{1.81}$$

1.12 Integral transforms

Integral transforms play a key role in the solution of partial differential equations. Integral transforms have the facility to convert a partial differential equation into an ordinary differential equation in the transformed domain. The solution of the reduced problem in the transform domain can be considerably simpler than the solution of the equivalent problem in the physical domain. The type of transform that can be employed in an analysis will invariably depend on the type of partial differential equation, the coordinate system employed to formulate the problem and the range of the variables identified as suitable for the application of a transform. While a variety of integral transforms are utilized for the solution of partial differential equations in science and engineering, we shall summarize here salient results applicable to the integral transforms identified with Pierre Simon Marquis de Laplace (1749-1827), Jean Baptiste Fourier (1768-1830), and Herrmann Hankel (1839-1874).

1.12.1 Laplace transform

(a) Definitions

Consider a function $F(t)$ which is defined for all positive values of the variable t. We can define a definite integral formed by multiplying $F(t)$ by e^{-st} and integrating between the limits 0 to ∞. The new function obtained through this operation is denoted by $f(s)$ i.e.

$$f(s) = \int_0^\infty e^{-st} F(t) dt \tag{1.82}$$

The new function $f(s)$ is referred to as the *Laplace transform* of $F(t)$. The function e^{-st} is referred to as the kernel of the transformation and $f(s)$ is the *image* of $F(t)$. In conventional abbreviation, the operation defined by (1.82) is given the notation

$$\mathcal{L}\{F(t)\} = \int_0^\infty e^{-st} F(t) dt \tag{1.83}$$

There are no restrictions placed on s and it can be complex valued.The *sufficient condition* for the existence of $\mathcal{L}\{F(t)$ are *piecewise continuity* of $F(t)$ and *exponential-order*. **Piecewise continuity**: Following conventional definitions, $F(t)$ is said to be piecewise continuous in an interval if (i) the interval can be represented as a finite number of sub-intervals in each of which $F(t)$ is continuous and (ii) the limits of $F(t)$ as t approaches the extremities of each sub-interval are finite.

Exponential-order: A function $F(t)$ is said to be exponential-order for $t > T$ if we can find constants M and α such that

$$\mid F(t) \mid \; \leq M e^{\alpha t} \quad \text{for} \quad t > T \tag{1.84}$$

Example 1.1

Find the Laplace transform of the functions $F(t) = C$; $F(t) = e^{\xi t}$; $F(t) = \cos \omega t$, for $t > 0$, where C, ξ and ω are constants.

Solution

$$\mathcal{L}\{C\} = C\int_0^\infty e^{-st}dt = \frac{C}{s} \quad ; \quad t > 0 \tag{1.85}$$

$$\mathcal{L}\{e^{\xi t}\} = \int_0^\infty e^{-(s-\xi)t}dt = \frac{1}{(s-\xi)} \quad ; \quad t > 0 \tag{1.86}$$

$$\mathcal{L}\{\cos(\omega t)\} = \int_0^\infty \cos(\omega t)e^{-st}dt = \frac{s}{(s^2+\omega^2)} \quad ; \quad t > 0 \tag{1.87}$$

We note that the Laplace transform is a linear operator, i.e.

$$\mathcal{L}\{\lambda_1 F_1(t) + \lambda_2 F_2(t)\} = \lambda_1\mathcal{L}\{F_1(t)\} + \lambda_2\mathcal{L}\{F_2(t)\} \tag{1.88}$$

where λ_1 and λ_2 are constants.

(b) Transform of derivatives

We can use an integration by parts procedure to determine the transform of the derivative of a function. If we denote the derivative by $F'(t)$ then

$$\mathcal{L}\{F'(t)\} = \int_0^\infty e^{-st}F'(t)dt = \left[e^{-st}F(t)\right]_0^\infty + s\int_0^\infty e^{-st}F(t)dt \tag{1.89}$$

If $F(t)$ is of order $e^{\alpha t}$ or less as t becomes large and if $s > \alpha$, then we have

$$\mathcal{L}\{F'(t)\} = sf(s) - F(0) \tag{1.90}$$

We can repeatedly apply this procedure to determine higher derivatives of $F(t)$; i.e.

$$\mathcal{L}\{F''(t)\} = s^2 f(s) - sF(0) - F'(0) \tag{1.91}$$

In general, we have

$$\mathcal{L}\{F^{(n)}(t)\} = s^{(n)}f(s) - s^{n-1}F(0) - s^{(n-2)}F'(0) \\ -s^{(n-3)}F''(0) - .... - sF^{(n-2)}(0) - F^{(n-1)}(0) \qquad (1.92)$$

(c) The inverse transform When a problem is solved in a transformed domain we should be able to revert back to the original variables by employing a transform inversion technique. This operation involves the determination of the function $F(t)$ given $f(s)$. Assuming that the inversion procedure exists, we can denote the inverse transform by the symbol $\mathcal{L}^{-1}$ and we can write

$$\mathcal{L}^{-1}\{f(s)\} = F(t) \qquad (1.93)$$

For example we note that

$$\mathcal{L}\{t^2\} = \frac{2!}{s^3} \qquad (1.94)$$

Hence it is evident that

$$\mathcal{L}^{-1}\left\{\frac{2!}{s^3}\right\} = t^2 \qquad (1.95)$$

Although a general formal procedure for Laplace transform inversion is available, it is possible to obtain the result for the inverse transform for most functions by reference to *Tables of Integral Transforms* or through symbolic mathematical manipulations offered by software such as MAPLE® and MATHEMATICA®. It should be noted that whenever a Laplace transform exists it is unique, whereas the same is not generally true for an inverse transform. It is possible to show that if two functions have the same Laplace transform, then they cannot differ from each other on an interval of positive length irrespective of how small the interval is. This is referred to as *Lerch's*

Theorem. The theorem states that if two functions have the same Laplace transform then they are the same for all practical purposes and we can assume that the inverse Laplace transform is essentially unique. As a corollary, if two continuous functions have the same Laplace transform then they are identical.

Example 1.2

$$F(t) = \begin{cases} t & \text{for } t \neq 0 \\ a & \text{for } t = 0 \end{cases} \tag{1.96}$$

where a is a constant, and compare the result with the Laplace transform for function

$$F^*(t) = t \tag{1.97}$$

Solution

First let us introduce the definitions of the *Heaviside step function* $H(t)$ and the *Dirac delta function* $\delta(t)$, (named after Paul A.M. Dirac (1902-1984)) which are also referred to as the *unit function* and the *unit impulse function*, respectively. We define the Heaviside step function (named after Oliver Heaviside (1850-1925))(Figure 1.7a)

$$H(t-t_0) = \begin{cases} 0 & \text{for } t < t_0 \\ 1 & \text{for } t \geq t_0 \end{cases} \tag{1.98}$$

and

$$F(t)H(t-t_0) = \begin{cases} 0 & \text{for } t < t_0 \\ F(t) & \text{for } t \geq t_0 \end{cases} \tag{1.99}$$

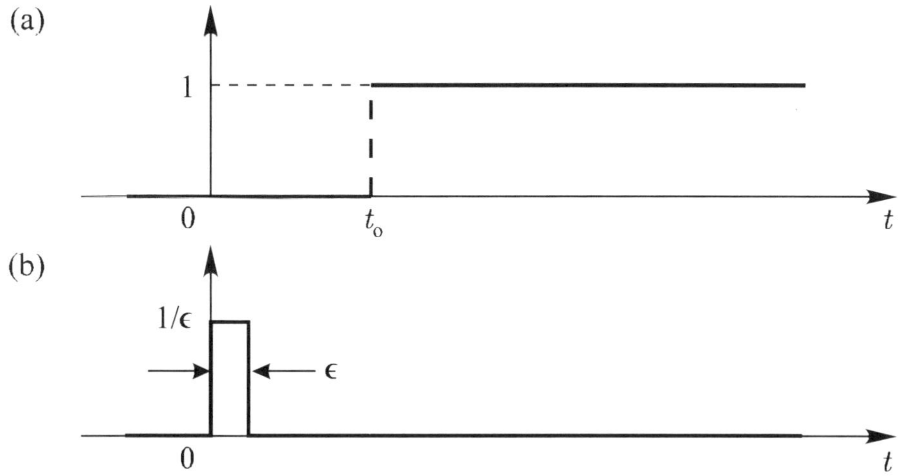

Figure 1.7: (a) the Heaviside step function $H(t - t_0)$ and (b) the Dirac delta function $\delta_\epsilon(t)$

The Laplace transform of the step function can be calculated using the result (1.83), i.e.

$$\mathcal{L}\{H(t - t_0)\} = \int_0^\infty e^{-st} H(t - t_0) dt = \int_{t_0}^\infty e^{-st} dt \tag{1.100}$$

Hence if $t_0 \geq 0$

$$\mathcal{L}\{H(t - t_0)\} = \frac{e^{-t_0 s}}{s} \tag{1.101}$$

The Dirac delta function has a useful physical significance in terms of an entity which is applied over a small extent either *spatially* or *temporally* but has a non zero-total value. An analogue of the Dirac delta function could be a force applied at a point on the surface of a medium, a point charge in electrostatics, a temperature pulse at a point, an acoustic impulse of short duration, etc. In relation to the present discussion, the Dirac delta function $\delta(t)$ is defined as

$$\delta(t) = 0 \qquad \text{for} \qquad t \neq 0 \tag{1.102}$$

$$\int_{-\infty}^{\infty} \delta(t)dt = 1 \tag{1.103}$$

These definitions imply that $\delta(t)$ is infinite at $t = 0$ although zero elsewhere. We can recast the definition of the Dirac delta function by considering the function to be distributed over an interval ϵ (Figure 1.7b) such that

$$\delta_\epsilon(t) = \begin{cases} 0 \text{ for } -\infty < t < 0 \\ 1/\epsilon \text{ for } \quad 0 < t < \epsilon \\ 0 \text{ for } \; \epsilon < t < \infty \end{cases} \tag{1.104}$$

Here again

$$\int_{-\infty}^{\infty} \delta_\epsilon(t)dt = \int_0^\epsilon \frac{dt}{\epsilon} = 1 \tag{1.105}$$

It may be noted that for a given function $F(t)$ which is continuous at $t = a$

$$\int_{-\infty}^{\infty} \delta(t-a)F(t)dt = F(a) \tag{1.106}$$

It is also possible to express the Heaviside step function in terms of the Dirac delta function as follows:

$$H(t - t_0) = \int_{-\infty}^{t} \delta(\tau - t_0)d\tau \tag{1.107}$$

since the interval of integrations contains t_0 only if $t > t_0$ and not if $t < t_0$. From (1.107) it is evident that

$$\delta(t - t_0) = H'(t - t_0) \tag{1.108}$$

which establishes the fact that $H'(t-t_0)$ is zero everywhere except at $t = t_0$, where the discontinuity in $H(t-t_0)$ occurs.

If $t_0 \geq 0$ we can define the Laplace transform of $\delta(t-t_0)$ as

$$\mathcal{L}\{\delta(t-t_0)\} = \int_0^\infty e^{-st}\delta(t-t_0)dt = e^{-t_0 s} \tag{1.109}$$

If $t_0 = 0$ then

$$\mathcal{L}^{-1}\{1\} = \delta(t). \tag{1.110}$$

Referring back to the problem posed in the example, we note that $F(t)$ in (1.96) can be written as

$$F(t) = t + a\ \delta(t) \tag{1.111}$$

From (1.88) we have

$$\mathcal{L}\{F(t)\} = \mathcal{L}\{t\} + a\mathcal{L}\{\delta(t)\} \tag{1.112}$$

or

$$\begin{aligned}\mathcal{L}\{F(t)\} &= \frac{1}{s^2} + a\mathcal{L}\{H(t) - H(t-\epsilon)\} \\ &= \frac{1}{s^2} + a\left[\frac{1}{s} - \frac{e^{-\epsilon s}}{s}\right]\end{aligned} \tag{1.113}$$

As $\epsilon \to 0$,

$$\mathcal{L}\{F(t)\} = \frac{1}{s^2} \tag{1.114}$$

This is identical to the Laplace transform of the function $F^*(t) = t$, which gives

$$\mathcal{L}\{F^*(t)\} = \mathcal{L}\{t\} = \frac{1}{s^2} \tag{1.115}$$

It is evident that, although the Laplace transforms given by (1.114) and (1.115) are identical, they are derived from two entirely different functions, which differ by the discontinuity at $x = 0$. Therefore, the function $F(t)$ defined by (1.96) cannot be recovered from the Laplace transform inversion of (1.114). Therefore, when discontinuous functions are encountered the inverse transform can represent more than one function.

(d) Partial fractions

When the Laplace transform $f(s)$ is a rational function (i.e. a quotient of two polynomials in s), transform inversion is best achieved by recourse to tabulated functions. On occasions, it is necessary to recast the transformed solutions into combination of tabulated functions. The transformed function can be decomposed into simpler functions by the use of partial fractions.

Example 1.3

Determine the inverse Laplace transform of

$$f(s) = \frac{3s+2}{s(s+2)} \tag{1.116}$$

Solution

From the theory of partial fractions we can express $f(s)$ as follows:

$$f(s) = \frac{1}{s} + \frac{2}{(s+2)} \tag{1.117}$$

Hence

$$\mathcal{L}^{-1}\{f(s)\} = \mathcal{L}^{-1}\left\{\frac{1}{s}\right\} + 2\mathcal{L}^{-1}\left\{\frac{1}{s+2}\right\} \tag{1.118}$$

or

$$\mathcal{L}^{-1}\left\{\frac{3s+2}{s(s+2)}\right\} = 1 + 2e^{-2t} \tag{1.119}$$

Other more complicated rational functions can similarly be decomposed into partial fractions for the application of the Laplace transform inversion procedure.

(e) Convolution theorem

The provides a useful procedure for inverting certain types of transformed results. Consider the functions $F(t)$ and $G(t)$ whose Laplace transforms are denoted by $f(s)$ and $g(s)$ respectively. i.e.

$$f(s) = \mathcal{L}\{F(t)\} \quad ; \quad g(s) = \mathcal{L}\{G(t)\} \tag{1.120}$$

We now introduce the "*shift*" or "*translation*" theorem:

THEOREM 1.1

If $f(s) = \mathcal{L}\{F(t)\}$ then

$$e^{-\beta s} f(s) = \mathcal{L}\{F_\beta(t)\} \tag{1.121}$$

where

$$F_\beta(t) = \begin{cases} 0 & \text{for} \quad 0 < t < \beta \\ F(t-\beta) & \text{for} \quad t > \beta \end{cases} \tag{1.122}$$

PROOF

We have

$$e^{-\beta s} f(s) = \int_0^\infty e^{-s(t+\beta)} F(t) dt \tag{1.123}$$

where β is a constant which is assumed to be greater than zero. Let $t+\beta = \tau$ such that the integral (1.123) becomes

$$e^{-\beta s} f(s) = \int_0^\infty e^{-s\tau} F(\tau - \beta) d\tau \tag{1.124}$$

since we have assumed that $t > 0$. Hence, if we define $F_\beta(t)$ as assumed by (1.122), we have proved the required preliminary result concerning the *shift* or *translation* theorem.

We can now apply the *shift* theorem to $g(s)$, i.e.

$$e^{-s\tau} g(s) = \mathcal{L}\{G(t-\tau)\} = \int_0^\infty e^{-st} G(t-\tau) dt \tag{1.125}$$

where it is implied that $G(t) = 0$ for $t < 0$. Then we can write

$$f(s)g(s) = g(s) \int_0^\infty F(\tau) e^{-s\tau} d\tau = \int_0^\infty F(\tau)[g(s)e^{-s\tau}] d\tau \tag{1.126}$$

Using the integral expression for $g(s)$ in (1.126) we have

$$f(s)g(s) = \int_0^\infty F(\tau)\left[\int_0^\infty e^{-st}G(t-\tau)dt\right]d\tau \tag{1.127}$$

We can express the result (1.127) in the form of a limit as follows:

$$f(s)g(s) = \lim_{T\to\infty}\int_0^\infty e^{-st}\left[\int_0^T F(\tau)G(t-\tau)d\tau\right]dt \tag{1.128}$$

Alternatively we can write

$$f(s)g(s) = \lim_{T\to\infty}\{I_1(T)+I_2(T)\} \tag{1.129}$$

where

$$I_1(t) = \int_0^T e^{-st}\int_0^T F(\tau)G(t-\tau)d\tau dt \tag{1.130}$$

$$I_2(t) = \int_T^\infty e^{-st}\int_0^T F(\tau)G(t-\tau)d\tau dt \tag{1.131}$$

As $T\to\infty, I_2(T)=0$. We also note that when $\tau > t,\ G(t-\tau)=0$, such that the upper limit of integration with respect to τ in (1.130) and (1.131) is replaced by t. Then we have

$$f(s)g(s) = \int_0^\infty e^{-st}\int_0^t F(\tau)G(t-\tau)d\tau dt \tag{1.132}$$

If we define

$$F(t) * G(t) = \int_0^t F(\tau)G(t-\tau)d\tau \tag{1.133}$$

we have

$$f(s)g(s) = \mathcal{L}\{F(t) * G(t)\} \tag{1.134}$$

or

$$\mathcal{L}^{-1}\{f(s)g(s)\} = F(t) * G(t) \tag{1.135}$$

This result is called the *convolution theorem* or the *Faltung Theorem.*

(f) The inversion integral

The transformed solutions to certain problems could contain results which are not amenable to Laplace transform inversion through results that are available from Tables. In this instance, it is necessary to develop an inversion procedure which can be applied universally. To develop such an inversion procedure we need to generalize the theory of the Laplace transformation to include values of s which are complex (e.g.: $s = x + iy$). The relevant results can be obtained by an extension of the integral formulae of Augustin Louis Cauchy (1789-1857).

From the theory of complex variables, Cauchy's integral formula states that if $f(z)$ is an *analytic function* within a closed region C and if z_0 is a point within C, then

$$f(z_0) = \frac{1}{2\pi i} \int_C \frac{f(z)dz}{(z - z_0)} \tag{1.136}$$

We note that for $f(t)$ to be an *analytic function*, it must be single-valued in C, and the derivative $f'(z)$, defined by

$$f'(z) = \operatorname*{Lim}_{\Delta z \to 0} \frac{f(z + \Delta z) - f(z)}{\Delta z} \tag{1.137}$$

exists, irrespective of the manner in which the limit is achieved. If the limit (1.137) exists at $z = z_0$ then the function is analytic at $z = z_0$. The function $f(z)$ must be continuous and single-valued in order for it to be analytic, although the converse is not necessarily true. A point where $f(z)$ is *not analytic* is called a *singular point* or a *singularity* of $f(z)$. If the function $f(z)$ exhibits a singularity at $z = a$, then the location $z = a$ is said to be a *pole* of the function. A *two-valued* function $f(z)$ can be separated into *two single valued functions* by separating them into branches. The branches can be isolated by providing an imaginary impenetrable *barrier* or a *cut*. Such barriers are called *branch cuts*.

To apply Cauchy's integral formulae we need to extend it according to the following theorem.

THEOREM 1.2

Let $f(z)$ be analytic for $\text{Re}(z) \geq \gamma$, where γ is a real constant other than zero. If $\text{Re}(z_0) > \gamma$ then

$$f(z_0) = -\frac{1}{2\pi i} \lim_{\beta \to \infty} \int_{\gamma - i\beta}^{\gamma + i\beta} \frac{f(z)dz}{(z - z_0)} \tag{1.138}$$

The location of $\text{Re}(z) = \gamma$ and z_0 are shown in Figure 1.8.

PROOF

The proof of the theorem is somewhat involved but basically focuses on choosing a rectangular contour S that encompasses z_0 and letting $\beta \to \infty$ in the final result for the equivalent of the integral (1.138). Invoking certain limiting arguments and inequalities, it can be shown that the contribution from the path S *vanishes* in the instance when S consists of the *horizontal* and *right hand vertical* portions of the path.

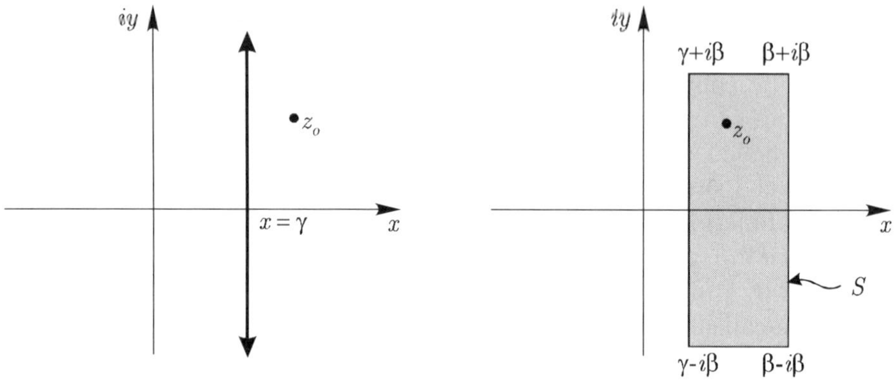

Figure 1.8: Location and the contour enclosing of z_0 in the complex plane.

In the previous theorem we replace z_0 by s such that

$$f(s) = \frac{1}{2\pi i} \operatorname*{Lim}_{\beta \to \infty} \int_{\gamma - i\beta}^{\gamma + i\beta} \frac{f(z)dz}{(s-z)} \tag{1.139}$$

where s is complex valued and given by $s = x + iy$. We note that $f(s)$ is assumed to be analytic in the half-plane $\text{Re}(s) > \gamma$. If we apply the inverse transform to either side of (1.139)

$$F(t) = \mathcal{L}^{-1}\{f(s)\} = \frac{1}{2\pi i}\mathcal{L}^{-1}\left\{ \operatorname*{Lim}_{\beta \to \infty} \int_{\gamma - i\beta}^{\gamma + i\beta} \frac{f(z)dz}{(s-z)} \right\} \tag{1.140}$$

or

$$F(t) = \frac{1}{2\pi i} \operatorname*{Lim}_{\beta \to \infty} \int_{\gamma - i\beta}^{\gamma + i\beta} f(z)\mathcal{L}^{-1}\left\{ \frac{1}{s-z} \right\} dz \tag{1.141}$$

We note that

$$\mathcal{L}^{-1}\left\{\frac{1}{(s-z)}\right\} = e^{zt} \tag{1.142}$$

Hence, we can write (1.141) as

$$F(t) = \frac{1}{2\pi i} \operatorname*{Lim}_{\beta\to\infty} \int_{\gamma-i\beta}^{\gamma+i\beta} e^{st} f(s) ds \tag{1.143}$$

(Note that in (1.143) the dummy integration variable z has been replaced by s.) The result (1.143) constitutes the inversion integral for the Laplace transform valid for any function $f(s)$. The path of the integration is referred to as the *Bromwich Contour, Laplace Transform Inversion* (named after T.J.I'Anson Bromwich (1875-1929).) Fundamentally, the Laplace transform inversion is now reduced to the application of contour integration in the complex plane. The specific aspects of the calculation will obviously depend on the nature of the function $f(s)$. Usually the function $f(s)$ is either.

(a) a single-valued function of s with a finite or infinite number of poles in the complex s-plane or

(b) a function of s having a number of branch points in the s-plane as well as a number of poles.

In the case of (a) a complete contour is formed from the Bromwich integral by adding a circle of radius R as shown in Figure 1.9.

The poles will be located entirely within the contour S such that from the *theory of residues*,

$$\int_C f(z)dz = 2\pi i \sum_1^n (\text{Res}) \tag{1.144}$$

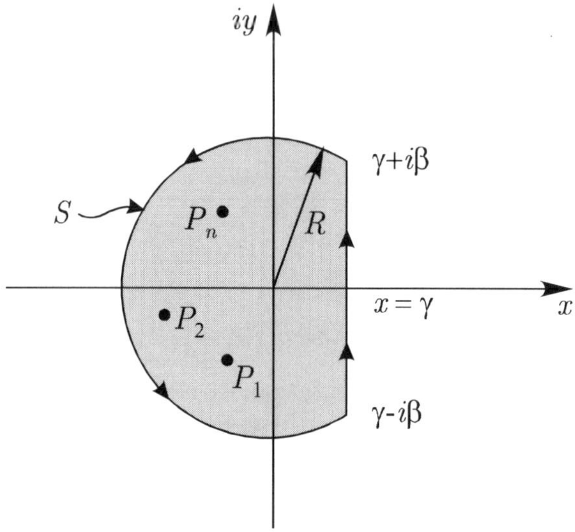

Figure 1.9: Bromwich contour enclosing poles

Applying this to (1.143) (see Figure 1.9) we have

$$\frac{1}{2\pi i}\int_S e^{st}f(s)ds+\frac{1}{2\pi i}\int_{\gamma-i\beta}^{\gamma+i\beta}e^{st}f(s)ds=\sum_1^n(\text{Res}) \tag{1.145}$$

By the correct application of inequalities and limits it can be shown that

$$\operatorname*{Lim}_{R\to\infty}\int_S e^{st}f(s)ds=0 \tag{1.146}$$

As a result we are left with only the *Bromwich integral* which is the inverse Laplace transform, i.e.

$$\mathcal{L}^{-1}\{f(s)\}=\sum_1^n(\text{Res}) \tag{1.147}$$

In the case of (b) the basic procedures are the same, except that $f(s)$ is no longer single-valued due to the presence of *branch points* and *branch cuts* and the contour cannot be closed in the usual way. The technique is to complete the contour but to configure it to *avoid* branch points and branch cuts. The technique is best illustrated by appeal to a specific example.

Example 1.4

Consider the Laplace transform

$$f(s) = \frac{e^{-a\sqrt{s}}}{s} \tag{1.148}$$

and outline a procedure for its inversion.

Solution

We note that $\sqrt{s}$ is not single-valued and a branch point exists at the origin. We can select the branch cut required to make $\sqrt{s}$ single-valued at any arbitrary orientation in the s-plane. It is, however, convenient to select the branch cut along the negative real axis as shown in Figure 1.10.

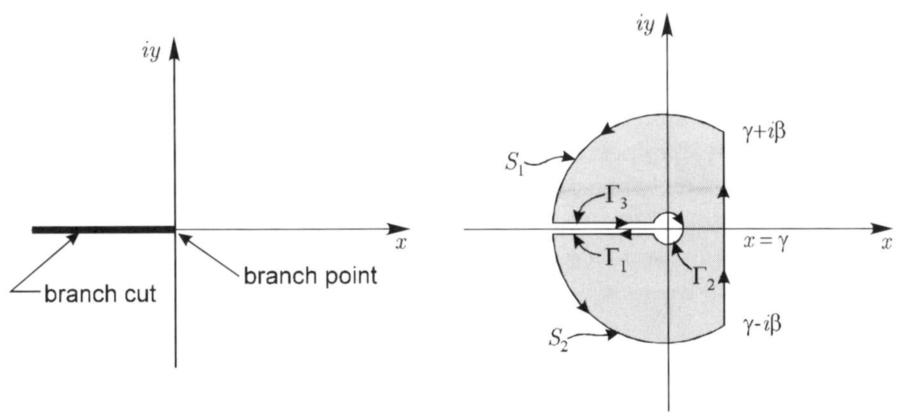

Figure 1.10: Bromwich contour for branch point at origin

The contour is configured to exclude the branch point $s = 0$ and the branch cut $-\infty < x \le 0$. The integration paths Γ_1 and Γ_3 are coincident with the x-axis but are shown separated for purposes of illustration. We have

$$\int_{\gamma-i\beta}^{\gamma+i\beta} + \int_{S_1} + \int_{\Gamma_3} + \int_{\Gamma_2} + \int_{\Gamma_1} + \int_{S_2} = 0 \tag{1.149}$$

By application of inequalities and limits it can be shown that

$$\int_{S_1} = \int_{S_2} = 0 \tag{1.150}$$

Comparing the results (1.143), (1.149) and (1.150) we obtain the inverse transform as

$$L^{-1}\{f(s)\} = -\int_{\Gamma_1} - \int_{\Gamma_2} - \int_{\Gamma_3} \tag{1.151}$$

The integral over Γ_2 will yield the residue at the origin and the integrals over Γ_1 and Γ_3 will be real integrals.

(g) Special Theorems involving Laplace transforms and inverse Laplace transforms

It is convenient to summarize a number of results involving Laplace transforms and inverse Laplace transforms which will be useful for purposes of future reference.

THEOREM I

If $\mathcal{L}\{F(t)\} = f(s)$ then

$$\mathcal{L}\{e^{at}F(t)\} = f(s-a) \tag{1.152}$$

Also, if $\mathcal{L}^{-1}\{f(s)\} = F(t)$, then

$$\mathcal{L}^{-1}\{f(s-a)\} = e^{at}F(t) \tag{1.153}$$

THEOREM II

If $\mathcal{L}\{F(t)\} = f(s)$, then

$$\mathcal{L}\{H(t-a)F(t-a)\} = e^{-as}f(s) \tag{1.154}$$

Also, if $\mathcal{L}^{-1}\{f(s)\} = F(t)$

$$\mathcal{L}^{-1}\{e^{-as}f(s)\} = H(t-a)F(t-a) \tag{1.155}$$

where $H(t-a)$ is the Heaviside step function.

THEOREM III

If $\mathcal{L}\{F(t)\} = f(s)$ then

$$\mathcal{L}\{F(at)\} = \frac{1}{a}f\left(\frac{s}{a}\right) \tag{1.156}$$

Also, if $\mathcal{L}^{-1}\{f(s)\} = F(t)$, then

$$\mathcal{L}^{-1}\left\{f\left(\frac{s}{a}\right)\right\} = aF(at) \tag{1.157}$$

THEOREM IV

If $\mathcal{L}\{F(t)\} = f(s)$, then if $n = 1, 2, 3,$

$$\mathcal{L}\{t^n F(t)\} = (-1)^n \frac{d^n f}{ds^n} = (-1)^n f^{(n)}(s) \tag{1.158}$$

Also, if $\mathcal{L}^{-1}\{f(s)\} = F(t)$, then

$$\mathcal{L}^{-1}\{f^{(n)}(s)\} = (-1)^n t^n F(t) \tag{1.159}$$

THEOREM V

Consider a periodic function $F(t)$ with period λ, such that $F(t+\lambda) = F(t)$, then

$$\mathcal{L}\{F(t)\} = \frac{\int_0^\lambda e^{-st} F(t) dt}{1 - e^{-s\lambda}} \tag{1.160}$$

THEOREM VI

If $\mathcal{L}\{F(t)\} = f(s)$, then

$$\mathcal{L}\{\int_0^t F(\xi) d\xi\} = \frac{f(s)}{s} \tag{1.161}$$

Also, if $\mathcal{L}^{-1}\{f(s)\} = F(t)$, then

$$\mathcal{L}^{-1}\left\{\frac{f(s)}{s}\right\} = \int_0^t F(\xi) d\xi \tag{1.162}$$

THEOREM VII

If $\operatorname*{Lim}_{t\to 0} [F(t)/t]$ exists and if $\mathcal{L}\{F(t)\} = f(s)$, then

$$\mathcal{L}\left\{\frac{F(t)}{t}\right\} = \int_s^\infty f(\xi) d\xi \tag{1.163}$$

THEOREM VIII

If $\mathcal{L}\{F(t)\} = f(s)$ and $\mathcal{L}\{G(t)\} = g(s)$, then

$$\mathcal{L}\{\int_0^t F(\xi)G(t-\xi)d\xi\} = f(s)g(s) \tag{1.164}$$

Also if $\mathcal{L}^{-1}\{f(s)\} = F(t)$ and $\mathcal{L}^{-1}\{g(s)\} = G(t)$, then

$$L^{-1}\{f(s)g(s)\} = \int_0^t F(\xi)G(t-\xi)d\xi \tag{1.165}$$

The integral in (1.165) is defined as the *convolution* of F and G or

$$F * G = \int_0^t F(\xi)G(t-\xi)d\xi \tag{1.166}$$

and the operation is commutative, i.e.

$$F * G = G * F \tag{1.167}$$

1.12.2 Fourier transforms

(a) Definitions

The Fourier transform $f(\xi)$ of a function $F(x)$, defined for all x, is given by

$$f(\xi) = \mathcal{F}\{F(x);\xi\} = \frac{1}{\sqrt{2\pi}}\int_{-\infty}^{\infty} e^{i\xi x}F(x)dx \tag{1.168}$$

where $e^{i\xi x}$ is referred to as the kernel function of the transformation. The result (1.168) is also referred to as the *exponential Fourier transform.*

If $F(x)$ is an *odd function*, (1.168) reduces to

$$f_s(\xi) = \mathcal{F}_s\{F(x);\xi\} = \sqrt{\frac{2}{\pi}}\int_0^\infty \sin(\xi x)F(x)dx \tag{1.169}$$

which is the *Fourier sine transform*. Similarly, if $F(x)$ is an *even function*, the exponential transform reduces to the *Fourier cosine transform*

$$f_c(\xi) = \mathcal{F}_c\{F(x);\xi\} = \sqrt{\frac{2}{\pi}}\int_0^\infty \cos(\xi x)F(x)dx \tag{1.170}$$

There is no general consensus regarding the choice of the normalizing constant; in the above we have selected $1/\sqrt{2\pi}$ as the normalizing constant; in some cases this is chosen as $1/2\pi$ or even as a constant incorporated within the argument of the exponential term.

(b) Fourier transform of derivatives

We can use an integration by parts procedure to determine the Fourier transform of dF/dx, i.e.

$$\begin{aligned}\mathcal{F}\left\{\frac{dF}{dx};\xi\right\} &= \frac{1}{\sqrt{2\pi}}\int_{-\infty}^{\infty} e^{i\xi x}F'(x)dx \\ &= \left[\frac{e^{i\xi x}}{\sqrt{2\pi}}F(x)\right]_{-\infty}^{\infty} - \frac{i\xi}{\sqrt{2\pi}}\int_{-\infty}^{\infty} e^{i\xi x}F(x)dx\end{aligned} \tag{1.171}$$

If we assume that $e^{i\xi x}F(x) \to 0$ as $x \pm \infty$ then

$$\mathcal{F}\left\{\frac{dF}{dx};\xi\right\} = -i\xi f(\xi) \tag{1.172}$$

The assumption, however, places restrictions on the behaviour of $F(x)$ for large x. A *sufficient* condition for the existence of the Fourier transform is that the integral

$$\int_{-\infty}^{\infty} \mid F(x) \mid dx$$

should converge. This, however, is not a *necessary* condition; $F(x)$ will *not* have a Fourier transform unless $F(x) \to 0$ as $\mid x \mid \to \infty$. This is a more stringent requirement than the exponential-order requirement (1.84) stipulated in connection with the Laplace transform. In general we can show that

$$\mathcal{F}\{F^{(n)}(x);\xi\} = (-i\xi)^{(n)} f(\xi) \tag{1.173}$$

It is of interest to note that the derivatives of the Fourier transform place restrictions on the behaviour of the function at $x = \pm\infty$, whereas the Laplace transform places conditions at both $t = 0$ and $t = \infty$. In both cases, appropriate boundary conditions are imposed on the function. The Fourier sine and Fourier cosine transforms have zeros as their lower limit and suitable initial conditions have to be incorporated in a manner similar to Laplace transforms, i.e.

$$f_s^{(n)}(\xi) = \mathcal{F}_s^{(n)}\{F(x);\xi\} = \sqrt{\frac{2}{\pi}} \int_0^{\infty} \sin(\xi x) F^{(n)}(x) dx \tag{1.174}$$

and

$$f_c^{(n)}(\xi) = \mathcal{F}_c^{(n)}\{F(x);\xi\} = \sqrt{\frac{2}{\pi}} \int_0^{\infty} \cos(\xi x) F^{(n)}(x) dx \tag{1.175}$$

Integration by parts of $f_c^{(n)}(\xi)$ gives

$$f_c^{(n)}(\xi) = \sqrt{\frac{2}{\pi}} \left[F^{(n-1)}(x)\cos(\xi x) \right]_0^\infty + \xi\sqrt{\frac{2}{\pi}} \int_0^\infty \sin(\xi x) F^{(n-1)}(x) dx \tag{1.176}$$

We assume that

$$\lim_{x\to 0} \sqrt{\frac{2}{\pi}}\, F^{(n-1)}(x) = a_{n-1} \quad ; \quad \lim_{x\to\infty} F^{(n-1)}(x) = 0 \tag{1.177}$$

Hence

$$f_c^{(n)}(\xi) = -a_{n-1} + \xi f_s^{(n-1)}(\xi) \tag{1.178}$$

Alternatively, integration of $f_s^{(n)}(\xi)$ by parts gives

$$f_s^{(n)}(\xi) = -\xi f_c^{(n-1)}(\xi) \tag{1.179}$$

From (1.178) and (1.179) we have

$$f_c^{(n)}(\xi) = -a_{n-1} - \xi^2 f_c^{(n-2)}(\xi) \tag{1.180}$$

We can treat results of this type as recurrence relationships to generate the following:

$$f_c^{(2n)}(\xi) = -\sum_{r=0}^{n-1} (-1)^r a_{2n-2r-1} (\xi)^{2r} + (-1)^n \xi^{(2n)} f_c(\xi) \tag{1.181}$$

$$f_c^{(2n+1)}(\xi) = -\sum_{r=0}^{n} (-1)^r a_{2n-2r} (\xi)^{2r} + (-1)^n \xi^{(2n+1)} f_s(\xi) \tag{1.182}$$

which give the Fourier cosine transform of the derivatives of $F(x)$ in terms of the Fourier cosine and sine transforms of $F(x)$.

Similarly, for the sine transforms of derivatives, we have

$$f_s^{(2n)}(\xi) = -\sum_{r=1}^{n}(-1)^r(\xi)^{2r-1}a_{2n-2r} + (-1)^{n+1}\xi^{(2n)}f_s(\xi) \tag{1.183}$$

$$f_s^{(2n+1)}(\xi) = -\sum_{r=1}^{n}(-1)^r(\xi)^{2r-1}a_{2n-2r+1} + (-1)^{n+1}\xi^{(2n+1)}f_c(\xi) \tag{1.184}$$

(c) The inverse transform

The inverse transform of the function $f(\xi)$ will give $F(x)$ or we can write

$$\mathcal{F}^{-1}\{f(\xi);x\} = F(x) \tag{1.185}$$

As in the case of Laplace transforms, the inversion procedure is particularly straight forward if $f(\xi)$ is identified as the transform of a known function $F(x)$. This can be done by reference to *Tables of Integral Transforms* or through the use of certain symbolic mathematical manipulation software such as MAPLE®and MATHEMATICA®. When such direct inversion procedures are not possible, more formal schemes have to be adopted to invert the transform. The general theorems and procedures used to achieve Fourier transform inversion are given in a number of texts on the subject and will not be covered here. We shall, however, summarize the relevant theorems.

THEOREM I

If $f(\xi)$ is the exponential Fourier transform of $F(x)$ defined by

$$f(\xi) = \mathcal{F}\{F(x);\xi\} = \frac{1}{\sqrt{2\pi}}\int_{-\infty}^{\infty} e^{i\xi x}F(x)dx \tag{1.186}$$

then $F(x)$ is given by

$$F(x) = \mathcal{F}^{-1}\{f(\xi); x\} = \frac{1}{\sqrt{2\pi}} \int_{-\infty}^{\infty} e^{-i\xi x} f(\xi) d\xi \tag{1.187}$$

THEOREM II

If $f_s(\xi)$ is the Fourier sine transform of $F(x)$ defined by

$$f_s(\xi) = \mathcal{F}_s\{F(x); \xi\} = \sqrt{\frac{2}{\pi}} \int_0^{\infty} \sin(\xi x) F(x) dx \tag{1.188}$$

Then

$$F(x) = \mathcal{F}_s^{-1}[f_s(\xi); x\} = \sqrt{\frac{2}{\pi}} \int_0^{\infty} \sin(\xi x) f_s(\xi) d\xi \tag{1.189}$$

THEOREM III

If $f_c(\xi)$ is the Fourier cosine transform of $F(x)$ defined by

$$f_c(\xi) = \mathcal{F}_c\{F(x); \xi\} = \sqrt{\frac{2}{\pi}} \int_0^{\infty} \cos(\xi x) F(x) dx \tag{1.190}$$

then

$$F(x) = \mathcal{F}_c^{-1}\{f_c(\xi); x\} = \sqrt{\frac{2}{\pi}} \int_0^{\infty} \cos(\xi x) f_c(\xi) d\xi \tag{1.191}$$

As is evident from Theorems II and III (equations (1.188) to (1.191)), the *Fourier sine and Fourier cosine transforms are self reciprocal.*

(d) A convolution theorem

Let $f(\xi)$ and $g(\xi)$ be the Fourier transforms of $F(x)$ and $G(x)$ respectively. Then, there is a convolution theorem which enables the inversion of a product of transforms $f(\xi)g(\xi)$ to be expressed as a convolution-type integral, i.e.

$$\mathcal{F}^{-1}\{f(\xi)g(\xi); x\} = F(x) * G(x) \tag{1.192}$$

The specific form of the integral can be obtained as follows: we have

$$\begin{aligned}\mathcal{F}^{-1}\{f(\xi)g(\xi); x\} &= \frac{1}{\sqrt{2\pi}} \int_{-\infty}^{\infty} f(\xi)g(\xi)e^{-i\xi x} d\xi \\ &= \frac{1}{2\pi} \int_{-\infty}^{\infty} f(\xi)e^{-i\xi x} d\xi \int_{-\infty}^{\infty} G(\eta)e^{i\eta\xi} d\eta \\ &= \frac{1}{2\pi} \int_{-\infty}^{\infty} G(\eta) d\eta \int_{-\infty}^{\infty} f(\xi)e^{-i\xi(x-\eta)} d\xi \end{aligned} \tag{1.193}$$

Also, we can write

$$F(x-\eta) = \frac{1}{\sqrt{2\pi}} \int_{-\infty}^{\infty} f(\xi)e^{-i\xi(x-\eta)} d\xi \tag{1.194}$$

Using (1.194) we can write (1.193) in the form

$$\mathcal{F}^{-1}\{f(\xi)g(\xi); x\} = \frac{1}{\sqrt{2\pi}} \int_{-\infty}^{\infty} G(\eta)F(x-\eta) d\eta \tag{1.195}$$

Thus we can write

$$F(x) * G(x) = \frac{1}{\sqrt{2\pi}} \int_{-\infty}^{\infty} G(\eta)F(x-\eta) d\eta \tag{1.196}$$

This is the *convolution* or *Faltung* integral form for the Fourier transform. Hence, the Fourier transform of the convolution of two functions is equal to the product of their Fourier transforms.

(e) Finite Fourier transform

It is possible to utilize the Fourier transform concepts developed previously for extended intervals to accommodate finite intervals. While a generalized approach to the formulation of the finite Fourier transform is possible, it is convenient to adopt a formulation based on Fourier series representations. The Fourier series representation for a function which is periodic in the interval $-\ell < x < \ell$ can be written as

$$F(x) = \frac{a_0}{2} + \sum_{n=1}^{\infty} \left\{ a_n \cos\left(\frac{n\pi x}{\ell}\right) + b_n \sin\left(\frac{n\pi x}{\ell}\right) \right\} \tag{1.197}$$

where

$$a_n = \frac{1}{\ell} \int_{-\ell}^{\ell} F(\xi) \cos\left(\frac{n\pi\xi}{\ell}\right) d\xi \tag{1.198}$$

$$b_n = \frac{1}{\ell} \int_{-\ell}^{\ell} F(\xi) \sin\left(\frac{n\pi\xi}{\ell}\right) d\xi \tag{1.199}$$

We note that if $F(x)$ is periodic in the interval $0 < x < \ell$ or *odd* in the interval $-\ell < x < \ell$, then the Fourier series representation of $F(x)$ takes the form

$$F(x) = \frac{2}{\ell} \sum_{n=1}^{\infty} \sin\left(\frac{n\pi x}{\ell}\right) \int_{0}^{\ell} F(\xi) \sin\left(\frac{n\pi\xi}{\ell}\right) d\xi \tag{1.200}$$

Similarly, if $F(x)$ is periodic in the interval $0 < x < \ell$ or *even* in the interval $-\ell < x < \ell$ then

$$F(x) = \frac{1}{\ell}\int_0^\ell F(\xi)d\xi + \frac{2}{\ell}\sum_{n=1}^{\infty}\cos\left(\frac{n\pi x}{\ell}\right)\int_0^\ell F(\xi)\cos\left(\frac{n\pi\xi}{\ell}\right)d\xi \quad (1.201)$$

We can now define the *finite Fourier transforms* of $F(x)$ as

$$\mathcal{F}_c(n) = \frac{1}{\ell}\int_0^\ell F(x)\cos\left(\frac{n\pi x}{\ell}\right)dx \quad (1.202)$$

$$\mathcal{F}_s(n) = \frac{1}{\ell}\int_0^\ell F(x)\sin\left(\frac{n\pi x}{\ell}\right)dx \quad (1.203)$$

where $\mathcal{F}_c(n)$ and $\mathcal{F}_s(n)$ are the finite Fourier cosine and sine transforms respectively.

(f) Finite Fourier transform of derivatives In the application of the finite Fourier transform to the solution of *partial differential equations* we invariably encounter integrals of the type

$$\int_0^\ell \frac{\partial^n F(x)}{\partial x^n}\sin\left(\frac{n\pi x}{\ell}\right)dx \quad , \quad \int_0^\ell \frac{\partial^n F(x)}{\partial x^n}\cos\left(\frac{n\pi x}{\ell}\right)dx$$

We therefore need to express integrals of this type in terms of the finite Fourier transform of $F(x)$. We proceed in exactly the same way as was done for deriving expressions for the derivatives of the exponential Fourier transform. Considering the integral involving the first derivative we have

$$\int_0^\ell \frac{\partial F(x)}{\partial x}\sin\left(\frac{n\pi x}{\ell}\right)dx = \left[F(x)\sin\left(\frac{n\pi x}{\ell}\right)\right]_0^\ell - \frac{n\pi}{\ell}\int_0^\ell F(x)\cos\left(\frac{n\pi x}{\ell}\right)dx \quad (1.204)$$

Since n is an integer, $\sin(n\pi x/\ell)$ vanishes at both $x = 0$ and $x = \ell$, such that

$$\int_0^\ell \frac{\partial F(x)}{\partial x}\sin\left(\frac{n\pi x}{\ell}\right)dx = -\frac{n\pi}{\ell}\mathcal{F}_C(n) \quad (1.205)$$

This result is true irrespective of the values of $F(0)$ and $F(a)$, provided they are *not infinite*. Similarly, integrating by parts, we have

$$\int_0^\ell \frac{\partial F(x)}{\partial x} \cos\left(\frac{n\pi x}{\ell}\right) dx = \left[F(x)\cos\left(\frac{n\pi x}{\ell}\right)\right]_0^\ell + \frac{n\pi}{\ell}\int_0^\ell F(x)\sin\left(\frac{n\pi x}{\ell}\right) dx \tag{1.206}$$

Since $\cos(n\pi) = (-1)^n$, where n is an integer

$$\int_0^\ell \frac{\partial F(x)}{\partial x} \cos\left(\frac{n\pi x}{\ell}\right) dx = (-1)^n F(\ell) - F(0) + \frac{n\pi}{\ell}\mathcal{F}_S(n) \tag{1.207}$$

If $F(x)$ *vanishes at both* $x = \ell$ *and* $x = 0$, then

$$\int_0^\ell \frac{\partial F(x)}{\partial x} \cos\left(\frac{n\pi x}{\ell}\right) dx = \frac{n\pi}{\ell}\mathcal{F}_S(n) \tag{1.208}$$

A repeated application of the above procedure will yield the following:

$$\int_0^\ell \frac{\partial^2 F(x)}{\partial x^2} \sin\left(\frac{n\pi x}{\ell}\right) dx = \left(\frac{n\pi}{\ell}\right)\left[(-1)^{n+1}F(\ell) + F(0)\right] - \frac{n^2\pi^2}{\ell^2}\mathcal{F}_S(n) \tag{1.209}$$

If $F(0) = F(\ell) = 0$, then (1.209) reduces to

$$\int_0^\ell \frac{\partial^2 F(x)}{\partial x^2} \sin\left(\frac{n\pi x}{\ell}\right) dx = -\frac{n^2\pi^2}{\ell^2}\mathcal{F}_S(n) \tag{1.210}$$

Similarly,

$$\begin{aligned}&\int_0^\ell \frac{\partial^2 F(x)}{\partial x^2}\cos\left(\frac{n\pi x}{\ell}\right)dx\\&= (-1)^n\left[\frac{\partial F}{\partial x}\right]_{x=\ell} - \left[\frac{\partial F}{\partial x}\right]_{x=0} - \frac{n^2\pi^2}{\ell^2}\mathcal{F}_C(n)\end{aligned} \tag{1.211}$$

Again, if $\partial F/\partial x$ vanishes at the end points of the interval $x = 0$ and $x = \ell$, then

$$\int_0^\ell \frac{\partial^2 F(x)}{\partial x^2}\cos\left(\frac{n\pi x}{\ell}\right)dx = -\frac{n^2\pi^2}{\ell^2}\mathcal{F}_C(n) \tag{1.212}$$

In a similar manner we can show that if $\partial^2 F/\partial x^2 = 0$ at the end points $x = 0$ and $x = \ell$ then

$$\begin{aligned}\int_0^\ell \frac{\partial^4 F(x)}{\partial x^4}\sin\left(\frac{n\pi x}{\ell}\right)dx &= -\frac{n^2\pi^2}{\ell^2}\int_0^\ell \frac{\partial^2 F(x)}{\partial x^2}\sin\left(\frac{n\pi x}{\ell}\right)dx\\&= \frac{n^4\pi^4}{\ell^4}\mathcal{F}_S(n)\end{aligned} \tag{1.213}$$

Similarly, if $\partial^3 F/\partial x^3 = \partial F/\partial x = 0$ when $x = 0$ and when $x = \ell$, we have

$$\int_0^\ell \frac{\partial^4 F(x)}{\partial x^4}\cos\left(\frac{n\pi x}{\ell}\right)dx = \frac{n^4\pi^4}{\ell^4}\mathcal{F}_C(n) \tag{1.214}$$

(g) The inverse of finite Fourier transforms

If $\mathcal{F}_c(n)$ is the finite Fourier cosine transform of $F(x)$ in $0 < x < \ell$, then

$$F(x) = \mathcal{F}_c(0) + 2\sum_{n=1}^{\infty}\mathcal{F}_c(n)\cos\left(\frac{n\pi x}{\ell}\right) \tag{1.215}$$

Similarly, if $\mathcal{F}_s(n)$ is the finite Fourier sine transform of $F(x)$ in $0 < x < \ell$, then

$$F(x) = 2\sum_{n=1}^{\infty} \mathcal{F}_s(n) \sin\left(\frac{n\pi x}{\ell}\right) \tag{1.216}$$

These representations are, in essence, the inverse transformations for (1.202) and (1.203) respectively.

(h) Multiple Fourier transforms The theory of Fourier transforms of functions of a single variable can be extended to functions of several variables. Consider for example the function $F(x,y)$ of the two independent variables x and y. If we consider the variable x, then $F(x,y)$ has the exponential Fourier transform

$$\overline{f}(\xi, y) = \frac{1}{\sqrt{2\pi}} \int_{-\infty}^{\infty} e^{i\xi x} F(x,y)dx \tag{1.217}$$

Now considering $\overline{f}(\xi, y)$ as a function of y, the exponential Fourier transform of $\overline{f}(\xi, y)$ is

$$f(\xi, \eta) = \frac{1}{\sqrt{2\pi}} \int_{-\infty}^{\infty} e^{i\eta y} \overline{f}(\xi, y)dy \tag{1.218}$$

Combining (1.217) and (1.218) we obtain

$$f(\xi, \eta) = \frac{1}{2\pi} \int_{-\infty}^{\infty} \int_{-\infty}^{\infty} F(x,y)e^{i(\xi x + \eta y)}dxdy \tag{1.219}$$

and we define $f(\xi, \eta)$ as the two-dimensional or double Fourier transform of $F(x,y)$.

Applying the inversion theorem to (1.217) we have

$$F(x,y) = \frac{1}{\sqrt{2\pi}} \int_{-\infty}^{\infty} \overline{f}(\xi, y) e^{-i\xi x} d\xi \tag{1.220}$$

Similarly, by applying the inversion theorem to (1.218) we obtain

$$\overline{f}(\xi, y) = \frac{1}{\sqrt{2\pi}} \int_{-\infty}^{\infty} f(\xi, \eta) e^{-i\eta y} d\eta \tag{1.221}$$

Combining (1.220) and (1.221) we obtain the inversion formula for the double Fourier transform of (1.219) as

$$F(x,y) = \frac{1}{2\pi} \int_{-\infty}^{\infty} \int_{-\infty}^{\infty} f(\xi, \eta) e^{-i(\xi x + \eta y)} d\xi d\eta \tag{1.222}$$

A further useful result concerns the application of the Faltung theorem to the product of two double Fourier transforms. Let $f(\xi, \eta)$ and $g(\xi, \eta)$ be two double Fourier transforms. The double Fourier transform of fg is

$$\begin{aligned}
&\frac{1}{2\pi} \int_{-\infty}^{\infty} \int_{-\infty}^{\infty} f(\xi, \eta) g(\xi, \eta) e^{-i(\xi x + \eta y)} d\xi d\eta \\
&= \frac{1}{2\pi} \int_{-\infty}^{\infty} \int_{-\infty}^{\infty} g(\xi, \eta) e^{-i(\xi x + \eta y)} d\xi d\eta \\
&\quad \left[\frac{1}{2\pi} \int_{-\infty}^{\infty} \int_{-\infty}^{\infty} F(u,v) e^{i(\xi u + \eta v)} du dv \right] \\
&= \frac{1}{2\pi} \int_{-\infty}^{\infty} \int_{-\infty}^{\infty} F(u,v) du dv \\
&\quad \left[\frac{1}{2\pi} \int_{-\infty}^{\infty} \int_{-\infty}^{\infty} g(\xi, \eta) e^{-i[\xi(x-u) + \eta(y-v)]} d\xi d\eta \right] \\
&= \frac{1}{2\pi} \int_{-\infty}^{\infty} \int_{-\infty}^{\infty} F(u,v) G(x-u, y-v) du dv
\end{aligned} \tag{1.223}$$

which gives the Faltung theorem

$$\int_{-\infty}^{\infty}\int_{-\infty}^{\infty} f(\xi,\eta)g(\xi,\eta)e^{-i(\zeta x+\eta y)}d\xi d\eta$$
$$= \int_{-\infty}^{\infty}\int_{-\infty}^{\infty} F(u,v)G((x-u),(y-v))dudv \tag{1.224}$$

(i) Finite double Fourier transform The concept of a finite double Fourier transform becomes useful when dealing with boundary value problems where the independent variables occupy finite intervals. Such an example is a two-dimensional domain where $0 \le x \le a$ and $0 \le y \le b$, and the values of the function are prescribed along the boundary of the region (i.e. Dirichlet conditions), and if the finite double Fourier transform is defined as

$$\mathcal{F}_S(m,n) = \int_0^a dx \int_0^b F(x,y)\sin\left(\frac{m\pi x}{a}\right)\sin\left(\frac{n\pi y}{b}\right)dy \tag{1.225}$$

then, at points within the rectangle where the function $F(x,y)$ is continuous

$$F(x,y) = \frac{4}{ab}\sum_{m=1}^{\infty}\sum_{n=1}^{\infty}\mathcal{F}_S(m,n)\sin\left(\frac{m\pi x}{a}\right)\sin\left(\frac{n\pi y}{b}\right) \tag{1.226}$$

Using integration by parts it can be shown that if $F(x,y)$ vanishes along the boundary of the rectangular region

$$\int_0^a\int_0^b \frac{\partial^2 F(x,y)}{\partial x^2}\sin\left(\frac{m\pi x}{a}\right)\sin\left(\frac{n\pi y}{b}\right)dxdy$$
$$= -\frac{\pi^2 m^2}{a^2}\mathcal{F}_S(m,n) \tag{1.227}$$

Similarly it can be shown that

$$\int_0^a \int_0^b \frac{\partial^2 F(x,y)}{\partial y^2} \sin\left(\frac{m\pi x}{a}\right) \sin\left(\frac{n\pi y}{b}\right) dx dy$$
$$= -\frac{\pi^2 n^2}{b^2} \mathcal{F}_S(m,n) \tag{1.228}$$

Combining (1.219) and (1.220) we obtain

$$\int_0^a \int_0^b \nabla^2 F \sin\left(\frac{m\pi x}{a}\right) \sin\left(\frac{n\pi y}{b}\right) dx dy \tag{1.229}$$
$$= -\pi^2 \left(\frac{m^2}{a^2} + \frac{n^2}{b^2}\right) \mathcal{F}_S(m,n)$$

and $\mathcal{F}_S(m,n)$ is defined by (1.225).

(j) The Fourier integral

If we consider (1.197), (1.198) and (1.199) we have, for the interval $-\ell < x < \ell$,

$$F(x) = \frac{1}{2\ell} \int_{-\ell}^{\ell} F(\xi) d\xi + \frac{1}{\ell} \sum_{n=1}^{\infty} \left\{ \int_{-\ell}^{\ell} F(\xi) \cos\left(\frac{n\pi\xi}{\ell}\right) \cos\left(\frac{n\pi x}{\ell}\right) d\xi \right.$$
$$\left. + \int_{-\ell}^{\ell} F(\xi) \sin\left(\frac{n\pi\xi}{\ell}\right) \sin\left(\frac{n\pi x}{\ell}\right) d\xi \right\} \tag{1.230}$$

Combining the sine and cosine integrals in (1.230) we can write

$$F(x) = \frac{1}{2\ell} \int_{-\ell}^{\ell} F(\xi) d\xi + \frac{1}{\ell} \sum_{n=1}^{\infty} \int_{-\ell}^{\ell} F(\xi) \cos\left[\frac{n\pi(\xi - x)}{\ell}\right] d\xi \tag{1.231}$$

If we denote $(n\pi/\ell)$ by α, and let $\ell \to \infty$ (keeping $\pi/\ell \to \delta\alpha$), the first term of (1.231) reduces to zero and the resulting equation takes the form

$$F(x) = \frac{1}{\pi} \sum_{n=1}^{\infty} \delta\alpha \int_{\pi/\delta\alpha}^{\pi/\delta\alpha} F(\xi) \cos\{n(\delta\alpha)(\xi - x)\} d\xi \tag{1.232}$$

or

$$F(x) = \frac{1}{\pi} \int_{-\pi/\delta\alpha}^{\pi/\delta\alpha} F(\xi) d\xi \sum_{n=1}^{\infty} \cos\{n(\delta\alpha)(\xi - x)\}(\delta\alpha) \tag{1.233}$$

We note that the limiting case of a series is an integral such that

$$\sum_{i=1}^{\infty} \Omega(\rho_i)\delta\rho_i \to \int_0^{\infty} \Omega(\rho) d\rho \tag{1.234}$$

A comparison of (1.234) and (1.233) indicates that $n(\delta\alpha)$ can be identified with ρ_i and $(\delta\alpha)$ can be identified with $\delta\rho_i$. Hence we can write (1.233) as

$$F(x) = \frac{1}{\pi} \int_{-\infty}^{\infty} F(\xi) d\xi \int_0^{\infty} \cos\{(\xi - x)\alpha\} d\alpha \tag{1.235}$$

The *Fourier integral representation* of the function $F(x)$ takes the form

$$F(x) = \frac{1}{\pi} \int_0^{\infty} d\alpha \int_{-\infty}^{\infty} F(\xi) \cos\{(\xi - x)\alpha\} d\xi \tag{1.236}$$

If the function $F(x)$ is defined in the range $0 < x < \infty$, the result (1.236) reduces to

$$F(x) = \frac{2}{\pi} \int_0^{\infty} \cos(\alpha x) d\alpha \int_0^{\infty} F(\xi) \cos(\alpha\xi) d\xi \tag{1.237}$$

1.12.3 Hankel transforms

(a) Definition

As indicated previously, the choice of a particular integral transform for the solution of a specific partial differential equation depends on the coordinate system within which the problem is formulated and the symmetries of the problem involved. When the problems examined are either two-dimensional with no angular dependence or three-dimensional problems which display symmetry about a particular axis, then it is convenient to employ the integral transforms proposed by Hermann Hankel (1839-1874). The Hankel transform of the function $F(r)$, of order ν, is defined as

$$\overline{F}(\xi) = \mathcal{H}_\nu\{F(r);\xi\} = \int_0^\infty rF(r)J_\nu(\xi r)d\xi \tag{1.238}$$

where $J_\nu(\xi r)$ is the Bessel function of the first kind of order ν (named after F.W. Bessel (1784-1846).

(b) Inverse transform

The Hankel inversion theorem is not only useful from the point of view of solution of partial differential equations but it also can be used to evaluate certain definite integrals. The Hankel inversion theorem can be written as

$$F(r) = \mathcal{H}_\nu^{-1}\{\overline{F}(\xi);r\} = \int_0^\infty \xi\overline{F}(\xi)J_\nu(\xi r)d\xi \tag{1.239}$$

i.e. the *Hankel transform is self-reciprocal.*

(c) The Hankel integral

It can be shown that, provided $\nu > -\frac{1}{2}$, an arbitrary function $F(r)$ can be expressed as

$$F(r) = \int_0^\infty \xi J_\nu(\xi r)d\xi \int_0^\infty \eta F(\eta)J_\nu(\eta\xi)d\eta \tag{1.240}$$

In order for (1.232) to hold, it is sufficient that

$$\int_0^\infty | F(r) | \, dr$$

converges and $F(r)$ be of bounded variation in the neighbourhood of $r = 0$. If $F(r)$ is discontinuous at $r = 0$, then the function is interpreted as

$$F(r) = \frac{1}{2}[F(r+0) + F(r-0)] \tag{1.241}$$

(d) Transform of derivatives

The Hankel transform of derivatives can be expressed as follows:

$$\mathcal{H}_\nu \left\{ r^{(\nu-1)} \frac{\partial}{\partial r}[r^{(1-\nu)} F(r)]; \xi \right\} = -\xi \mathcal{H}_{\nu-1} \{F(r); \xi\} \tag{1.242}$$

$$\mathcal{H}_\nu \left\{ r^{-(\nu+1)} \frac{\partial}{\partial r}[r^{(1+\nu)} F(r)]; \xi \right\} = \xi \mathcal{H}_{\nu+1} \{F(r); \xi\} \tag{1.243}$$

$$\mathcal{H}_\nu \left\{ \tilde{\nabla}_\nu^2 F(r); \xi \right\} = -\xi^2 \mathcal{H}_\nu \{F(r); \xi\} \tag{1.244}$$

where

$$\tilde{\nabla}_\nu^2 = \left(\frac{\partial^2}{\partial r^2} + \frac{1}{r}\frac{\partial}{\partial r} - \frac{\nu^2}{r^2} \right) \tag{1.245}$$

(e) Finite Hankel transforms

A function $g(x)$ is said to satisfy the conditions proposed by P.G.J.Dirichlet (1805-1859), in the interval (a, b), if (i) $g(x)$ has only a finite number of maxima and minima in (a, b), and (ii) $g(x)$ has only a finite number of finite discontinuities in (a, b) and no infinite discontinuities. Hence any function which is continuous in (a, b) and has only a finite number of maxima or minima over the range (a, b) will satisfy Dirichlet's conditions. For example the function $x/(a^2 + x^2)$ where a is a constant satisfies Dirichlet's conditions for $x \in (-\infty, \infty)$. The function $1/(a^2 - x^2)$ does not satisfy Dirichlet's conditions on any interval which includes $x = a$ since the function is singular at $x = a$.

If a function $g(x)$ satisfies Dirichlet's conditions then the series

$$\frac{2}{a^2} \sum A_i \frac{J_\mu(x\xi_i)}{\left[J'_\mu(a\xi_i)\right]^2}$$

in which A_i denotes the definite integral

$$A_i = \int_0^a x g(x) J_\mu(x\xi_i) dx \tag{1.246}$$

and the summation is over the positive zeros of the function $J_\mu(a\xi_i)$, converges to the sum $[g(x+0) + g(x-0)]/2$. The result can be written as an inversion theorem for the finite Hankel transform

$$\widetilde{g}_\mu(\xi_i) = \mathcal{H}^f_\mu[g(x); \xi_i] = \int_0^a x g(x) J_\mu(x\xi_i) dx \tag{1.247}$$

Formal theorems associated with the finite Hankel transforms can be stated as follows:

THEOREM I:

If $g(x)$ is a function which satisfies Dirichlet's conditions in the interval $(0, a)$ and if its finite Hankel transform in the range is defined to be

$$\widetilde{g}^f_\mu(\xi_i) = \mathcal{H}^f_\mu \{g(x); \xi_i\} = \int_0^a x g(x) J_\mu(x\xi_i) dx \tag{1.248}$$

where ξ_i is a root of the transcendental equation

$$J_\mu(a\xi_i) = 0 \tag{1.249}$$

then, at any point of the interval $(0, a)$ at which the function $g(x)$ is continuous

$$g(x) = \frac{2}{a^2} \sum_{i=1,2,} \widetilde{g}_f(\xi_i) \frac{J_\mu(x\xi_i)}{[J'_\mu(a\xi_i)]^2} \tag{1.250}$$

where the summation is taken over all the positive roots of (1.249).

In many boundary value problems which lend themselves to the application of finite Fourier transforms the specific case $\mu = 0$ features prominently. It can be shown that

$$J'_0(x) = -J_1(x) \tag{1.251}$$

such that when $\mu = 0$, the result (1.250) reduces to

$$g(x) = \frac{2}{a^2} \sum_{i=1,2,} \widetilde{g}_f(\xi_i) \frac{J_0(x\xi_i)}{[J_1(a\xi_i)]^2} \tag{1.252}$$

THEOREM II:

If $g(x)$ satisfies Dirichlet's conditions in the closed interval of $(0, a)$ and if its finite Hankel transform is defined as

$$\widetilde{g}^f_\mu(\xi_i) = \mathcal{H}^f_\mu \{g(x); \xi_i\} = \int_0^a x g(x) J_\mu(x\xi_i) dx \tag{1.253}$$

where ξ_i is a root of the transcendental equation

$$\xi_i J'_\mu(\xi_i a) + \lambda J_\mu(\xi_i a) = 0 \tag{1.254}$$

then at each point of the interval $(0, a)$ in which $g(x)$ is continuous

$$g(x) = \frac{2}{a^2} \sum_{i=1,2,} \frac{\xi_i^2 \widetilde{g}^f_\mu(\xi_i)}{\left[\lambda^2 + \left(\xi_i^2 - \frac{\mu^2}{a^2}\right)\right]} \frac{J_\mu(x\xi_i)}{\left[J_\mu(a\xi_i)\right]^2} \tag{1.255}$$

where the sum is taken over the positive roots of (1.254) with $\lambda = const.$.

Again, in the special case when $\mu = 0$.

$$g(x) = \frac{2}{a^2} \sum_{i=1,2,} \frac{\xi_i^2 \widetilde{g}^f_\mu(\xi_i)}{\left(\lambda^2 + \xi_i^2\right)} \frac{J_0(x\xi_i)}{\left[J_0(a\xi_i)\right]^2} \tag{1.256}$$

where the sum is taken over all the positive roots of the equation

$$\lambda J_0(\xi_i a) = \xi_i J_1(\xi_i a) \tag{1.257}$$

Other generalization of these results will be considered as an when they are required for the solution of specific boundary value or initial boundary value problems which could be examined by the application of finite Hankel transforms.

(f) Orthogonality Properties of Bessel Functions

If λ and μ are different constants if can be shown that

$$\int_0^1 rJ_n(\lambda r)J_n(\mu r)dr = \frac{\mu J_n(\lambda)J_n'(\mu) - \lambda J_n(\mu)J_n'(\lambda)}{(\lambda^2 - \mu^2)} \tag{1.258}$$

and

$$\int_0^1 rJ_n^2(\lambda r)dr = \frac{1}{2}\left[(J_n'(\lambda))^2 + \left(1 - \frac{n^2}{\lambda^2}\right)J_n^2(\lambda)\right] \tag{1.259}$$

In general if λ and μ can be derived as roots of a characteristic equation

$$\phi_1 J_n(r) + \phi_2 rJ_n'(r) = 0 \tag{1.260}$$

where ϕ_1 and ϕ_2 are constants, then

$$\int_0^1 rJ_n(\lambda r)J_n(\mu r)dr = 0 \tag{1.261}$$

which implies that $\sqrt{r}J_n(\lambda r)$ and $\sqrt{r}J_n(\mu r)$ are *orthogonal* in the interval $(0, 1)$.

Expansions in series involving Bessel functions

If a function $F(r)$ satisfies the Dirichlet conditions then at every point of continuity of $F(r)$ in the interval $0 < r < 1$, there exists a Bessel series expansion of the form

$$F(r) = \sum_{m=1,}^{\infty} A_m J_n(\lambda_m r) \tag{1.262}$$

where $\lambda_1, \lambda_2, \lambda_3, \ldots$ are the positive roots of the transcendental equation with $\phi_1/\phi_2 \geq 0$; $\phi_2 \neq 0$. Also

$$A_m = \frac{2\lambda_m^2}{[\lambda_m^2 - n^2 + (\phi_1/\phi_2)^2]J_n^2(\lambda_m)} \int_0^1 rJ_n(\lambda_m r)F(r)dr \tag{1.263}$$

At any discontinuity, the series in (1.262) converges to $(1/2)[F(r+0)+F(r-0)]$ which can be used to represent the function $F(r)$ in (1.262).

When $\phi_2 = 0$ such that $\lambda_1, \lambda_2, \lambda_3 \ldots$ are the roots of

$$J_n(r) = 0 \tag{1.264}$$

we have

$$A_m = \frac{2}{[J_{n+1}(\lambda_m)]^2} \int_0^1 rJ_n(\lambda_m r)F(r)dr \tag{1.265}$$

Alternatively, if $\phi_1 = 0$ and $n = 0$ the series (1.262) has a leading constant term

$$A_1 = 2\int_0^1 rF(r)dr \tag{1.266}$$

1.13 PROBLEM SET 1

1.1 Obtain from first principles an expression for ∇ and ∇^2 for the system of orthogonal coordinates (u, v, w) which are related to the rectangular Cartesian coordinates as follows:

$$x = \frac{1}{2}\left(u^2 - v^2\right) \quad ; \quad y = uv \quad ; \quad z = w$$

1.2 Obtain the flux of the vector $\mathbf{F} = \mathbf{r}/r^3$, through any sphere $r = a$.

1.3 Consider the function $f = \ln\left(x^2 + y^2\right)$ and $\mathbf{F} = z\mathbf{k}$. Evaluate the following

a) $\mathrm{grad} f$,

b) $\mathbf{F} \cdot \mathrm{grad}\ f$,

c) $\mathbf{F} \times \mathrm{grad}\ f$,

d) $\mathrm{div}\ \mathbf{F}$

e) $\nabla^2 f$,

f) $\mathrm{curl}\left(\mathbf{F} \times \mathrm{grad}\ f\right)$

1.4 For a particular two-dimensional flow pattern in a fluid, the velocity vector $\mathbf{V}$, with components V_x, V_y and V_z, is expressed in terms of a scalar potential φ by the relationship $\mathbf{V} = \mathrm{grad}\ \varphi$. If

$$\varphi = A \ln\left(x^2 + y^2\right) + B \tan^{-1}\left(y/x\right)$$

show that the flow is incompressible.

1.5 If $\mathbf{v} = (\mathbf{i}v_x + \mathbf{j}v_y + \mathbf{k}v_z)$, then show that

$$\mathrm{curl\ curl}\ \mathbf{v} = \mathrm{grad\ div}\ \mathbf{v} - \nabla^2 \mathbf{v}$$

1.6 If $\mathbf{u}$ and $\mathbf{v}$ are vectors referred to a system of rectangular Cartesian coordinates, then show that

a) $\mathrm{div}\left(\mathbf{u} \times \mathbf{v}\right) = \left(\mathrm{curl}\ \mathbf{u}\right) \cdot \mathbf{v} - \left(\mathrm{curl}\ \mathbf{v}\right) \cdot \mathbf{u}$

b) $\mathrm{curl}\ \left(\mathbf{u} \times \mathbf{v}\right) = \left(\mathbf{v} \cdot \mathrm{grad}\right) \mathbf{u} - \left(\mathbf{u} \cdot \mathrm{grad}\right) \mathbf{v} + \mathbf{u}\ \mathrm{div}\ \mathbf{v} - \mathbf{v}\ \mathrm{div}\ \mathbf{u}$

1.7 Show that

a) for any scalar function φ
$\mathrm{curl\ grad}\ \varphi = 0$

b) for any vector $\mathbf{v}$
$\mathrm{div\ curl}\ \mathbf{v} = 0,$ and

c) div $(\varphi \mathbf{v}) = \varphi$ div $\mathbf{v} + \mathbf{v} \cdot$ grad φ

1.8 The vector $\mathbf{v}$ has components x, y, z and magnitude $v^2 = \left(x^2 + y^2 + z^2\right)$. Determine

a) grad $\mathbf{v}$

b) grad (v^{-n})

c) div $(v^n \mathbf{v})$

d) curl $(v^{-n} \mathbf{v})$

e) $\nabla^2 (1/v)$

where n is a positive integer.

1.9 Obtain from first principles, the Laplace transforms of the following functions:

(i) $F(t) = 3t^4 - 2t^{3/2} + 6$

(ii) $F(t) = \dfrac{1}{\sqrt{t}}$

(iii) $F(t) = H(t - a)$

(iv) $F(t) = (1 + t)^2$

(v) $F(t) = \begin{cases} 0 & \text{for} \quad 0 \le t < 1 \\ 1 & \text{for} \quad 1 \le t < 2 \\ 0 & \text{for} \quad t > 2 \end{cases}$

(vi) $F(t) = \dfrac{c^{\alpha t}}{\sqrt{t}}$

(vii) $F(t) = t^2 \sin 2t$

(viii) $F(t) = \begin{cases} \sin t & \text{for} \quad 0 \le \pi \\ 0 & \text{for} \quad \pi \le t < 2\pi \end{cases}$

(ix) $F(t) = \begin{cases} \sin t & \text{for} \quad t < \pi \\ t & \text{for} \quad t > \pi \end{cases}$

(x) $F(t) = \dfrac{1 - e^{-t}}{t}$

1.10 Obtain the inverse Laplace transforms of the following functions:

(i) $f(s) = \dfrac{3s + 4}{(s^2 - 4s + 13)}$

(ii) $f(s) = \dfrac{1}{s(1 + s)^{\frac{1}{2}}}$

(iii) $f(s) = \dfrac{se^{-s}}{(s^2 + 9)}$

(iv) $f(s) = \dfrac{e^{-3s}}{\sqrt{s - 3}}$

(v) $f(s) = \dfrac{1}{(s^2 + \omega^2)}$

(vi) $f(s) = \dfrac{1}{s(s^2 + 16)}$

1.11 Using Laplace transforms obtain the solution of the integral equation

$$F(t) = t^2 + \int_0^t F(\xi) \sin(t - \xi) d\xi$$

1.12 Use the method of residues to evaluate the inverse Laplace transforms of the following functions:

(i) $f(s) = \dfrac{1}{(s - 1)}$

(ii) $f(s) = \dfrac{1}{(s - 1)^2}$

(iii) $f(s) = \dfrac{1}{(s^2 - 1)}$

(iv) $f(s) = \dfrac{1}{(s^2 + 4)^2(s - 2)}$

(v) $f(s) = \dfrac{e^{\lambda s}}{(s-3)^2}$

(vi) $f(s) = \dfrac{s}{(s^2+1)^3}$

1.13 Use the complex inversion formula to show that the inverse Laplace transform of

$$f(s) = \frac{e^{-\lambda\sqrt{s}}}{s}$$

is given by

$$F(t) = 1 - \frac{1}{\pi}\int_0^\infty \frac{e^{-\xi t}}{\xi}\sin(\lambda\sqrt{\xi})d\xi = \text{erfc}\left(\frac{\lambda}{2\sqrt{t}}\right)$$

1.14 Evaluate the inverse Laplace transform of the following functions, using the complex inversion formula

(i) $f(s) = \dfrac{s}{(s^2+\omega^2)}$

(ii) $f(s) = \dfrac{1}{s\cosh(s)}$

1.15 Obtain the Fourier transforms of the following functions:

(i) $F(x) = \begin{cases} 1 & \text{for} \quad |x| < L \\ 0 & \text{for} \quad |x| > L \end{cases}$

(ii) $F(x) = \begin{cases} (L-x) & \text{for} \quad |x| < L \\ 0 & \text{for} \quad |x| > L \end{cases}$

(iii) $F(x) = \begin{cases} e^{-x} & \text{for} \quad x \geq 0 \\ 0 & \text{for} \quad x < 0 \end{cases}$

(iv) $F(x) = \begin{cases} e^{-x} & \text{for} \quad x \geq 0 \\ -e^{-x} & \text{for} \quad x < 0 \end{cases}$

(v) $F(x) = \dfrac{1}{(x^2+b^2)}$ where b is a constant

(vi) $F(x) = e^{-\lambda|x-a|}$

(vii) $F(x) = \sin(\omega x^2)$

(viii) $F(x) = \dfrac{\sin \lambda x}{x}$

1.16 Evaluate the integrals:

(i) $F(x) = \displaystyle\int_0^\infty \frac{\cos(\xi x) d\xi}{(\xi^2 + a^2)}$

(ii) $F(x) = \displaystyle\int_0^\infty \frac{\sin(\xi x) d\xi}{(\xi^2 + a^2)}$

1.17 Determine a generalized expression for $F(x)$ by solving the integral equation

$$\int_{-\infty}^{\infty} \frac{F(\xi) d\xi}{[(x-\xi)^2 + \lambda^2]} = \frac{x}{(x^2 + \mu^2)} \quad ; \quad 0 < \lambda < \mu$$

1.18 A function is defined as follows, over the interval $x \in (-2, 2)$;

$$f(x) = \begin{cases} -x \; ; & -2 \le x \le 0 \\ x \; ; & 0 \le x \le 2 \end{cases}$$

Obtain the coefficients of the Fourier series for $f(x)$.

1.19 Obtain the Fourier series for

$$(i) \;\; f(x) = x \;\; ; \;\; (ii) \;\; f(x) = x^2$$

on the interval $(-\pi, \pi)$.

1.20 A convergence theorem for Fourier series states that if $f(x)$ is piecewise smooth on $x \in (-L, L)$ (i.e. a finite number of segments of $f(x)$ and $f\,'(x)$ over subsets of $x \in (-L, L)$), then the Fourier series for $f(x)$ converges at each point in $x \in (-L, L)$ to

$\frac{1}{2}\left[f(x^+) + f(x^-)\right]$

which implies that, at each x between $-L$ and L, the Fourier series converges to the average of the *left* and *right* limits of x.

Develop the Fourier series for

$$f(x) = \begin{cases} 0 & ; \ -2 \leq x \leq 0 \\ (2-x) & ; \ 0 \leq x \leq 2 \end{cases}$$

1.21 A series of digitally filtered signals have the following forms

$$(i)\ \ f(x) = \begin{cases} 0 & ; \ -\pi \leq \theta \leq 0 \\ \sin\theta & ; \ 0 \leq \theta \leq \pi \end{cases}$$

$$(ii)\ \ f(x) = \begin{cases} 1 & ; \ -2 < x < 0 \\ 0 & ; \ 0 < x < 2 \end{cases}$$

$$(iii)\ \ f(x) = \begin{cases} 1 & ; \ 0 \leq x \leq \pi/2 \\ 2 & ; \ \pi/2 < x < \pi \end{cases}$$

Obtain the coefficients of the Fourier series for $f(x)$.

1.22 A stretched string with uniform tension T is held fixed at $x = 0$ and $x = L$. A weight W is attached to the location $x = a$, which causes the string to deflect. Assuming that the deflections are small in comparison to a and $(L - a)$, derive a Fourier series expression for the deflection of the string.

1.23 The boundary of a stretched membrane occupies the region $x \in (-\ell, \ell)$. This boundary is subjected to a displacement of the following form

$$f(x) = \begin{cases} 0 & ; \ -\ell \leq x \leq -\ell/2 \\ \Delta_0 \left[\left(\dfrac{2x}{\ell} \right)^2 - 1 \right] & ; \ -\ell/2 \leq x \leq \ell/2 \\ 0 & ; \ \ell/2 \leq x \leq \ell \end{cases}$$

where Δ_0 is the maximum deflection. Derive a Fourier series expression for the boundary deflection.

1.24 A convergence theorem for Fourier integrals states if $f(x)$ is piecewise continuous on every interval $(-a, a)$, and in addition if

$\int_{-\infty}^{\infty} f(x)dx$

is convergent, then the Fourier integral of f converges to

$\frac{1}{2}\left[f(x) + f(-x)\right]$

at each x where f has a *left* and *right* derivative. Also if f is continuous with *left* and *right* derivaties of x, then the Fourier integral converges to $f(x)$.

Determine the Fourier integral coefficients of

$$f(x) = \begin{cases} 0 \ ; & x < -a \\ 1 \ ; & -a \leq x \leq a \\ 0 \ ; & x > a \end{cases}$$

and obtain the values to which the integral converges over the ranges $-a \leq x < a$; $x = -a$; $x = a$; $x < -a$; $x > a$.

Chapter 2

General concepts in partial differential equations

Differential calculus was developed independently by Newton (1642-1727) and Leibniz (1646-1716). Both mathematicians employed equations in a differential form in their work but the term "*differential equations*" is generally attributed to Leibniz. In the 18th century, the differential equation was accepted as a powerful tool in the formulation of problems in physics and mechanics. A number of eminent mathematicians applied differential equations to formulate and solve problems in a variety of areas including mechanics, hydrodynamics and astronomy. These scientists and engineers include Jakob Bernoulli (1654-1705), Johann Bernoulli (1667-1748), Daniel Bernoulli (1700-1782), Leonhard Euler (1707-1783), Jean d'Alembert (1717-1783), Joseph Lagrange (1736-1813) and Pierre Simon de Laplace (1749-1827). Although the applications dominated the development of differential equations, questions pertaining to existence of a solution was assured only by physical intuition. Augustin Cauchy (1789-1857) obtained the first existence theorems. These were followed by contributions from J.F. Pfaff (1765-1825), Joseph Fourier (1768-1830), Siméon Poisson (1781-1840) and Carl Jacobi (1804-1851). Modern advances in the theory and applications of differential equations continue with the milestone contributions of Sophus Lie (1842-1899). After three centuries, the applications of partial differential equations is so extensive that only a few can master the mathematical developments and physical applications in their entirety.

2.1 Fundamental concepts

A partial differential equation establishes a relationship between an unknown function of several variables and its partial derivatives. This is in contrast to an ordinary differential equation where the relationship exists between an unknown function of a single variable and its total derivatives. Consider a partial differential equation in which the independent variables are denoted by $x, y, z, \ldots,$ and the dependent variables are defined by $u, v, w, \ldots,$ etc. Direct functionality is often written as

$$u = u\,(x, y, z) \tag{2.1}$$

which, in this particular case, designates u as a function of the independent variables x, y and z. Partial derivatives are often denoted by the simplified expressions such as

$$u_x = \frac{\partial u}{\partial x} \quad ; \quad u_y = \frac{\partial u}{\partial y} \quad ; \quad u_{xx} = \frac{\partial^2 u}{\partial x^2} \quad ; \quad u_{xy} = \frac{\partial^2 u}{\partial x \partial y}$$

Using the above definitions and (2.1) we can represent a partial differential equation in the general form

$$F\,(x,\ y,\ u,\ u_x,\ u_y,\ u_{xx},\ u_{yy},\ u_{xy}, \ldots) = 0 \tag{2.2}$$

where F is a function of the indicated quantities and at least one partial derivative exists. For example, the following are partial differential equations:

$$\frac{\partial^2 u}{\partial x^2} + \frac{\partial^2 u}{\partial y^2} = 0$$

$$\frac{\partial u}{\partial x} = u + x^2 + y^2$$

$$\frac{\partial^3 u}{\partial x^3} = \frac{\partial^2 u}{\partial y^2} + u^2 \tag{2.3}$$

$$\left(\frac{\partial u}{\partial x}\right)^2 + \left(\frac{\partial u}{\partial y}\right)^2 = e^u$$

2.1.1 The order of a partial differential equation

The order is defined by the highest derivative in the equations.

e.g.

$$\begin{aligned}
\frac{\partial u}{\partial x} - b\frac{\partial u}{\partial y} &= 0 \quad ; \quad \text{First} - \text{order} \\
\frac{\partial^2 u}{\partial x^2} + \frac{\partial u}{\partial y} &= 0 \quad ; \quad \text{Second} - \text{order} \\
\frac{\partial^4 u}{\partial x^4} + 2\frac{\partial^4 u}{\partial x^3 dy} + \frac{\partial^4 u}{\partial y^4} &= 0 \quad ; \quad \text{Fourth} - \text{order} \\
\frac{\partial^2 u}{\partial x^2} + \lambda\left(\frac{\partial u}{\partial x}\right)^3 &= 0 \quad ; \quad \text{Second} - \text{order}
\end{aligned} \tag{2.4}$$

When several inter-dependent partial differential equations are encountered, "the order" is established by combining all the equations into a single form.

e.g. Find the order of the coupled partial differential equations

$$\frac{\partial u}{\partial x} + \frac{\partial v}{\partial y} = \frac{\partial u}{\partial z} \quad ; \quad u = \frac{\partial w}{\partial x} \quad ; \quad v = \frac{\partial w}{\partial y} \tag{2.5}$$

These equations can be reduced to the form

$$\frac{\partial^2 w}{\partial x^2} + \frac{\partial^2 w}{\partial y^2} - \frac{\partial^2 w}{\partial x \partial z} \tag{2.6}$$

Hence the system of equations (2.5) is of the second-order.

2.1.2 The linearity of a partial differential equation

A partial differential equation is said to be linear if it is of the first degree in the field variable and its partial derivatives. For example,

$$\frac{\partial F}{\partial x} = \frac{\partial F}{\partial y} \qquad \text{is linear} \tag{2.7}$$

$$\frac{\partial F}{\partial x} = \left(\frac{\partial F}{\partial y}\right)^2 \qquad \text{is non} - \text{linear} \tag{2.8}$$

A further definition of the linearity of a partial differential equation can be provided by considering an operator equation of the form

$$L\left(u\right) = 0 \tag{2.9}$$

where the operator L is defined as

$$L = \frac{\partial}{\partial x} + y\frac{\partial}{\partial y} \tag{2.10}$$

Let us assume that the function u admits representations

$$u = \alpha\left(x, y\right) + \beta\left(x, y\right) \quad ; \quad u = c\gamma\left(x, y\right) \tag{2.11}$$

where α, β, γ are arbitrary functions and c is a constant. It is clear that

$$L(u) = L(\alpha + \beta) = L(\alpha) + L(\beta) \tag{2.12}$$

and

$$L(u) = cL(\gamma) \ \cdot \tag{2.13}$$

Now consider the non-linear partial differential equation

$$L^*(u) = 0 \tag{2.14}$$

where the non-linear operator takes the form

$$L^* = \frac{\partial}{\partial t} - \frac{\partial^2}{\partial x^2} + u^2 \tag{2.15}$$

From (2.11) and (2.15) it is evident that

$$L^*(u) = L^*(\alpha) + L^*(\beta) + 2\alpha\beta \tag{2.16}$$

and

$$L^*(c\gamma) = cL^*(\gamma) - c\gamma^2(1-c) \tag{2.17}$$

In general, for linear partial differential equations if $u(x, y, z, t)$, $v(x, y, z, t)$,

etc. are solutions of a linear operator equation of the type (2.9), then a linear combination of these solutions is also a solution of the same operator equation.
In the solution of partial differential equations, the property of linearity plays an important part. Consider for example the first-order equation

$$A\left(x,y,\frac{\partial u}{\partial x},\frac{\partial u}{\partial y}\right)\frac{\partial u}{\partial x}+B\left(x,y,\frac{\partial u}{\partial x},\frac{\partial u}{\partial y}\right)\frac{\partial u}{\partial y}=C\left(x,y,\frac{\partial u}{\partial x},\frac{\partial u}{\partial y}\right) \tag{2.18}$$

a) In the case of (2.18), if the coefficients A and B are constants or functions of the independent variables only, [i.e. $A = A(x,y)$, $B = B(x,y)$], then the partial differential equation is *linear*.
b) If the coefficients A and B are also functions of the dependent variables [i.e. $A = A(x,y,u)$, $B = B(x,y,u)$], then the partial differential equation is *quasi-linear*.
c) If the coefficients A and B are functions of the first derivatives [i.e. $A = A(x,y,u,\ \partial u/\partial x, \partial u/\partial y)$, $B = B(x,y,u,\partial u/\partial x,\partial u/\partial y)$], then the partial differential equation is *non-linear*.

In general; when the *coefficients* of an n th-order partial differential equation depends on (i) the n th-order derivatives, the partial differential equation is *non-linear*, (ii) the m th-order derivatives ($n > m$), the partial differential equation is *quasi-linear*. This type of classification or identification of the category of the partial differential equation is important since many analytical properties of linear and quasi-linear partial differential equations are known and, as a general rule, each non-linear partial differential equation must be considered individually.

2.1.3 Homogeneity of a partial differential equation

A linear partial differential equation is considered to be homogeneous if every term in the equation contains either the field variable or one of its derivatives. If we define a linear operator of the form

$$L\left(\frac{\partial}{\partial x},\frac{\partial}{\partial y},\frac{\partial^2}{\partial x^2},\frac{\partial^2}{\partial y^2},\ldots,\frac{\partial}{\partial t},\frac{\partial^2}{\partial t^2}\cdots\right)$$

Then a linear homogeneous partial differential equation can be written in the form

$$L(u) = 0 \tag{2.19}$$

and a linear non-homogeneous equation can be written in the form

$$L(u) = f(x, y, z, t) \tag{2.20}$$

where $f(x, y, z, t)$ is a non-zero function.

2.2 Well-posed problems

In general, there can be infinitely many solutions to partial differential equations which characterize physical systems. In order to select a single function that represents the solution to a physical problem, it is necessary to impose certain auxilliary conditions which are specific to the physical system that is being modelled. These auxiliary conditions can take two forms, namely *boundary conditions* and *initial conditions*.

2.2.1 Boundary conditions

These are conditions that must be satisfied at the boundary S of the region V. Consider the operator form of the PDE (2.9). The boundary conditions which must be imposed on u must be governed by the physical characteristics of the specific problem under consideration. Some general attributes of typical boundary conditions are characterized by the following three types of boundary conditions.

(a) *Dirichlet Boundary Condition* : $u = g$ on $\mathbf{x} \in S$
(named after P.G.J. Dirichlet (1805 – 1859)) (2.21)

(b) *Neumann Boundary Condition* : $\dfrac{\partial u}{\partial n} = h \quad \text{on} \quad \mathbf{x} \in S$
(named after Carl Gottfried Neumann (1832 – 1925)) (2.22)

(c) *Mixed (or Robin) Condition* : $\lambda u + \mu \dfrac{\partial u}{\partial n} = k \quad \text{on} \quad \mathbf{x} \in S$
(named after V.G. Robin (1855 – 1897)) (2.23)

where g, h, k, λ and μ are functions prescribed on S. These boundary conditions can also be prescribed on surfaces $S_1, S_2, \ldots$, etc. which form subsets of S.

Example 2.1

The two-dimensional flow of fluid through a porous granular medium is governed by Laplace's equation for an unknown potential φ, i.e.

$$\nabla^2 \varphi = 0 \quad \text{in} \quad \mathbf{x} \in V \tag{2.24}$$

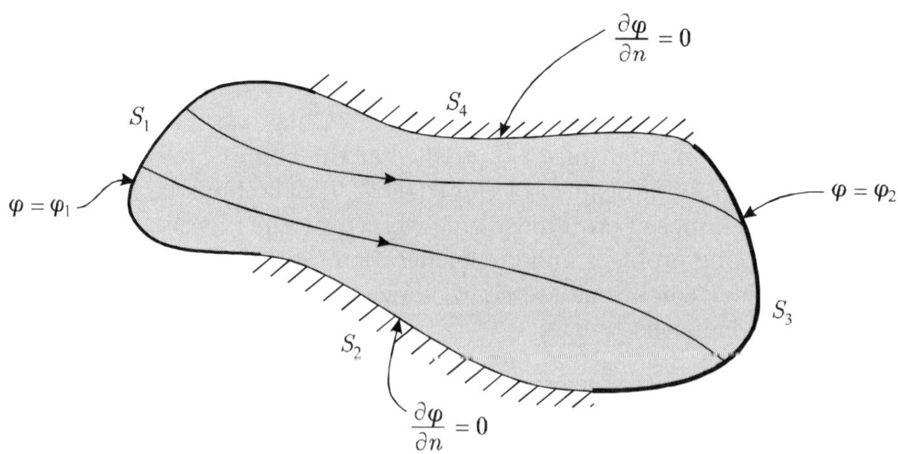

Figure 2.1: Mixed boundary conditions for the potential problem

The boundary conditions specified on S (Fig.2.1) can be as follows:

$$
\begin{aligned}
\varphi &= \varphi_1 \quad \text{on} \quad \mathbf{x} \in S_1 \\
\frac{\partial \varphi}{\partial n} &= 0 \quad \text{on} \quad \mathbf{x} \in S_2 \\
\varphi &= \varphi_2 \quad \text{on} \quad \mathbf{x} \in S_3 \\
\frac{\partial \varphi}{\partial n} &= 0 \quad \text{on} \quad \mathbf{x} \in S_4
\end{aligned}
\tag{2.25}
$$

where $S = S_1 \cup S_2 \cup S_3 \cup S_4$ and $S_1 \cap S_2 \cap S_3 \cap S_4 = 0$.

Example 2.2

The flexural or bending deflections $w(x, y)$ of an elastic plate which is subjected to a transverse load $p(x, y)$ is given by

$$
D\nabla^2\nabla^2 w(x, y) = p(x, y) \qquad \text{for} \quad \mathbf{x} \in \mathbb{D} \tag{2.26}
$$

where $\mathbb{D}$ is the plate region and D is the "flexural rigidity" of the plate. The plate is *fixed* on a part of the boundary and *supported* without transverse movement (Fig.2.2).

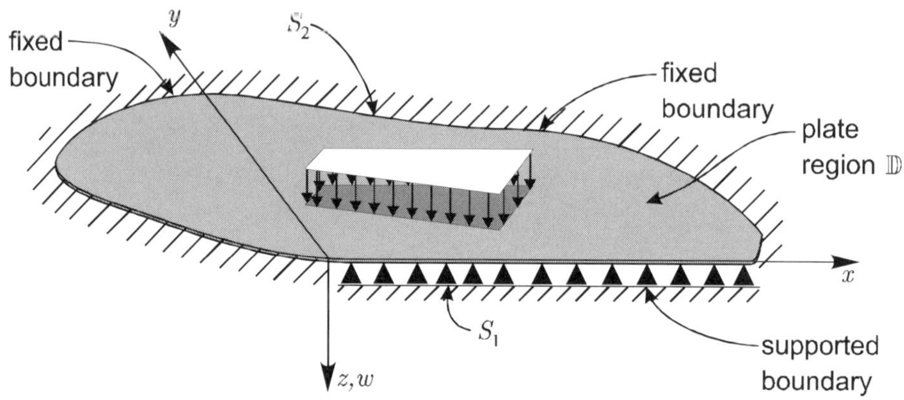

Figure 2.2: Mixed boundary conditions for a plate problem.

$$w = 0 \;\; ; \quad \left(\frac{\partial^2 w}{\partial y^2} + \nu \frac{\partial^2 w}{\partial x^2} \right) = 0 \; ; \quad \text{on} \quad \mathbf{x} \in S_1 \tag{2.27}$$

$$w = 0 \;\; ; \quad \frac{\partial w}{\partial n} = 0 \; ; \quad \text{on} \quad \mathbf{x} \in S_2 \tag{2.28}$$

where ν is a material constant.

2.2.2 Initial conditions

Initial conditions are prescribed conditions that must be satisfied throughout V at the instant when the processes described by the partial differential equation commences. A typical initial condition can prescribe a combination of the independent variable and its time derivatives.

Example 2.3

The two-dimensional form of the partial differential equation governing heat conduction in a solid is given by

$$\nabla^2 T = c\frac{\partial T}{\partial t} \quad \text{for} \quad \mathbf{x} \in V \tag{2.29}$$

The boundary condition governing the problem can be of the form

$$T(x, y, t) = T_1^*(x, y) \qquad \text{on} \qquad \mathbf{x} \in S_1 \tag{2.30}$$

$$\frac{\partial T}{\partial n}(x, y, t) = T_2^*(x, y) \qquad \text{on} \qquad \mathbf{x} \in S_2 \tag{2.31}$$

The initial condition governing the problem can be of the form

$$T(x, y, 0) = T_0 \qquad \text{on} \qquad \mathbf{x} \in V \tag{2.32}$$

2.2.3 Well-posed problems

The set of prescribed *boundary conditions* and *initial conditions* along with the coefficient functions and any inhomogeneous terms in the linear partial differential equation constitute the "data" in the problem modelled by the partial differential equation. The problem is said to be "well-posed" in the sense of Hadamard (J. Hadamard, 1865-1963) if the solution *depends continuously* on the "*data*" and if small changes in the "data" produce correspondingly small changes in the solution.

The problem itself is said to be "well-posed" if

a) a solution *exists* (the purpose of an existence theorem is to establish conditions under which, for the "data" prescribed, at least one solution exists),
b) the solution is *unique* (the purpose of a uniqueness theorem is to assure us that if for the prescribed data a solution exists it is unique) and
c) the solution *depends continuously* on the "data".

If these conditions are not satisfied the resulting problem is said to be "*ill posed*".

2.3 PROBLEM SET 2

2.1 If $u(x, y)$, $u(x, t)$, $u(x, y, z)$, etc., are continuously differentiable and non-zero functions of the independent variables, classify the following partial differential equations with respect to order, linearity and homogeneity

(i) $\dfrac{\partial u}{\partial x} - x\dfrac{\partial u}{\partial y} = 0$

(ii) $\left(\dfrac{\partial}{\partial x} - \dfrac{\partial}{\partial y}\right) u^2 = u\dfrac{x}{y}$

(iii) $\dfrac{\partial^2 u}{\partial x^2} - \dfrac{\partial u}{\partial t}\dfrac{\partial u}{\partial x} = xt\ u$

(iv) $\dfrac{\partial^3 u}{\partial x^3} - \dfrac{\partial^3 u}{\partial x \partial t^2} = \sin u$

(v) $\left(\dfrac{\partial}{\partial x} + \dfrac{\partial}{\partial t}\right)\left(\dfrac{\partial}{\partial x} - \dfrac{\partial}{\partial t}\right) u = xy$

(vi) $\nabla^2 u = 2$

(vii) $(\nabla^2 u^2) = u + 2\left[\left(\dfrac{\partial u}{\partial x}\right)^2 + \left(\dfrac{\partial u}{\partial y}\right)^2 + \left(\dfrac{\partial u}{\partial z}\right)^2\right]$

(viii) $\nabla^2 u = \frac{1}{c^2}\left(\frac{\partial u}{\partial t}\right)^2$; c is a constant

and

$$\nabla^2 = \frac{\partial^2}{\partial x^2} + \frac{\partial^2}{\partial y^2} + \frac{\partial^2}{\partial z^2}$$

2.2 Classify the following partial differential equations as to order, linearity and homogeneity.

(i) $\frac{\partial^2 u}{\partial x^2} - \lambda \frac{\partial u}{\partial x} + \frac{\partial u}{\partial x} = 0$

(ii) $\frac{\partial^2 u}{\partial x^2} + 2u\frac{\partial u}{\partial y} = 3xy$

(iii) $\left(\frac{\partial^4 u}{\partial y^4}\right)^{\frac{1}{2}} + 3\frac{\partial^2 u}{\partial x \partial y} + xu = 0$

(iv) $u^2 + 3\frac{\partial u}{\partial x}\frac{\partial u}{\partial y} - 6 = 0$

(v) $\left(\frac{\partial u}{\partial x}\right)^2 - \frac{\partial u}{\partial y} = 0$

(vi) $\sin^2(xy)\,\frac{\partial^2 u}{\partial x^2} - 3\cos(xy)\,\frac{\partial^2 u}{\partial x \partial y} = 0$

2.3 Obtain the partial differential equations, for which the general solutions take the forms

(i) $u(x,y) = f(x+y)$

(ii) $u(x,y) = f\left(x^2+y^2\right)$

(iii) $u(x,y) = f(x+ct) + g(x-ct)$

(iv) $u(x,y) = f\left(x+y^3\right) + g\left(x-y^3\right)$

(v) $u(x,y) = f\left(x+\sqrt{y}\right) + g\left(x-\sqrt{y}\right)$

(vi) $u(x,y) = f\left(x^2 + xy + y^2\right)$

where () represent the arguments of the functions f and g.

2.4 Find the value of λ in the partial differential equation

$$\frac{\partial^2 u}{\partial x^2} - \lambda \frac{\partial u}{\partial x} + \frac{\partial u}{\partial t} = 0$$

such that

$$u = \left[\cos\left(x + 2t\right) + i \sin\left(x + 2t\right)\right] \sin x$$

is a solution of the partial differential equation $\left(i = \sqrt{-1}\right)$

2.5 Obtain the general solution of

$$2\frac{\partial^2 u}{\partial x^2} - 5\frac{\partial^2 u}{\partial x \partial y} - 3\frac{\partial^2 u}{\partial y^2} = 0$$

by assuming a solution of the form $u = f(\Omega x + y)$ where Ω is a constant.

2.6 Assume that the solution of the partial differential equation

$$\frac{\partial^2 u}{\partial x^2} - \lambda^2 \frac{\partial u}{\partial t} = -Ae^{-\lambda x}$$

admits a representation

$$u\left(x,t\right) = w\left(x,t\right) + \varphi\left(x\right)$$

Determine the function $\varphi\left(x\right)$ such that the partial differential equation is reduced to a homogeneous form.

2.7 Determine the explicit forms of the solution for the following boundary value problems

a) $$\frac{\partial^2 u}{\partial y^2} = 2xy$$

with boundary conditions

i. $u\left(x,0\right) = x^2$

ii. $\left(\frac{\partial u}{\partial y}\right)_{y=0} = x$

b) $\frac{\partial^2 u}{\partial x \partial y} = c$

with boundary conditions

i. $u(x, y) = 0 \quad \text{on} \quad (x - y) = 0$

ii. $\frac{\partial u}{\partial x}(x, y) = 0 \quad \text{on} \quad (x - y) = 0$

2.8 Use a trial and error technique to obtain polynomial solutions $u(x, t)$ of degree 2 and 3 in x such that the governing partial differential equation

$$\frac{\partial^2 u}{\partial t^2} = \frac{\partial^2 u}{\partial x^2}$$

is satisfied.

Chapter 3

Partial differential equations of the first-order

Although the major focus of these introductory volumes is on linear partial differential equations of second- and higher-order, it is instructive to examine some aspects of first-order partial differential equations. These can arise in a number of engineering applications including the study of waves in shallow water, traffic flow, gas dynamics, isothermal plug flow reactors in chemical engineering and in the analysis of the thermal efficiency of heat exchangers.

3.1 General concepts

We focus attention on the quasi-linear first-order partial differential equation which has the form

$$A(x, y, u)\frac{\partial u}{\partial x} + B(x, y, u)\frac{\partial u}{\partial y} = C(x, y, u) \tag{3.1}$$

where x and y are the independent variables. The solution of (3.1) is a surface in the (x, y, u) space. Let us consider a point $P(x, y, u)$ on the solution surface $u = u(x, y)$ and we move along a direction given by the vector $\mathbf{t} = \{A(x, y, u), B(x, y, u), C(x, y, u)\}$. However, at any point on the surface, the direction of the normal is given by (Figure 3.1) $\mathbf{n} = \{\partial u/\partial x, \partial u/\partial y, -1\}$.

From these it is clear that **n** and **t** are orthogonal. Therefore any solution surface must be tangent to a vector with components $\{A, B, C\}$ and since this vector is tangential we do not leave the solution surface. Also, the total differential du is given by

$$du = \frac{\partial u}{\partial x} \cdot dx + \frac{\partial u}{\partial y} \cdot dy \tag{3.2}$$

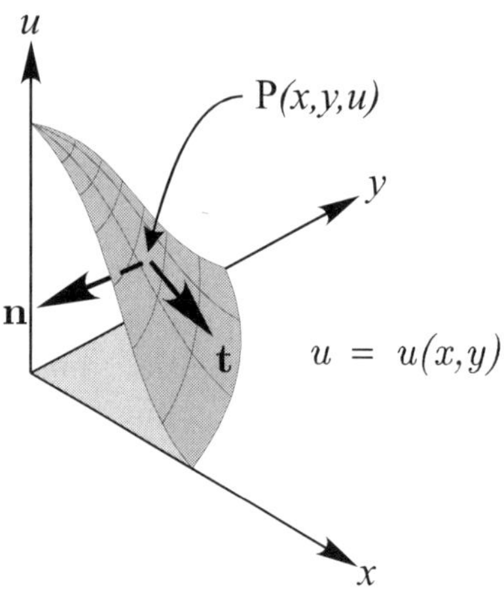

Figure 3.1: Solution surface for $u\,(x, y)$.

Consequently, from (3.1) and (3.2)

$$\{A, B, C\} \equiv \{dx, dy, du\} \tag{3.3}$$

The solution to equation (3.1) can be obtained by using the following theorem.

THEOREM 3.1

The general solution of the quasi-linear PDE

$$A(x,y,u)\frac{\partial u}{\partial x}+B(x,y,u)\frac{\partial u}{\partial y}=C(x,y,u) \tag{3.4}$$

can be written in the form

$$\mathcal{F}(F,G)=0 \tag{3.5}$$

where $\mathcal{F}$ is an arbitrary function and $F(x,y,u)=k_1$, and $G(x,y,u)=k_2$ form a solution of the equations

$$\frac{dx}{A(x,y,u)}=\frac{dy}{B(x,y,u)}=\frac{du}{C(x,y,u)} \tag{3.6}$$

PROOF

The equations (3.6) consist of a set of two independent ODE's (i.e. a two parameter family of curves in space). Also, one set can be written as

$$\frac{dy}{dx}=\frac{B(x,y,u)}{A(x,y,u)} \tag{3.7}$$

which is referred to as a "characteristic curve".

(i) If $A=A(x,y)$ and $B=B(x,y)$ then (3.7) is a function in the (x,y) space. This is referred to as a base curve.

(ii) When A and B are constants (3.7) defines a set of parallel lines in the (x,y) space.

In both cases (i) and (ii) (3.7) can be evaluated without a knowledge of $u(x,y)$. In the quasi-linear case (3.7) cannot be evaluated until $u(x,y)$ is known.

Returning to the equations (3.1) and (3.2) we can write

$$\begin{bmatrix} A & B \\ dx & dy \end{bmatrix} \begin{bmatrix} \partial u/\partial x \\ \partial u/\partial y \end{bmatrix} = \begin{bmatrix} C \\ du \end{bmatrix} \tag{3.8}$$

Both equations must hold on the solution surface and one can interpret each equation as a plane element, where they intersect on a line along which $\partial u/\partial x$ and $\partial u/\partial y$ may exist; i.e. $\partial u/\partial x$ and $\partial u/\partial y$ are themselves indeterminate along the intersection line but they are related to or determinate to each other since (3.8) must hold.

Using a principle in linear algebra, if a square coefficients matrix of a set of n simultaneous equations has a vanishing determinant, a necessary condition for the existence of finite solutions is that when the right hand side is substituted for any column, the determinants of the resulting coefficients matrices must also vanish, i.e.

$$\begin{vmatrix} A & B \\ dx & dy \end{vmatrix} = \begin{vmatrix} A & C \\ dx & du \end{vmatrix} = \begin{vmatrix} C & B \\ du & dy \end{vmatrix} = 0 \tag{3.9}$$

Evaluating the determinants we have

$$\frac{dx}{A(x,y,u)} = \frac{dy}{B(x,y,u)} = \frac{du}{C(x,y,u)} \tag{3.10}$$

3.2 Examples involving first-order equations

In this section we shall examine some basic first-order equations and their characteristic curves.

Example 3.1

Examine the characteristic curves for the simplest possible first-order partial differential equation

$$\frac{\partial u}{\partial x} = 0 \tag{3.11}$$

Solution

The general solution of equation (3.11) is

$$u = f(y) \tag{3.12}$$

where $f(y)$ is a completely arbitrary function of y. The characteristic curves in the x, y plane are straight lines (Figure 3.2).

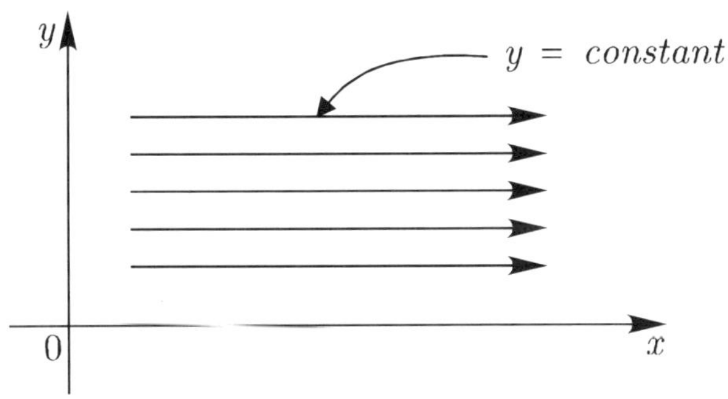

Figure 3.2: Characteristic curves for $\partial u/\partial x = 0$.

Example 3.2

Consider the first-order linear homogeneous partial differential equation

$$\frac{\partial u}{\partial x} + \frac{\partial u}{\partial y} = 0 \tag{3.13}$$

Examine the characteristic curves associated with this equation.

Solution

Comparing this with (3.1) we have $A = 1$ and $B = 1$; and (3.10) gives

$$dx = dy \tag{3.14}$$

Integrating (3.14) we have

$$x - y = \text{const} \tag{3.15}$$

Consequently, the general form of the solution is

$$u = F\,(x - y) \tag{3.16}$$

where F is an arbitrary function. Therefore if $y = x$ or $y = x + \Delta$, then $u = \text{const}$. Hence there is a characteristic direction in the x, y plane along which u is constant (Figure 3.3).

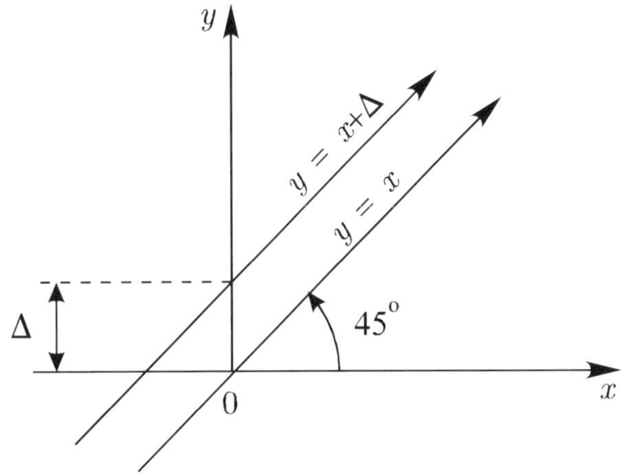

Figure 3.3: Characteristic curves for $\partial u/\partial x + \partial u/\partial y = 0$.

The general solution (3.16) can be made specific by imposing additional conditions.

e.g. If $u\,(x, y) = \eta x$, where η is a constant, when $y = 0$, the specific solution for (3.16) is

$$u\,(x, y) = \eta\,(x - y) \tag{3.17}$$

Example 3.3

Generalize the results presented in connection with Example 3.2 to include the case

$$A_0 \frac{\partial u}{\partial x} + B_0 \frac{\partial u}{\partial y} = 0 \tag{3.18}$$

where A_0 and B_0 are both *constants.*

Solution

We can prove that the characteristic curves are still straight lines except that they are inclined at an angle $\alpha = \tan^{-1}(B_0/A_0)$ to the x-axis (Figure 3.4). Avoiding details, the general solution of (3.18) takes the form

$$u(x,y) = \widetilde{F}\left(\frac{x}{A_0} - \frac{y}{B_0}\right) \tag{3.19}$$

where $\widetilde{F}$ is an arbitrary function. Again specific forms for $\widetilde{F}$ can be obtained by assigning additional conditions.

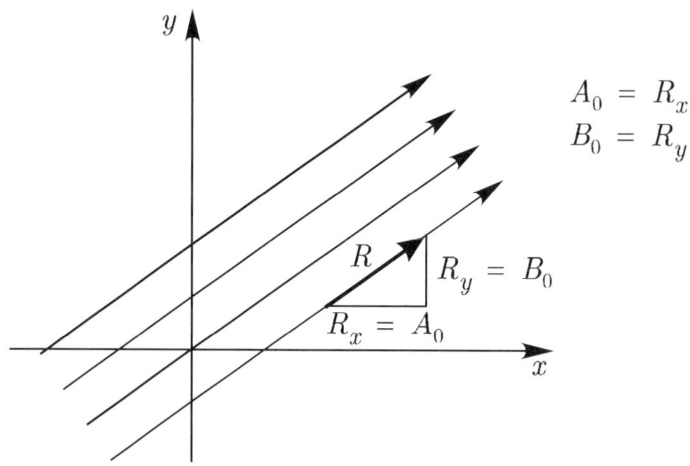

Figure 3.4: Characteristic curves for $A_0\,(\partial u/\partial x) + B_0\,(\partial u/\partial y) = 0$.

Example 3.4

Consider the first-order partial differential equation (3.18) which can be further generalized to the following form

$$A(x,y)\frac{\partial u}{\partial x} + B(x,y)\frac{\partial u}{\partial y} = 0 \tag{3.20}$$

Examine the characteristic curves associated with the partial differential equation

Solution

Considering (3.7) we have

$$\frac{dy}{dx} = \frac{B(x,y)}{A(x,y)} \tag{3.21}$$

Since $A(x,y)$ and $B(x,y)$ are arbitrary functions, the orientations of the tangent vector $\mathbf{R}$ to the characteristic lines are still parallel to the x, y plane (i.e. independent of u) but has a variable orientation in the x, y plane (Figure 3.5).

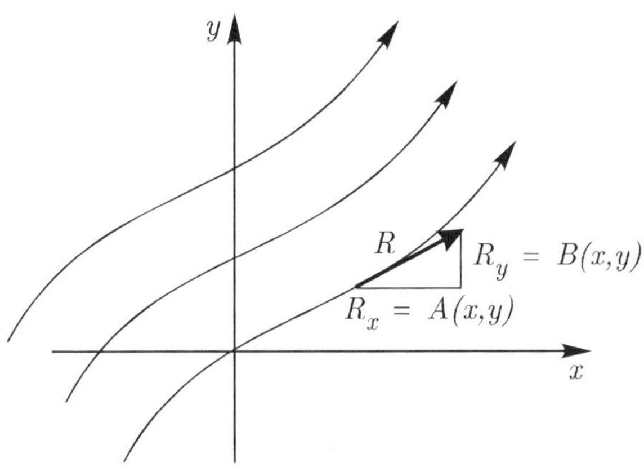

Figure 3.5: Characteristic curves $A(x,y)(\partial u/\partial x) + B(x,y)(\partial u/\partial y) = 0$.

We can rewrite (3.21) as

$$\frac{dx}{A(x,y)} = \frac{dy}{B(x,y)} = ds \tag{3.22}$$

Using (3.2) and (3.22) we have

$$du = \frac{\partial u}{\partial x}dx + \frac{\partial u}{\partial y}dy = ds\left[A(x,y)\frac{\partial u}{\partial x} + B(x,y)\frac{\partial u}{\partial y}\right] = 0 \tag{3.23}$$

The result (3.23) confirms that along a characteristic curve $du = 0$, irrespective of $A(x,y)$ and $B(x,y)$.

Example 3.5

Examine a specific case of the general first-order homogeneous partial differential equation (3.20);

$$\frac{1}{x}\frac{\partial u}{\partial x} + \frac{1}{y}\frac{\partial u}{\partial y} = 0 \tag{3.24}$$

Obtain a solution such that $u(x,0) = \mu x^4$ where μ is a constant.

Solution

A comparison with (3.20) gives

$$A(x,y) = \frac{1}{x} \quad ; \quad B(x,y) = \frac{1}{y} \tag{3.25}$$

and using (3.25) in (3.21) and integrating we have

$$\int \frac{dx}{A(x,y)} = \int \frac{dy}{B(x,y)}$$

or

$$\int x \, dx = \int y \, dy$$

or

$$x^2 - y^2 = -C_0 \tag{3.26}$$

where C_0 is a constant. Therefore the most general functional form of the solution of (3.24) is

$$u(x, y) = F^* \left(x^2 - y^2\right) \tag{3.27}$$

Substituting (3.27) in (3.24) it is evident that the latter is satisfied for all choices of F^*. Again, a specific form of (3.27) can be obtained by imposing additional conditions on $u(x, y)$. e.g. If $u(x, 0) = \mu x^4$ is it evident that the specific solution of $u(x, y)$ is

$$u(x, y) = \mu \left(x^2 - y^2\right)^2 \tag{3.28}$$

where μ is a constant.

• • •

Let us now consider the inhomogeneous first-order partial differential equation posed by a reduced form of (3.1). i.e.

$$A(x, y) \frac{\partial u}{\partial x} + B(x, y) \frac{\partial u}{\partial y} = C(x, y) \tag{3.29}$$

The characteristic curves occupy the (x, y, u) space except that a characteristic line does not lie in a plane $u = \text{const}$. Consider (3.10) and (3.22)

$$\frac{dx}{A(x,y)} = \frac{dy}{B(x,y)} = \frac{du}{C(x,y)} = ds \tag{3.30}$$

Of these, two equations can be taken to be independent, i.e.

$$\frac{dx}{A(x,y)} = \frac{dy}{B(x,y)} \quad ; \quad \frac{dy}{B(x,y)} = \frac{du}{C(x,y)} \tag{3.31}$$

and as stated previously, if these equations are integrable, they can be evaluated in the form

$$F(x,y) = C_1 \ \ ; \ \ G(x,y) = C_2 \tag{3.32}$$

and the general solution of the partial differential equation (3.29) can be written in the form

$$\Omega(F,G) = 0 \tag{3.33}$$

Example 3.6

As a specific case of (3.29), consider the non-homogeneous partial differential equation

$$\frac{1}{x}\frac{\partial u}{\partial x} + \frac{1}{y}\frac{\partial u}{\partial y} = \frac{1}{y} \tag{3.34}$$

Develop a solution such that $u = \frac{1}{2}\left(3 - x^2\right)$ when $y = 1$.

Solution

Comparing (3.29) and (3.34) we have

$$A(x,y) = \frac{1}{x} \quad ; \quad B(x,y) = \frac{1}{y} \quad ; \quad C(x,y) = \frac{1}{y} \tag{3.35}$$

The associated ordinary differential equations corresponding to (3.31) are

$$\frac{dx}{(1/x)} = \frac{dy}{(1/y)} \quad ; \quad \frac{dy}{(1/y)} = \frac{du}{(1/y)} \tag{3.36}$$

Integrating (3.36) we have

$$F = \left(x^2 - y^2\right) = C_1 \quad ; \quad G = (u - y) = C_2 \tag{3.37}$$

and the general form of the function is

$$\Omega\left[\left(x^2 - y^2\right), (u - y)\right] = 0 \tag{3.38}$$

Hence for $y = 1$, (3.38) gives

$$\Omega\left[\left(x^2 - 1\right), (u - 1)\right] = 2u - 3 + x^2 = 2(u - 1) + \left(x^2 - 1\right) \tag{3.39}$$

Therefore the general relationship equivalent to (3.33) is obtained by replacing $\left(x^2 - 1\right)$ by $\left(x^2 - y^2\right)$ and $(u - 1)$ by $(u - y)$ in (3.39), i.e.

$$\Omega\left[(u - y), \left(x^2 - y^2\right)\right] = 2(u - y) + \left(x^2 - y^2\right) \tag{3.40}$$

which gives

$$u(x,y) = \frac{1}{2}\left[2y - \left(x^2 - y^2\right)\right] \tag{3.41}$$

It can be verified by substitution that (3.41) satisfies the first-order partial differential equation (3.34) subject to the additional condition (3.39) when $y = 1$.

3.3 Advective transport in reactor column

In the preceding, we have examined the various types of first-order partial differential equations in a very abstract sense without reference to any engineering application. We now apply the general theory concerning the first-order non-homogeneous partial differential equation to a problem which has applications in both *environmental engineering* and

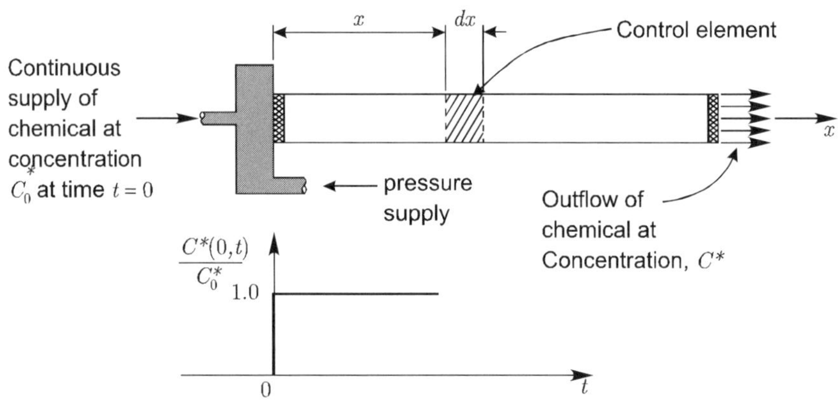

Figure 3.6: Advective transport in column.

chemical engineering. The problem deals with the so called "*advective transport*" of a species (such as a chemical, biological medium, effluent, etc.) along a reactor column. The reactor column is basically a tube with a circular cross sectional area which contains either a reactive or non-reactive porous solid (Figure 3.6). A chemical species is introduced into the flow regime and the "*concentration*" of the chemical or species is maintained constant at the start of the experiment. The objective of the analysis is to determine the variation of the concentration of the species along the tube as a function of time.

3.3.1 Governing equation - one dimensional case

Before attempting to formulate the governing equation, it is necessary to define the parameters governing the problem. The porous medium is characterized by continuous flow paths within the void space. A measure of the void space is the scalar quantity *porosity* (n^*) . This is defined by

$$n^* = \frac{\text{volume of voids}}{\text{total volume}} = \frac{V_v}{V} \tag{3.42}$$

where V_v is the volume of voids and V is the total volume.

Fluid flow which results in advection of the species takes place in the void space. The average local velocity of flow in the void space at any cross section is defined by $\overline{v}$. We define an average advective velocity over the entire cross section as $\widehat{v}$. In the case of one dimensional flow, both these velocities would be aligned in the x-direction (Figure 3.6). For conservation of mass we require.

$$\widehat{v}A = \overline{v}A_v \tag{3.43}$$

For most porous media with an *isotropic* or *direction independent* fabric or structure,

$$\frac{A_v}{A} \simeq \frac{V_v}{V} = n^* \tag{3.44}$$

We define the concentration $C(x,t)$ as the mass of the species per *unit volume of the fluid* conveying the species. We can also define a concentration $\widehat{C}(x,t)$ as the mass of the species per *unit total volume of the entire porous medium.* If the porous medium is *saturated*, the volume of the fluid is identically equal to the volume of the voids. From these definitions we have,

$$C(x,t) = \frac{\text{mass of species}}{\text{volume of fluid}} = \frac{\text{mass of species}}{\text{volume of voids}} = \frac{\text{mass of species}}{V_v}$$

or

$$\frac{\text{mass of species}}{\text{total volume}} = \frac{C(x,t)V_v}{V} = n^* C(x,t) = \widehat{C}(x,t) \tag{3.45}$$

Consider the one-dimensional flow in the tube, where, prior to the introduction of the species, a pressure gradient induces a uniform average fluid velocity of $\widehat{v}$. The cross sectional area of the tube is a_0. The mass of species transported in the tube by "*advection*", per unit cross sectional area per unit time is $\widehat{v}n^* C(x,t)$.

The total mass of the species which enters the control element of cross sectional area a_0 and length dx is

$$F_x = a_0 \widehat{v} n^* C(x,t) \tag{3.46}$$

The total mass of the species leaving the control element

$$= F_x + \frac{\partial F_x}{\partial x} \cdot dx \tag{3.47}$$

The conservation of mass equation for the control element can be stated as follows:

$$\begin{aligned} \begin{bmatrix} \text{net rate of change} \\ \text{of mass of species} \\ \text{within the element} \end{bmatrix} &= \begin{bmatrix} \text{flux of species} \\ \text{out of the element} \end{bmatrix} \\ &- \begin{bmatrix} \text{flux of species} \\ \text{into the element} \end{bmatrix} \\ &\pm \begin{bmatrix} \text{loss or gain of the} \\ \text{mass of the species} \\ \text{to reactions} \end{bmatrix} \end{aligned} \tag{3.48}$$

From (3.48) we have

$$-\frac{\partial}{\partial t}\left(n^{*}Ca_{0}dx\right)=\frac{\partial}{\partial x}\left(\widehat{v}n^{*}Ca_{0}\right)dx\pm\xi n^{*}Ca_{0}dx \tag{3.49}$$

where ξ is the rate of accumulation (+) or loss (-) of chemicals within the control element due to reactions. We can reduce (3.49) to the form

$$\frac{\partial C}{\partial t}+\widehat{v}\frac{\partial C}{\partial x}=\pm\xi C \tag{3.50}$$

For the simple one-dimensional advective flow problem, which includes a reaction term, the governing first-order differential equation is identical in form to (3.1). The boundary and initial conditions governing (3.50) can be written as follows:

$$C=0 \quad ; \quad t=0 \quad ; \quad x\geq 0 \tag{3.51}$$

$$C=C_{0} \quad ; \quad t>0 \quad ; \quad x=0 \tag{3.52}$$

In the instance where there is loss of the chemical concentration due to reactive processes we take the negative component of the right hand side of (3.50). Furthermore we introduce a change of variables such that

$$\eta=\frac{C}{C_{0}} \quad ; \quad \theta=\xi t \quad ; \quad \tau=\frac{x\xi}{\widehat{v}} \tag{3.53}$$

Consequently we can write

$$\frac{\partial\eta}{\partial\theta}+\frac{\partial\eta}{\partial\tau}=-\eta \tag{3.54}$$

with the appropriate boundary conditions being

$$\eta = 0 \ ; \quad \theta = 0 \ ; \quad \tau \geq 0 \tag{3.55}$$

$$\eta = 1 \ ; \quad \theta > 0 \ ; \quad \tau = 0 \tag{3.56}$$

The characteristic equations are

$$\frac{d\theta}{1} = \frac{d\tau}{1} = -\frac{d\eta}{\eta} \tag{3.57}$$

which gives

$$\frac{d\tau}{d\theta} = 1 \ ; \quad \theta \geq 0 \ ; \quad \tau \geq 0 \tag{3.58}$$

$$\frac{d\eta}{d\tau} = -\eta \ ; \begin{array}{l} \eta = 0 \ ; \quad \theta = 0 \ ; \quad \tau \geq 0 \\ \eta = 1 \ ; \quad \theta > 0 \ ; \quad \tau = 0 \end{array} \tag{3.59}$$

These linear ordinary differential equations can be integrated to give the solution

$$\frac{C}{C_0} = 0 \ ; \quad x > \widehat{v}t \tag{3.60}$$

and

$$\frac{C}{C_0} = e^{-x\xi/\widehat{v} \ \ ; \ \ x<\widehat{v}t} \tag{3.61}$$

In this simple example we have considered only "advective transport" processes of the species in the porous medium. There is a further mechanism called "diffusive transport" which accounts for transport by virtue of a gradient of the concentration of the species at a point. These processes can occur simultaneously and the degree to which one process is dominant will depend on the velocity of flow and the mechanical properties which govern the diffusion processes. These will be discussed in a later chapter.

3.3.2 Governing equation - generalized formulation

The formulation of the advective transport problem presented in the previous section is useful from the point of view of illustrating the distinct factors that should be taken into consideration in the development of the final governing equation in one-dimension. It is also possible to present a generalized formulation of the advective transport problem applicable to three dimensions.

The fundamental definitions required for the three dimensional formulation will now be briefly reviewed. Consider an arbitrary region V of a porous medium with surface S. (Figure 3.7)

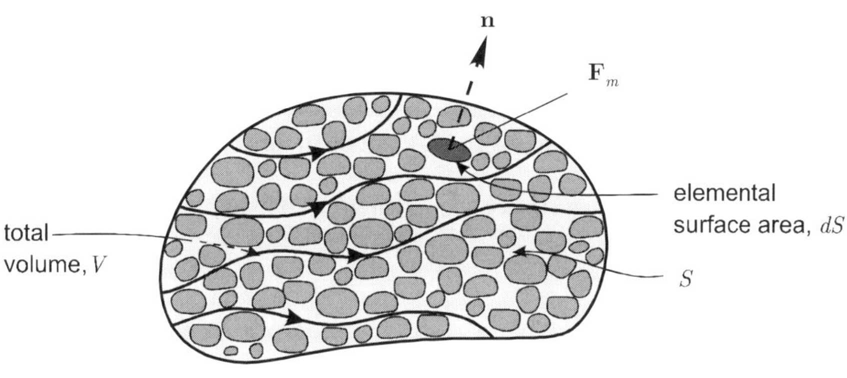

Figure 3.7: Advective transport in a porous medium.

The porosity measure n^* defined by (3.42) and (3.44) is equally applicable to the three dimensional case. The concentration of the species per unit volume of the fluid is defined by $C(\mathbf{x}, t)$; indicating that the concentration is a function of the position vector $\mathbf{x}$. Similarly the concentration of the species per unit total volume of the porous medium is defined by $\widehat{C}(\mathbf{x}, t)$. Again at any given location $\mathbf{x}$ at time t

$$\widehat{C}(\mathbf{x},t) = n^* C(\mathbf{x},t) \tag{3.62}$$

The advective velocity averaged over the entire cross section of the porous medium is defined by the vector $\widehat{\mathbf{v}}(\mathbf{x},t)$. The *flux* is defined as the mass of the species either *entering* or *leaving* a unit area in unit time. The flux vector, of mass being transported by advection, per unit total area per unit time is given by

$$\mathbf{F}_a = \widehat{\mathbf{v}}\widehat{C}(\mathbf{x},t) = n^* \widehat{\mathbf{v}} C(\mathbf{x},t) \tag{3.63}$$

The total mass of the species transported *into* the region V of surface area S is given by

$$m_i = -\int\int_S \mathbf{F}_a \mathbf{n}\, dS \tag{3.64}$$

where $\mathbf{n}$ is the outward unit normal to the elemental surface are dS.

Using the *divergence theorem* we can rewrite (3.64) as

$$m_i = -\int\int\int_V \nabla \cdot \{n^* \widehat{\mathbf{v}} C\}\, dV \tag{3.65}$$

where the dependency on $\mathbf{x}$ and t is suppressed for convenience of presentation.

The species is assumed to accumulate in the fluid in the void space of the medium. The rate of accumulation of species is given by

$$m_a = \frac{d}{dt}\int\int\int_V n^* C dV = \int\int\int_V n^* \frac{\partial C}{\partial t}\, dV \tag{3.66}$$

The rate of mass production due to generation (+) or decay (-) of the species within the medium due to chemical reactions is given by

$$m_p = \pm \int\int\int_V n^* \xi C \, dV \tag{3.67}$$

where ξ is a specific generation term referred to the fluid volume (i.e. generation per *unit volume of the fluid*) For conservation of the mass of the species

$$m_a = m_i \pm m_p \tag{3.68}$$

Substituting (3.65), (3.66) and (3.67) in (3.68) we have

$$\begin{aligned}\int\int\int_V n^* \frac{\partial C}{\partial t} \, dV = & -\int\int\int_V \nabla.\{n^* \hat{\mathbf{v}} C\} \, dV \\ & \pm \int\int\int_V n^* \xi C \, dV\end{aligned} \tag{3.69}$$

or

$$\int\int\int_V n^* \left[\frac{\partial C}{\partial t} + \nabla.\{\hat{\mathbf{v}} C\} \mp \xi C\right] dV = 0 \tag{3.70}$$

Since V is arbitrary we require the integrand be zero. Since $n^* \neq 0$, we have

$$\frac{\partial C}{\partial t} + \nabla.(\hat{\mathbf{v}} C) = \pm \xi C \tag{3.71}$$

In arriving at (3.71) we have implicitly made use of the *Dubois-Reymond Lemma,* for the existence of a local equation which is also valid for $\widehat{C}$.

Lemma

The lemma due to Paul Dubois-Reymond (1831-1889) states that if $F(\mathbf{x}, t)$ is a continuous function in an arbitrary domain V and if for every smooth subdomain $V' \subset V$, $\int_{V'} F(\mathbf{x}, t) dV = 0$, then, $F(\mathbf{x}, t) = 0$, in V.

⊙ ⊙ ⊙

We shall consistently employ this lemma to develop the governing partial differential equations as local equations.

Equation (3.71) represents the generalized three-dimensional form of the differential equation governing advective transport in a porous medium in the presence of *decay* or *generation* phenomena. In the case of *one-dimensional* advective flow such as in a reactor column,

$$C = C(x, t) \tag{3.72}$$

and

$$\hat{\mathbf{v}} = v_0 \mathbf{i} \tag{3.73}$$

where v_0 is a constant. Using (3.72) and (3.73) in (3.71) we have

$$\frac{\partial C}{\partial t} + v_0(\mathbf{i}) \left\{ \mathbf{i} \frac{\partial C}{\partial x} + \mathbf{j} \frac{\partial C}{\partial y} + \mathbf{k} \frac{\partial C}{\partial z} \right\} = \pm \xi C \tag{3.74}$$

or

$$\frac{\partial C}{\partial t} + v_0 \frac{\partial C}{\partial x} = \pm \xi C \tag{3.75}$$

which agrees with the result (3.50), $\widehat{v} = v_0$, derived from elementary considerations.

We shall now consider an alternative procedure for the solution of the one-dimensional advective transport problem which is based on the application of *Laplace transforms*. We first examine the case where the species generation (or species decay) term is absent. The partial differential equation governing the time-dependent distribution concentration $C(x,t)$ is given by

$$\frac{\partial C}{\partial t} + v_0 \frac{\partial C}{\partial x} = 0 \tag{3.76}$$

As a requirement for well-posedness we need to specify both a *boundary condition* and an *initial condition*. We shall assume that the species is supplied at one end ($x = 0$) of the reactor column at a *constant* concentration C_0. i.e.

$$C(0,t) = C_0 H(t) \tag{3.77}$$

where $H(t)$ is the Heaviside step function.

The governing initial condition assumes that prior to the start of the advective transport of the species, the reactor column is void of any concentration, i.e.

$$C(x,0) = 0 \tag{3.78}$$

We now define the Laplace transform (1.82) of the concentration $C(x,t)$ with respect to t by

$$\mathcal{L}\{C(x,t)\} = c(x,s) = \int_0^\infty C(x,t) e^{-st} dt \tag{3.79}$$

From (1.92)

$$\mathcal{L}\left\{\frac{\partial C}{\partial t}\right\} = sc(x,s) \tag{3.80}$$

Using (3.79) and (3.80) we can convert (3.76) to the form

$$sc(x,s) = -v_0 \frac{dc(x,s)}{dx} \tag{3.81}$$

or

$$\frac{dc(x,s)}{c} = -\frac{s}{v_0} dx \tag{3.82}$$

Integrating (3.82) we can write

$$c(x,s) = Ae^{-sx/v_0} \tag{3.83}$$

where A is a constant of integration. This constant needs to be determined by making use of the boundary condition (3.77) applicable to the problem. Since $c(x, s)$ is in the Laplace transform domain we need to formulate the boundary condition also in the Laplace transform domain, i.e.

$$\mathcal{L}\{C(0,t)\} = c(0,s) = \frac{C_0}{s} \tag{3.84}$$

Using (3.84) we can evaluate A. i.e.

$$c(0,s) = \frac{C_0}{s} = A \tag{3.85}$$

The solution of the problem in the Laplace transform domain is given by

$$c(x,s) = C_0 \frac{e^{-sx/v_0}}{s} \tag{3.86}$$

We can now invert the Laplace transform to generate $C(x,t)$. i.e.

$$C(x,t) = C_0\mathcal{L}^{-1}\left\{\frac{e^{-sx/v_0}}{s}\right\} \tag{3.87}$$

From (1.85) and (1.155) (or from *Tables of Inverses of Laplace Transforms)* we have

$$C(x,t) = C_0 H\left(t - \frac{x}{v_0}\right) \tag{3.88}$$

where $H\left(t - \frac{x}{v_0}\right)$ is the Heaviside step function defined by

$$H\left(t - \frac{x}{v_0}\right) = \begin{cases} 1 & \left(t - \frac{x}{v_0}\right) > 0 \\ 0 & \left(t - \frac{x}{v_0}\right) < 0 \end{cases} \tag{3.89}$$

Hence

$$\frac{C(x,t)}{C_0} = \begin{cases} 1 & x < v_0 t \\ 0 & x > v_0 t \end{cases} \tag{3.90}$$

The solution clearly illustrates the characteristic features of the advective transport problem. (i) Since there is no *decay* or *generation* of the species, the concentration level will remain constant within the region where the species has migrated. (ii) The concentration is zero at or beyond the migration front. This sharp discontinuity stems from the purely advective processes in the porous medium. If diffusive processes are incorporated, the migration from will have a smooth profile. The solution of the simple advective transport problem can be represented in the C, x, t space. The resulting representation is shown in Figure 3.8. As is evident, the front is established on the characteristic line $x = v_0 t$.

Let us now examine the case where the decay or generation term of (3.75) is taken into account. For purposes of comparison with the results given in section 3.3.1. we shall restrict the analysis to the case where there is decay of the species. The basic one-dimensional advective transport is governed by the equation

$$\frac{\partial C}{\partial t} + v_0 \frac{\partial C}{\partial x} = -\xi C \tag{3.91}$$

The boundary condition governing the problem is still (3.77) and the initial condition is (3.78). We define the Laplace transform of $C(x,t)$ with respect to t as defined by (3.79). Applying the Laplace transform to (3.91) we have

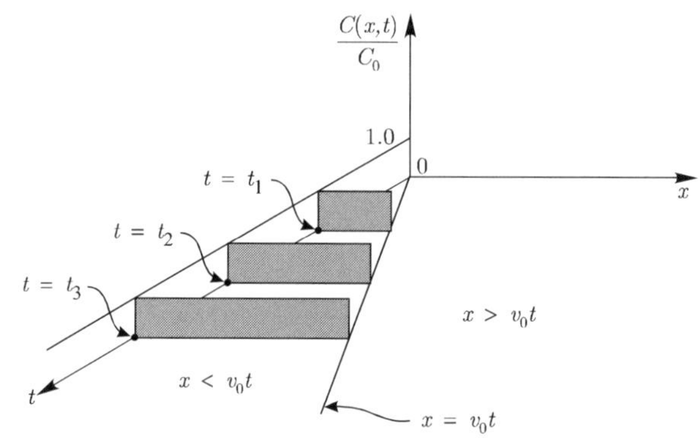

Figure 3.8: Time-dependent advective movement of a species in a porous medium.

$$sc(x,s) + v_0 \frac{dc}{dx} = -\ \xi c(x,s) \tag{3.92}$$

We can reduce (3.92) to the form

$$\frac{dc}{c} = -\frac{(s+\xi)}{v_0}dx \tag{3.93}$$

Integrating (3.93) we have

$$c(x,s) = Ae^{-\frac{(s+\xi)x}{v_0}} \tag{3.94}$$

The constant of integration A can be determined by making use of the boundary condition (3.78) and its transformed equivalent (3.84). We thus obtain

$$c(x,s) = C_0\frac{e^{-\frac{(s+\xi)x}{v_0}}}{s} \tag{3.95}$$

Inverting the Laplace transform we can recover $C(x,t)$ i.e.

$$C(x,t) = c_0 e^{-\frac{\xi x}{v_0}}\, H\left(t - \frac{x}{v_0}\right) \tag{3.96}$$

or

$$\frac{C(x,t)}{C_0} = \begin{cases} e^{-\xi x/v_0} & x \le v_0 t \\ 0 & x > v_0 t \end{cases} \tag{3.97}$$

Here again, the solution demonstrates both the influence of the advection term and the role of the decay term in attenuating the migration of the species along the direction of flow. The Figure 3.9 illustrates decay dependent advective transport profiles in the C, x, t space.

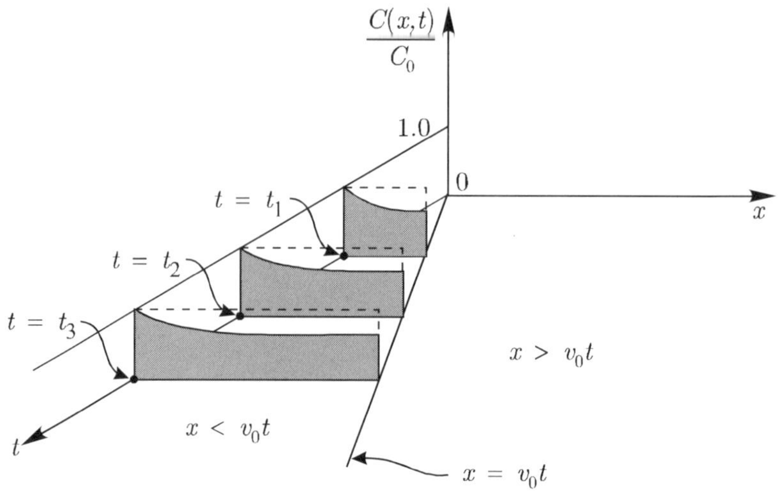

Figure 3.9: Time-dependent advective movement of a species in a porous medium; influence of decay.

3.4 A heat exchanger problem

Heat exchangers are basically devices that are used to efficiently transfer heat between a warmer material and a colder material. The most rudimentary form of a heat exchanger consists of a tube which is immersed in a material region of temperature T_R. The material region could simply be a water bath, or in the case of geothermal heat pumps, the ground. The schematic view of a heat exchanger of this type is shown in Figure 3.10. The temperature of the fluid flowing through the tube at any distance z along the tube at any time t is denoted by $T(z,t)$. When water moves through the tube, at a uniform velocity V, heat is transferred through the wall to the fluid if $T_R > T(z,t)$. For the purposes of the elementary model, we assume that during this heat transfer process, there is no change in the ambient temperature T_R. Let us now consider the heat balance over the sections between z and $z + \Delta z$, over the time interval $(t, t + \Delta t)$.

The increase in heat in time Δt, within the element

$$= \pi a^2 (\Delta z) [T(z, t + \Delta t) - T(z,t)] C_P$$

where C_P is the heat capacity of the moving fluid per unit volume.

Heat supplied to the element at time Δt by the fluid moving at velocity V

$$= V\pi a^2 (\Delta t)\, C_P \left[T(z,t) - T(z+\Delta z, t)\right]$$

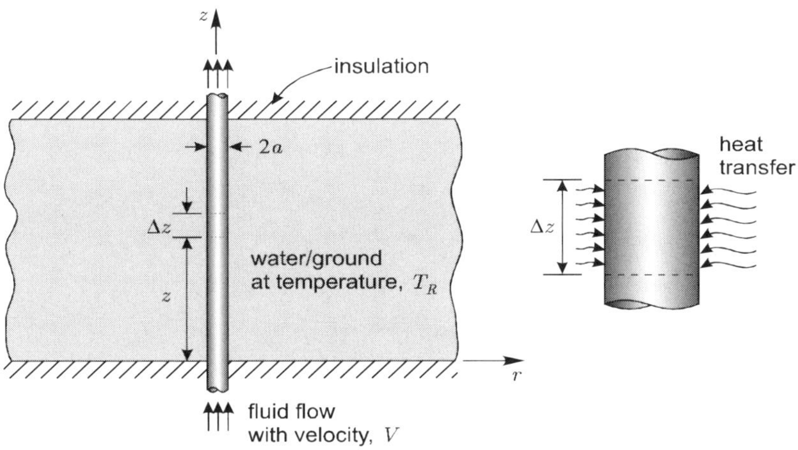

Figure 3.10: The tubular heat exchanger.

Heat supplied to the element for the increment of time Δt due to heat exchange with the reservoir $= U(2\pi a)\,\Delta z\,(T_R - T)\,\Delta t$ where U is an overall heat transfer coefficient at the boundary of the tube.

If heat energy is conserved, then we obtain

$$\begin{aligned} \pi a^2 \Delta z C_P \left[T(z, t+\Delta t) - T(z,t)\right] &= V\pi a^2 (\Delta t)\, C_P \left[T(z,t) - T(z+dz,t)\right] \\ &\quad + U(2\pi a)\,\Delta z\,(T_R - T)\,\Delta t \end{aligned} \tag{3.98}$$

Dividing by $\Delta z\ \Delta t$ and passing to the limit we obtain

$$\frac{\partial T}{\partial t}+V\frac{\partial T}{\partial z}-\frac{2U}{aC_P}\left(T_R-T\right) \tag{3.99}$$

By introducing the substitutions

$$T^*=T-T_R \quad ; \quad \zeta=\frac{2U}{aC_P}$$

the equation (3.99) gives

$$\frac{\partial T^*}{\partial t}+V\frac{\partial T^*}{\partial z}=-\zeta T^* \tag{3.100}$$

which is a first-order partial differential equation for $T^*\left(z,t\right)$. To make the problem well-posed we need to specify the inlet temperature and the initial temperature distribution along the tube. i.e.

$$T\left(0,t\right)=T_e\left(t\right) \quad ; \quad T\left(z,0\right)=T_s\left(t\right) \tag{3.101}$$

The analogy between the one-dimensional advective transport in a reactor column discussed in Section 3.3 and the tubular heat exchanger problem is very clear. The solution to the problem of the tubular heat exchanger can be obtained using exactly the same procedures described in Section 3.3.

3.5 PROBLEM SET 3

3.1 Determine the *first-order* partial differential equations for which the general solutions are as follows:

(i) $u\left(x,y\right)=f\left(\frac{xy}{u^2}\right)$

(ii) $u\left(x,y\right)=xy+f\left(x^2+y^2\right)$

(iii) $u(x,y) = f\left(\frac{xy}{u}\right)$

(iv) $u(x,y) = \left(\frac{y^2}{x}\right) + f(xy)$

(v) $u(x,y) = f(\ln r) + g(e^{r^2})$

(vi) $u(x,y) = e^{f(2\ln r)}$

(vii) $u(x,y) = x^n f\left(\frac{x}{y}\right)$

where

$r^2 = (x^2 + y^2)$

and $f()$ and $g()$ are continuously differentiable non-zero functions and the terms within the brackets represent the respective arguments.

3.2 Find the general form of the solution of the following *first-order* partial differential equations

(i) $\frac{\partial u}{\partial x} + y\frac{\partial u}{\partial y} = 0$

(ii) $\frac{1}{2x}\frac{\partial u}{\partial x} + \frac{\partial u}{\partial y} = 0$

(iii) $x\frac{\partial u}{\partial x} + y\frac{\partial u}{\partial y} = y^2$

3.3 (i) Obtain the solution of the following *first-order* partial differential equation

$$x\frac{\partial u}{\partial x} + y\frac{\partial u}{\partial y} = 3y^3$$

subject to the condition

$$u(x,1) = \cosh^2 x$$

(ii) Verify your solution by back substitution into the original PDE.

3.4 (i) Obtain the solution of *first-order* partial differential equation

$$\frac{\partial u}{\partial x} + \left\{ \frac{\cos y}{\cos x} \right\}^2 (\sin x) \left(\frac{\partial u}{\partial y} \right) = 0$$

such that

$$u(0, y) = (\sec^2 y - 2 \tan y)$$

(ii) Obtain the *first-order* partial differential equation for which

$$u(x, y) = x^m f\left(\frac{y}{x}\right)$$

is a solution, where f is an arbitrary function with continuous derivatives.

3.5 Obtain the three specific forms of the solution to the *first-order* partial differential equation

$$\frac{\partial u}{\partial x} + \frac{\partial u}{\partial y} = 0$$

such that

1. $u(x, 0) = \alpha x^2$
2. $u(x, 0) = \beta x^3$
3. $u(x, 0) = \gamma e^{\alpha x}$

where α, β and γ are constants.

3.6 (i) Derive from first principles, the general equation governing advective transport of a chemical species in a porous medium

$$\frac{\partial C}{\partial t} + \nabla . (\widehat{\mathbf{v}} C) = \pm \xi C$$

where C is the concentration of the species, $\widehat{\mathbf{v}}$ is the averaged advective velocity vector and ξ is a generation (+) or decay (-) factor.

(ii) Consider the one dimensional case of the above advective transport problem for a column of cross sectional area 'a' which occupies the region $0 \leq x < \infty$. The advective velocity $\widehat{\mathbf{v}}$ is assumed to be constant equal to $\widehat{v}\mathbf{i}$. A chemical species is injected at $x = 0$ with a time-dependent concentration of the form

$$C(0,t) = C_0\left[1 + \sin \omega t\right]$$

where C_0 and ω are constants. Derive an expression for the time-dependent distribution of the concentration $C(x,t)$ along the column if there is a decay factor ξ and the column is free of any chemical at the start of the injection process.

$$\left[\text{Given: } \mathcal{L}^{-1}\left\{e^{-\lambda s} f(s)\right\} = H(t-\lambda)\, F(t-\lambda)\right]$$

3.7 The Figure 3.11 shows the longitudinal and cross section of a small stream into which an effluent is discharged at a concentration which is time-dependent. The velocity of flow in the stream, averaged over the cross section, is v_0. The effluent is moved by advective transport and concentration at any cross section downstream of the discharge location is $C(x,t)$. The equation governing advective flow induced transport is

$$\frac{\partial C}{\partial t} + v_0 \frac{\partial C}{\partial x} = -\zeta C$$

where ζ is a decay term due to natural oxygenation. If the effluent is discharged at a concentration

$$C(0,t) = C_0 e^{-\eta t}$$

and prior to the discharge of the effluent the stream is effluent free, determine the time-dependent variation of the effluent concentration along the stream.

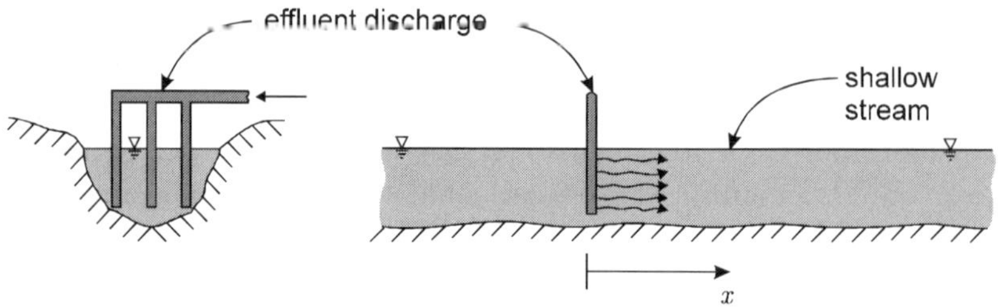

Figure 3.11: Effluent transport by advection

Chapter 4

Partial differential equations of the second-order

Partial differential equations of the second- and higher-order are encountered in a number of areas of importance to engineering and mathematical physics. As the processes that are accounted for in the development of the governing differential equations increase in complexity, the differential equations themselves tend to acquire a higher-order. The theory of second-order partial differential equations has found extensive applications in the study of problems in fluid mechanics, flow in porous media, heat conduction in solids, diffusive transport of chemicals in porous media, wave propagation in strings and membranes, and in mechanics of solids. Higher-order partial differential equations have been developed particularly in relation to studies in mechanics of solids. These include the formulation of the biharmonic equation for the solution of problems in plane elasticity and in the flow of viscous fluids, the fourth-order partial differential equations associated with the flexural mechanics of prismatic plates, the sixth-order partial differential equations used to describe the mechanics of shell structures and other fourth-order partial differential equations which describe the mechanics of plastic solids. In this chapter we shall restrict attention to the examination of second-order partial differential equations.

4.1 Classification of second-order partial differential equations

The objectives of a classification system of differential equations are basically threefold. Firstly, to identify the general character of the differential equation which may not be self evident from the representation provided. Secondly, to identify a method of solution of the partial differential equation which is specific to the category of equation under consideration. Finally, to establish *a priori* the particular attributes of the solution by appeal to expected behaviour derived from physical or mathematical considerations of a general nature.

We will consider the general form of a second-order linear partial differential equation in two independent variables, i.e.

$$\begin{aligned} A\left(x,y\right)\frac{\partial^2 u}{\partial x^2} + 2B\left(x,y\right)\frac{\partial^2 u}{\partial x \partial y} + C\left(x,y\right)\frac{\partial^2 u}{\partial y^2} + D\left(x,y\right)\frac{\partial u}{\partial x} \\ + E\left(x,y\right)\frac{\partial u}{\partial y} + F\left(x,y\right)u = G\left(x,y\right) \end{aligned} \tag{4.1}$$

In (4.1) the coefficients $A, B, \ldots, G$ are functions of the independent variables (x, y). The most convenient means of classifying (4.1) would be to restrict attention to terms of the coefficients $A\left(x,y\right)$, $B\left(x,y\right)$ and $C\left(x,y\right)$ of the second-order terms only. The character of the differential equation is entirely dependent on the relative values of these coefficients. As a lead into the discussion, we will restrict attention to the situation where A, B and C are all constants such that

$$A\frac{\partial^2 u}{\partial x^2} + 2B\frac{\partial^2 u}{\partial x \partial y} + C\frac{\partial^2 u}{\partial y^2} + \text{lowerorder terms} = 0 \tag{4.2}$$

The classification can be further aided by the identification of the *discriminant* $\left[B^2 - AC\right]$ which represents an inherent property of the partial differential equation. We classify the partial differential equation according to the sign of the *discriminant*, i.e.

(a) $B^2 - AC < 0$: the differential equation is of the *elliptic*-type
(b) $B^2 - AC = 0$: the differential equation is of the *parabolic*-type
(c) $B^2 - AC > 0$: the differential equation is of the *hyperbolic*-type

The choice for this terminology becomes somewhat evident if we relate the classification of the second-order partial differential equations to the classification of *conics* in conventional analytic geometry. The general algebraic equation of the second degree in two variables x and y can be written as

$$Ax^2 + 2Bxy + Cy^2 + \text{lowerorder terms} = 0 \tag{4.3}$$

The above result represents a conic and is analogous to (4.2). If we were to consider a transformation of coordinates whereby the new orthogonal coordinate system $(\widetilde{x}, \widetilde{y})$ is obtained by rotating the (x, y) coordinate system by an angle α, it can be shown that (4.3) takes the form

$$\widetilde{A}\widetilde{x}^2 + 2\widetilde{B}\widetilde{x}\widetilde{y} + \widetilde{C}\widetilde{y}^2 + \text{lowerorder terms} = 0 \tag{4.4}$$

where

$$\widetilde{A} = \widetilde{A}\,(A, B, C, \ldots, \alpha)$$

$$\widetilde{B} = \widetilde{B}\,(A, B, C, \ldots, \alpha)$$

etc. Furthermore it can be shown that the discriminant is an invariant under a rotation of axes.

$$B^2 - AC = \widetilde{B}^2 - \widetilde{A}\widetilde{C} \tag{4.5}$$

The discriminant represents an inherent property of the conic and can be used to classify conics. We can identify three cases as follows:

$$\begin{aligned} &Case\ (i)\ \widetilde{B}^2 - \widetilde{A}\widetilde{C} < 0; \\ &Case\ (ii)\ \widetilde{B}^2 - \widetilde{A}\widetilde{C} = 0; \\ &Case\ (iii)\ \widetilde{B}^2 - \widetilde{A}\widetilde{C} > 0 \end{aligned} \tag{4.6}$$

The rationale for transforming coordinates is to simplify the equation (4.3) and the most obvious simplification is the elimination of the second term, which, when referred to (4.4), implies that $\widetilde{B} = 0$. We can now interpret this simplification in relation to (4.6).

For Case (i)

If $\widetilde{B} = 0$ then $-\widetilde{A}\widetilde{C} < 0$ or $\widetilde{A}\widetilde{C} > 0$. Hence, $\widetilde{A}$ and $\widetilde{C}$ must be of the same sign. The equation (4.4) can be written as

$$\mid \widetilde{A} \mid \widetilde{x}^2 + \mid \widetilde{C} \mid \widetilde{y}^2 + \text{lowerorder terms} = 0 \tag{4.7}$$

which represents an ellipse. Thus (4.4) represents an ellipse if $\left(\widetilde{B}^2 - \widetilde{A}\widetilde{C}\right) < 0$.

For Case (ii)

If $\widetilde{B} = 0$ then $\widetilde{A}\widetilde{C} = 0$. Therefore either $\widetilde{A} = 0$ or $\widetilde{C} = 0$ but not both if the equation is to remain a quadratic. If $\widetilde{C} = 0$, then (4.4) reduces to

$$\widetilde{A}\widetilde{x}^2 + \text{terms of lowerorder} = 0 \tag{4.8}$$

which represents a parabola. Thus (4.4) represents a parabola if $\left(\widetilde{B}^2 - \widetilde{A}\widetilde{C}\right) = 0$.

For Case (iii)

If $\widetilde{B} = 0$ then $-\widetilde{A}\widetilde{C} > 0$ or $\widetilde{A}\widetilde{C} < 0$. Consequently $\widetilde{A}$ and $\widetilde{C}$ must be opposite in sign and equation (4.4) gives

$$\mid \widetilde{A} \mid \widetilde{x}^2 - \mid \widetilde{C} \mid \widetilde{y}^2 + \text{terms of lowerorder} = 0 \tag{4.9}$$

which represents a hyperbola. Thus (4.4) represents a hyperbola if $\left(\widetilde{B}^2 - \widetilde{A}\widetilde{C}\right) > 0$.

4.2 Reduction to canonical forms

Partial differential equations which are encountered in science and engineering can present themselves in a variety of forms. In some instances the type of category of the partial differential equation is easily recognizable by a comparison with the standard forms. In other instances, the category is not so evident. In developing either a method of solution, or assessing the well-posed nature of the problem or establishing a priori the general nature of the solution it is highly desirable to ascertain the "*canonical*", namely, the "*simplest*" form of the partial differential equation. This becomes quite important when attention is focused on the development of numerical schemes for the solution of the partial differential equation where the method of solution will invariably depend on the character (*elliptic* or *parabolic* or *hyperbolic*) of the partial differential equation.

In order to obtain the canonical form of a more complicated partial differential equation we proceed by introducing a transformation of the independent variables x and y; such that

$$\xi = \xi(x, y) \quad ; \quad \eta = \eta(x, y) \tag{4.10}$$

and that both ξ and η possess continuous derivatives up to the second-order. We obtain

$$\frac{\partial u}{\partial x} = \frac{\partial u}{\partial \xi}\frac{\partial \xi}{\partial x} + \frac{\partial u}{\partial \eta}\frac{\partial \eta}{\partial x} \tag{4.11}$$

$$\frac{\partial u}{\partial y} = \frac{\partial u}{\partial \xi}\frac{\partial \xi}{\partial y} + \frac{\partial u}{\partial \eta}\frac{\partial \eta}{\partial y} \tag{4.12}$$

$$\begin{aligned}\frac{\partial^2 u}{\partial x^2} &= \frac{\partial^2 u}{\partial \xi^2}\left(\frac{\partial \xi}{\partial x}\right)^2 + 2\frac{\partial^2 u}{\partial \xi \partial \eta}\frac{\partial \xi}{\partial x}\frac{\partial \eta}{\partial x} + \frac{\partial^2 u}{\partial \eta^2}\left(\frac{\partial \eta}{\partial x}\right)^2 \\ &\quad + \frac{\partial u}{\partial \xi}\left(\frac{\partial^2 \xi}{\partial x^2}\right) + \frac{\partial u}{\partial \eta}\left(\frac{\partial^2 \eta}{\partial x^2}\right)\end{aligned} \tag{4.13}$$

$$\begin{aligned}\frac{\partial^2 u}{\partial x \partial y} &= \frac{\partial^2 u}{\partial \xi^2}\frac{\partial \xi}{\partial x}\frac{\partial \xi}{\partial y} + \frac{\partial^2 u}{\partial \xi \partial \eta}\frac{\partial \xi}{\partial x}\frac{\partial \eta}{\partial y} + \frac{\partial^2 u}{\partial \xi \partial \eta}\frac{\partial \xi}{\partial y}\frac{\partial \eta}{\partial x} \\ &\quad + \frac{\partial^2 u}{\partial \eta^2}\frac{\partial \eta}{\partial x}\frac{\partial \eta}{\partial y} + \frac{\partial u}{\partial \xi}\left(\frac{\partial^2 \xi}{\partial x \partial y}\right) + \frac{\partial u}{\partial \eta}\left(\frac{\partial^2 \eta}{\partial x \partial y}\right)\end{aligned} \tag{4.14}$$

$$\begin{aligned}\frac{\partial^2 u}{\partial y^2} &= \frac{\partial^2 u}{\partial \xi^2}\left(\frac{\partial \xi}{\partial y}\right)^2 + 2\frac{\partial^2 u}{\partial \xi \partial \eta}\frac{\partial \xi}{\partial y}\frac{\partial \eta}{\partial y} + \frac{\partial^2 u}{\partial \eta^2}\left(\frac{\partial \eta}{\partial y}\right)^2 \\ &\quad + \frac{\partial u}{\partial \xi}\left(\frac{\partial^2 \xi}{\partial y^2}\right) + \frac{\partial u}{\partial \eta}\left(\frac{\partial^2 \eta}{\partial y^2}\right)\end{aligned} \tag{4.15}$$

By substituting (4.13) to (4.15) in (4.2) and neglecting terms of order less than the second we obtain

$$\begin{aligned}& A\frac{\partial^2 u}{\partial x^2} + 2B\frac{\partial^2 u}{\partial x \partial y} + C\frac{\partial^2 u}{\partial y^2} + \text{lowerorder terms} \\ &\quad = \frac{\partial^2 u}{\partial \xi^2}\left[A\left(\frac{\partial \xi}{\partial x}\right)^2 + 2B\frac{\partial \xi}{\partial x}\frac{\partial \xi}{\partial y} + C\left(\frac{\partial \xi}{\partial y}\right)^2\right] + \\ &\quad + 2\frac{\partial^2 u}{\partial \xi \partial \eta}\left[A\frac{\partial \xi}{\partial x}\frac{\partial \eta}{\partial x} + B\left(\frac{\partial \xi}{\partial x}\frac{\partial \eta}{\partial y} + \frac{\partial \xi}{\partial y}\frac{\partial \eta}{\partial x}\right) + C\frac{\partial \xi}{\partial y}\frac{\partial \eta}{\partial y}\right] \\ &\quad + \frac{\partial^2 u}{\partial \eta^2}\left[A\left(\frac{\partial \eta}{\partial x}\right)^2 + 2B\frac{\partial \eta}{\partial x}\frac{\partial \eta}{\partial y} + C\left(\frac{\partial \eta}{\partial y}\right)^2\right] \\ &\quad + \text{lowerorder terms}\end{aligned} \tag{4.16}$$

If we denote the terms within the square brackets of the right hand side by

$\widetilde{A}$, $\widetilde{B}$, and $\widetilde{C}$, we obtain the transformed equivalent of (4.2) as

$$\widetilde{A}\frac{\partial^2 u}{\partial \xi^2} + 2\widetilde{B}\frac{\partial^2 u}{\partial \xi \partial \eta} + \widetilde{C}\frac{\partial^2 u}{\partial \eta^2} + \text{lowerorder terms} = 0 \tag{4.17}$$

If we were to examine the discriminant of the transformed equation $\left(\widetilde{B}^2 - \widetilde{A}\tilde{C}\right)$ we can show that

$$\left(\widetilde{B}^2 - \widetilde{A}\widetilde{C}\right) = \left(B^2 - AC\right)\left[\frac{\partial \xi}{\partial x}\frac{\partial \eta}{\partial y} - \frac{\partial \xi}{\partial y}\frac{\partial \eta}{\partial x}\right]^2 \tag{4.18}$$

where, the second term in square brackets in (4.18) can be identified as the Jacobian (named after C.G.J. Jacobi (1804-1851))of the transformation

$$J = \begin{vmatrix} \dfrac{\partial \xi}{\partial x} & \dfrac{\partial \xi}{\partial y} \\ \dfrac{\partial \eta}{\partial x} & \dfrac{\partial \eta}{\partial y} \end{vmatrix} \tag{4.19}$$

If we choose a rotation of the axes such that $\mid J \mid \equiv 1$, then the equation (4.17) is identical to (4.2). In the case of a more general transformation, J^2 is not zero, but will introduce at least a change of scale. As discussed previously, we can classify the second-order partial differential equations as

$$\begin{aligned} &\text{(i) elliptic equations if } \widetilde{B}^2 - \widetilde{A}\widetilde{C} < 0 \\ &\text{(ii) parabolic equations if } \widetilde{B}^2 - \widetilde{A}\widetilde{C} = 0 \\ &\text{(iii) hyperbolic equations if } \widetilde{B}^2 - \widetilde{A}\widetilde{C} > 0 \end{aligned} \tag{4.20}$$

The complete transformed equation can be recast in the general form

$$\widetilde{A}\frac{\partial^2 u}{\partial \xi^2} + 2\widetilde{B}\frac{\partial^2 u}{\partial \xi \partial \eta} + \widetilde{C}\frac{\partial^2 u}{\partial \eta^2} + \text{lowerorder terms} = 0 \tag{4.21}$$

where $\widetilde{A}$, $\widetilde{B}$, $\widetilde{C}$ are defined by

$$\widetilde{A} = A\left(\frac{\partial \xi}{\partial x}\right)^2 + 2B\frac{\partial \xi}{\partial x}\frac{\partial \xi}{\partial y} + C\left(\frac{\partial \xi}{\partial y}\right)^2 \tag{4.22}$$

$$\widetilde{B} = A\frac{\partial \xi}{\partial x}\frac{\partial \eta}{\partial x} + B\left(\frac{\partial \xi}{\partial x}\frac{\partial \eta}{\partial y} + \frac{\partial \xi}{\partial y}\frac{\partial \eta}{\partial x}\right) + C\frac{\partial \xi}{\partial y}\frac{\partial \eta}{\partial y} \tag{4.23}$$

$$\widetilde{C} = A\left(\frac{\partial \eta}{\partial x}\right)^2 + 2B\frac{\partial \eta}{\partial x}\frac{\partial \eta}{\partial y} + C\left(\frac{\partial \eta}{\partial y}\right)^2 \tag{4.24}$$

etc. The terms of lower-order are not central to the discussions involving classification of second-order partial differential equations.

In the preceding we have shown that the basic form of the second-order partial differential equation given by (4.2) can be transformed to an equation with identical form (4.21) and the associated coefficients are defined by (4.22) to (4.24). By introducing the substitutions

$$\alpha\left(x,y\right) = \frac{\partial \xi/\partial x}{\partial \xi/\partial y}, \text{ and } \beta\left(x,y\right) = \frac{\partial \eta/\partial x}{\partial \eta/\partial y} \tag{4.25}$$

we can rewrite the expressions for $\widetilde{A}$, $\widetilde{B}$, and $\widetilde{C}$ as follows:

$$\widetilde{A} = \left[A\alpha^2 + 2B\alpha + C\right]\left(\frac{\partial \xi}{\partial y}\right)^2 \tag{4.26}$$

$$\widetilde{B} = \left[A\alpha\beta + B\left(\alpha+\beta\right) + C\right]\left(\frac{\partial \xi}{\partial y}\right)\left(\frac{\partial \eta}{\partial y}\right) \tag{4.27}$$

$$\widetilde{C} = \left[A\beta^2 + 2B\beta + C\right]\left(\frac{\partial \eta}{\partial y}\right)^2 \tag{4.28}$$

If we are to reduce the governing equation and therefore the transformed version of the governing equation to a "*canonical*" form, it is necessary to obtain those functions ξ and η which will render $\widetilde{A}$, $\widetilde{B}$, $\widetilde{C}$ etc., zero. Let us first consider the case when $\widetilde{A} = 0$. From (4.26) we note that for $\widetilde{A}$ to be zero

$$A\alpha^2 + 2B\alpha + C = 0 \tag{4.29}$$

the roots of which are

$$\alpha = \frac{-B \pm \sqrt{B^2 - AC}}{A} \tag{4.30}$$

An inspection of (4.28) will also reveal that if β has roots similar to α, given by (4.30), we have $\widetilde{C} \equiv 0$. Therefore, without loss of generality, we can assume that

$$\alpha = \frac{1}{A}\left[-B + \sqrt{B^2 - AC}\right] \tag{4.31}$$

$$\beta = \frac{1}{A}\left[-B - \sqrt{B^2 - AC}\right] \tag{4.32}$$

Therefore, if we can find the values of α and β as defined by (4.31) and (4.32), the values of $\widetilde{A}$ and $\widetilde{C}$ are identically zero and the partial differential equation reduces to a very simple form

$$\frac{\partial^2 u}{\partial \xi \partial \eta} + \frac{1}{2\tilde{B}} F\left(\xi, \eta, u, \frac{\partial u}{\partial \xi}, \frac{\partial u}{\partial \eta}\right) = 0 \tag{4.33}$$

$$\alpha\beta = \frac{C}{A} \;\; ; \;\; (\alpha + \beta) = -\frac{2B}{A} \tag{4.34}$$

and F is identified as the grouping of terms of lower-order. Referring to the developments in Chapter 3 and noting that for ξ=const., and η=const.,

$$\frac{dy}{dx} = -\frac{\partial\xi/\partial x}{\partial\xi/\partial y} \quad \text{and} \quad \frac{dy}{dx} = -\frac{\partial\eta/\partial x}{\partial\eta/\partial y} \tag{4.35}$$

respectively, we obtain from (4.25) and (4.35) the following ordinary differential equations

$$\frac{dy}{dx} + \alpha(x,y) = 0 \tag{4.36}$$

$$\frac{dy}{dx} + \beta(x,y) = 0 \tag{4.37}$$

The functions $\xi(x,y)$ and $\eta(x,y)$ can be taken as C_1 and C_2 (see eq.(3.32)) obtained by integrating (4.36) and (4.37). These equations are known as the *characteristic equations.* They are ordinary differential equations for families of curves in the x, y plane along which ξ = constant and η = constant. The integration of (4.36) and (4.37) yields the *characteristic curves.*

Let us now interpret the three categories of second-order partial differential equations by examining in sequence the value of the discriminant as interpreted through the parameters α and β.

(i) Elliptic Case.

In this instance $\left(B^2 - AC\right) < 0$; consequently the roots α and β are complex conjugate i.e.

$$\alpha = \frac{1}{A}\left[-B + i\sqrt{AC - B^2}\right] \tag{4.38}$$

$$\beta = \frac{1}{A}\left[-B - i\sqrt{AC - B^2}\right] \tag{4.39}$$

We can, for convenience introduce transformations which will allow us to deal with real numbers, i.e.

$$\mu = \xi + \eta \ \ ; \ \ \lambda = i\,(\xi - \eta) \tag{4.40}$$

The canonical form of the equation is (4.33); with the aid of the transformations (4.40) we can now show that

$$\frac{\partial^2 u}{\partial \xi \partial \eta} = \frac{\partial^2 u}{\partial \mu^2} + \frac{\partial^2 u}{\partial \lambda^2} \tag{4.41}$$

Hence, the canonical form of the elliptic equation is

$$\frac{\partial^2 u}{\partial \mu^2} + \frac{\partial^2 u}{\partial \lambda^2} + \frac{\widetilde{F}}{2\widetilde{B}} \left(\mu, \lambda, \frac{\partial u}{\partial \mu}, \frac{\partial u}{\partial \lambda} \right) = 0 \tag{4.42}$$

If the lower-order terms in (4.42) are absent it is clear that the canonical form of an elliptic partial differential equation must be of the form

$$\frac{\partial^2 u}{\partial \mu^2} + \frac{\partial^2 u}{\partial \lambda^2} = 0 \tag{4.43}$$

(ii) Parabolic Case.

In this case $\left(B^2 - AC\right) = 0$, as such (4.31) and (4.32) yield the same root, i.e.

$$\alpha = -\frac{B}{A} \tag{4.44}$$

which gives both $\widetilde{A} = 0$ and $\widetilde{C} = 0$ in (4.26) and (4.28) respectively. The function η can be set equal to any value which is independent of ξ. Hence, the canonical form of a partial differential equation of the parabolic-type can be written in the form

$$\frac{\partial^2 u}{\partial \eta^2}+\frac{1}{\widetilde{C}}\widetilde{F}\left(\xi,\eta,u,\frac{\partial u}{\partial \xi},\frac{\partial u}{\partial \eta}\right)=0 \tag{4.45}$$

Using (4.25) and (4.28) we have

$$\widetilde{C}=A\left(\frac{\partial \eta}{\partial x}\right)^2+2B\left(\frac{\partial \eta}{\partial x}\right)\left(\frac{\partial \eta}{\partial y}\right)+C\left(\frac{\partial \eta}{\partial y}\right)^2 \tag{4.46}$$

(iii) Hyperbolic Case. In this case $\left(B^2-AC\right)>0$ and the roots given by (4.31) and (4.32) are real and the canonical form of the equation is exactly that given by (4.33) where ξ and η are obtained from the solution of the differential equations (4.36) and (4.37).

4.3 Applications of the procedures

The general theory and procedures indicated in the previous section will become a little more clear through their application to some specific cases of second-order partial differential equations.

Example 4.1

Reduce the partial differential equation to a canonical form.

$$\frac{\partial^2 u}{\partial x^2}+2\sqrt{3}\frac{\partial^2 u}{\partial x \partial y}+\frac{\partial^2 u}{\partial y^2}=0 \tag{4.47}$$

Solution

If we compare (4.47) with (4.2) we have

$$A = 1 \quad ; \quad B = \sqrt{3} \quad ; \quad C = 1$$

or

$$B^2 - AC = 2 > 0$$

Hence the partial differential equation (4.47) is *hyperbolic*. From (4.31) and (4.32) we obtain

$$\alpha = \frac{1}{A}\left[-B + \sqrt{B^2 - AC}\right] = \left(-\sqrt{3} + \sqrt{2}\right) \tag{4.48}$$

$$\beta = \frac{1}{A}\left[-B - \sqrt{B^2 - AC}\right] = \left(-\sqrt{3} - \sqrt{2}\right) \tag{4.49}$$

The characteristic equations are therefore

$$\frac{dy}{dx} = \left(\sqrt{3} - \sqrt{2}\right) \quad ; \quad \frac{dy}{dx} = \left(\sqrt{3} + \sqrt{2}\right) \tag{4.50}$$

Integrating the equations (4.50) we have

$$y = \left(\sqrt{3} - \sqrt{2}\right)x + C_1 \quad ; \quad y = \left(\sqrt{3} + \sqrt{2}\right)x + C_2 \tag{4.51}$$

Therefore, the characteristic coordinates are given by

$$y - \left(\sqrt{3} - \sqrt{2}\right)x = C_1 = \xi$$

$$y - \left(\sqrt{3} + \sqrt{2}\right) x = C_2 = \eta \tag{4.52}$$

Substituting (4.52) into (4.22) to (4.24) we have

$$\widetilde{A} = 0 \;\; ; \;\; \widetilde{B} = -4 + 2\sqrt{6} \;\; ; \;\; \widetilde{C} = 0$$

Hence the canonical form of the equation (4.47), in terms of the characteristic coordinates, is

$$\frac{\partial^2 u}{\partial \xi \partial \eta} = 0 \tag{4.53}$$

Example 4.2

Classify the following partial differential equation and reduce it to its canonical form.

$$\frac{\partial^2 u}{\partial x^2} + 2\frac{\partial^2 u}{\partial x \partial y} + \frac{\partial^2 u}{\partial y^2} + \frac{\partial u}{\partial x} = 0 \tag{4.54}$$

Solution

By comparison with (4.2) we obtain

$$A = 1 \;\; ; \;\; B = 1 \;\; ; \;\; C = 1$$

or

$$B^2 - AC = 0$$

Hence the partial differential equation is *parabolic* and from (4.31) and (4.32) we have

$$\alpha = \beta = -\frac{B}{A} = -1 \tag{4.55}$$

The characteristic equations reduce to a single equation

$$\frac{dy}{dx} - 1 = 0 \quad \text{or} \quad y - x = C_1 = \xi \tag{4.56}$$

The value of η can be taken as any function of x and y independent of ξ provided the Jacobian (4.19) is non-zero, e.g. Let us assume that

$$\eta = y + x \tag{4.57}$$

Using (4.56) and (4.57) in (4.19) we have

$$J = \begin{vmatrix} \dfrac{\partial \xi}{\partial x} & \dfrac{\partial \xi}{\partial y} \\ \dfrac{\partial \eta}{\partial x} & \dfrac{\partial \eta}{\partial y} \end{vmatrix} = \begin{vmatrix} -1 & 1 \\ 1 & 1 \end{vmatrix} = -2$$

Substituting (4.56) and (4.57) into (4.22) to (4.24) we obtain

$$\tilde{A} = \tilde{B} = 0 \quad \text{and} \quad \tilde{C} = 4$$

Also, considering (4.11), (4.56) and (4.57) we have

$$\frac{\partial u}{\partial x} = \frac{\partial u}{\partial \xi}\frac{\partial \xi}{\partial x} + \frac{\partial u}{\partial \eta}\frac{\partial \eta}{\partial x} = -\frac{\partial u}{\partial \xi} + \frac{\partial u}{\partial \eta} \tag{4.58}$$

The canonical form for (4.54) can now be written as

$$\frac{\partial^2 u}{\partial \eta^2} + \frac{1}{4}\left(-\frac{\partial u}{\partial \xi} + \frac{\partial u}{\partial \eta}\right) = 0 \tag{4.59}$$

Example 4.3

Consider the second-order partial differential equation of the form

$$\frac{\partial^2 u}{\partial x^2} + 2\frac{\partial^2 u}{\partial x \partial y} + 5\frac{\partial^2 u}{\partial y^2} = x\frac{\partial u}{\partial x} \tag{4.60}$$

Obtain the characteristic equations, the characteristic coordinates and the canonical form of the differential equation.

Solution

From (4.2) and (4.60) we have

$$A = 1 \;\; ; \;\; B = 1 \;\; ; \;\; C = 5$$

The discriminant

$$B^2 - AC = -4 < 0$$

Hence the partial differential equation is *elliptic*. From (4.31) and (4.32), the roots of the characteristic equation are

$$\alpha = [-1 + 2i] \;\; ; \;\; \beta = [-1 - 2i] \tag{4.61}$$

The ordinary differential equations for determining the characteristic coordinates take the forms

$$\frac{dy}{dx} + (-1 + 2i) = 0 \;\; ; \;\; \frac{dy}{dx} + (-1 - 2i) = 0 \tag{4.62}$$

Integrating (4.62) we have

$$\begin{aligned} &y - (1 - 2i)\, x = C_1 = \xi \\ &y - (1 + 2i)\, x = C_2 = \eta \end{aligned} \tag{4.63}$$

As noted previously, we can introduce transformations, which will allow us to deal with real numbers instead of the complex quantities (see e.g. (4.40)). We can introduce new coordinates λ and μ such that

$$\lambda = i\,(\xi - \eta) = -4x \;\; ; \;\; \mu = \xi + \eta = 2\,(y - x) \tag{4.64}$$

It can be shown that

$$\frac{\partial}{\partial \xi} = \frac{\partial}{\partial \mu} + i\frac{\partial}{\partial \lambda} \;\; ; \;\; \frac{\partial}{\partial \eta} = \frac{\partial}{\partial \mu} - i\frac{\partial}{\partial \lambda} \;\; ; \;\; \frac{\partial^2}{\partial \xi \partial \eta} = \frac{\partial^2}{\partial \mu^2} + \frac{\partial^2}{\partial \lambda^2} \tag{4.65}$$

Also it can be shown that

$$x\frac{\partial u}{\partial x} = \frac{\lambda}{2}\left(\frac{\partial u}{\partial \mu} + 2\frac{\partial u}{\partial \lambda}\right) \tag{4.66}$$

Substituting (4.61) in (4.26) to (4.28) it can be shown that

$$\widetilde{A} = \widetilde{C} = 0 \ ; \ \widetilde{B} = 8$$

Consequently, the canonical form of the partial differential equation is

$$\frac{\partial^2 u}{\partial \lambda^2} + \frac{\partial^2 u}{\partial \mu^2} = \frac{\lambda}{16}\left(\frac{\partial u}{\partial \mu} + 2\frac{\partial u}{\partial \lambda}\right) \tag{4.67}$$

• • •

Up to now we have examined the category of second-order partial differential equations where the parameters A, B and C, which form the discriminant, are constants. As a further example consider the problem where A, B and C are not necessarily constants.

Example 4.4

Obtain the characteristic coordinates and the canonical form of the partial differential equation

$$x^2\frac{\partial^2 u}{\partial x^2} + 2xy\frac{\partial^2 u}{\partial x \partial y} + y^2\frac{\partial^2 u}{\partial y^2} = x\frac{\partial u}{\partial x} + y\frac{\partial u}{\partial y} \tag{4.68}$$

Solution

Considering (4.1) and (4.68) we have

$$A(x,y) = x^2 \ ; \ B(x,y) = xy \ ; \ C(x,y) = y^2 \tag{4.69}$$

The discriminant

$$B^2 - AC = 0$$

Hence the partial differential equation is *parabolic* for all x and y. From (4.31) and (4.32), the roots of the characteristic equation are

$$\alpha = \beta = -\frac{B}{A} = -\frac{y}{x} \tag{4.70}$$

The characteristic equations reduce to a single equation

$$\frac{dy}{dx} - \frac{y}{x} = 0 \quad \text{or} \quad \ln y = \ln x + \ln C_1$$

Hence

$$\frac{y}{x} = C_1 = \xi \tag{4.71}$$

Again, the value of η can be taken as any function of x and y, independent of ξ provided the Jacobian (4.19) is non-zero, e.g.

$$\eta = y \tag{4.72}$$

Substituting (4.71) and (4.72) into (4.22) to (4.24) we have

$$\widetilde{A} = 0 \ ; \quad \widetilde{B} = 0 \ ; \quad \widetilde{C} = y^2 = \eta^2 \tag{4.73}$$

Therefore the canonical form of the partial differential equation (4.68) can be written as

$$\widetilde{C}\frac{\partial^2 u}{\partial \eta^2} = x\frac{\partial u}{\partial x} + y\frac{\partial u}{\partial y} \tag{4.74}$$

Using (4.71) and (4.72), the right hand side of (4.74) can be reduced to the form

$$x\frac{\partial u}{\partial x}+y\frac{\partial u}{\partial y}=\eta\frac{\partial u}{\partial \eta} \tag{4.75}$$

Using (4.73) and (4.75) we can reduce (4.74) to the form

$$\frac{\partial^2 u}{\partial \eta^2}=\frac{1}{\eta}\frac{\partial u}{\partial \eta} \tag{4.76}$$

which is the canonical form of the partial differential equation (4.68) and is essentially a *first-order partial differential equation* for $\partial u/\partial \eta$.

• • •

The classification and identification of the three basic classes of partial differential equations have been presented in a very general sense without appeal to any engineering problems. In the subsequent chapters we will examine in detail the engineering situations which yield the separate classes of *elliptic*, *parabolic* and *hyperbolic* equations. The applications of these three classes of partial differential equations to problems in engineering and science is sufficiently well developed to warrant the subject matter being presented under the three separate categories.

As a separate exercise perform the calculations for the reduction to a canonical form by making use of the *chain rule* and *product rule* differentiations given by (4.13), (4.14) and (4.15), and verify that the final result corresponding to $\xi=(y/x)$ and $\eta=y$ is identical to the result (4.76).

4.4 Classification of second-order pdes for n independent variables

The procedures discussed in the previous sections focused primarily on the classification procedures for second-order partial differential equations with *two-independent variables.* The classification concepts become more meaningful, in terms of concepts based on *analytic geometry* when there are only two independent variables. When partial differential equations are formulated in a four dimensional space, three independent variables could be spatial dimensions and the fourth independent variable could be designated as time. The classification of second-order partial differential equations in 'n' independent variables is a generalization of the classification concepts developed for two independent variables. The classification aspects become important in formalizing the category of the partial differential equations of the second-order even in the four dimensional case.

Consider a generalized form of the second-order partial differential equation n independent variables, which has a form

$$\sum_{i=1,}^{n} \sum_{j=1}^{n} A_{ij} \frac{\partial^2 u}{\partial x_i \partial x_j} + \sum_{i=1}^{n} B_i \frac{\partial u}{\partial x_i} + Cu = D \tag{4.77}$$

where A_{ij}, B_i, C and D could be functions of the independent variables $x_1, x_2, ..., x_n$.

An important observation in the development of the theory of partial differential equations is that the most important properties of solutions of linear partial differential equations depend only on the form of the highest order terms appearing in the equation. These terms from what is usually referred to as the *principal part* of the equation. For example in equation (4.77), the principal part of the equation involves only the second-order terms with multiplying constants or functions A_{ij}.

If we can construct a "*coefficients matrix*" involving A_{ij}, we can classify the partial differential equation according to the character of $[A_{ij}]$, the coefficients matrix of the *principal part.* Since the operator $\partial^2/\partial x_i \partial x_j$ is symmetric i.e.

$$\frac{\partial^2 u}{\partial x_i \partial x_j} = \frac{\partial^2 u}{\partial x_j \partial x_i} \tag{4.78}$$

we can assume that $[A_{ij}]$ is symmetric. i.e.

$$A_{ij} = A_{ji} \qquad \text{or } [A_{ij}] = [A_{ij}] \;;\; \begin{matrix}(i = 1, 2, \ldots n \\ j = 1, 2, \ldots . n)\end{matrix} \tag{4.79}$$

If we consider the *principal part* of the partial differential equation in two independent variables, the coefficients matrix $[A_{ij}]$ can be constructed as follows

$$A_{ij} = \begin{bmatrix} A_{11}\dfrac{\partial^2 u}{\partial x_1^2} & A_{12}\dfrac{\partial^2 u}{\partial x_1 \partial x_2} \\ \\ A_{12}\dfrac{\partial^2 u}{\partial x_2 \partial x_1} & A_{22}\dfrac{\partial^2 u}{\partial x_2^2} \end{bmatrix} \tag{4.80}$$

We can write

$$\begin{aligned}\sum_{i=1}^{2}\sum_{j=1}^{2} A_{ij}\frac{\partial^2 u}{\partial x_i \partial x_j} &= A_{11}\frac{\partial^2 u}{\partial x_1^2} + A_{12}\frac{\partial^2 u}{\partial x_1 \partial x_2} + A_{21}\frac{\partial^2 u}{\partial x_2 \partial x_1} \\ &\quad + A_{22}\frac{\partial^2 u}{\partial x_2^2}\end{aligned} \tag{4.81}$$

If we compare (4.81) with

$$A\frac{\partial^2 u}{\partial x^2} + B\frac{\partial^2 u}{\partial x \partial y} + B\frac{\partial^2 u}{\partial y \partial x} + C\frac{\partial^2 u}{\partial y^2} = \text{Lower-order terms} \tag{4.82}$$

we have

$$A_{11} = A \quad ; \quad A_{12} = A_{21} = B \quad ; \quad A_{22} = C \tag{4.83}$$

$$x_1 = x \quad ; \quad x_2 = y$$

The "coefficients matrix" of the principal part is therefore

$$[A_{ij}] = \begin{bmatrix} A & B \\ B & C \end{bmatrix} = [A_{ij}]^T \tag{4.84}$$

The eigen-values of the coefficient's matrix of the principal part is given by the determinantal equation

$$\begin{vmatrix} (A-\lambda) & B \\ B & (C-\lambda) \end{vmatrix} = 0 \quad ; \quad \text{or } |A_{ij} - \lambda\delta_{ij}| = 0 \tag{4.85}$$

where δ_{ij} is the Kronecker delta defined by

$$\delta_{ij} = \begin{cases} 1 & i = j \\ 0 & i \neq j \end{cases} \tag{4.86}$$

Expanding (4.85) we have

$$\lambda^2 - (A+C)\lambda - (B^2 - AC) = 0 \tag{4.87}$$

The roots of (4.87) are

$$\lambda_1 = \frac{(A+C) + [(A-C)^2 + (2B)^2]^{\frac{1}{2}}}{2} \tag{4.88}$$

$$\lambda_2 = \frac{(A+C) - [(A-C)^2 + (2B)^2]^{\frac{1}{2}}}{2} \tag{4.89}$$

Note that both λ_1 and λ_2 are real, which is also proved by the fact that the coefficients matrix of the principal part is *symmetric*. If we multiply the expression of the eigenvalues given by (4.88) and (4.89), we have

$$\lambda_1\lambda_2 = (B^2 - AC) = -\Delta \tag{4.90}$$

where Δ is the discriminant. The conclusions relating to the sign of Δ can be interpreted in relation to the *sign of the eigen-values λ_1 and λ_2*

(i) $\Delta > 0$: implies that λ_1 and λ_2 are *non-zero* and are of opposite sign
(ii) $\Delta = 0$: implies that *one* of λ_1 and λ_2 is *zero*
(iii) $\Delta < 0$: implies that λ_1 and λ_2 are non-zero but have the same sign.

This classification concept based on the *signs of the eigen-values of the coefficients matrix of the principal part* is another way of classifying a second-order partial differential equation.

In the case of a second-order partial differential equation of 'n' independent variables, the *coefficients matrix of the principal part* is

$$[A_{ij}] = \begin{bmatrix} A_{11} & A_{13} & A_{13}\cdots\cdots\cdots\cdots A_{1n} \\ A_{21} & A_{22} & A_{23}\cdots\cdots\cdots\cdots A_{2n} \\ A_{31} & A_{32} & A_{33}\cdots\cdots\cdots\cdots A_{3n} \\ & & \\ & & \\ A_{n1} & A_{n2} & A_{n3}\cdots\cdots\cdots\cdots A_{nn} \end{bmatrix} \tag{4.91}$$

The eigen-values of the symmetric coefficients matrix is obtained from the determinantal equation

$$|A_{ij} - \lambda\delta_{ij}| = 0 \tag{4.92}$$

From a theorem in linear algebra we can show that if

$$A_{ij} = A_{ji} \tag{4.93}$$

the eigenvalues λ_n are all *real*. Invoking a further theorem in linear algebra it can be shown that for a specific location $\mathbf{x}_0$ (in n dimensions), the number of positive, zero and negative eigenvalues of $[A_{ij}]$ remain invariant under smooth non-singular transformation of coordinates. These properties of the λ_n eigenvalues of $[A_{ij}]$ can now be used to classify the second-order partial differential equation referred to an n-dimensional space.

We define $\lambda_1, \lambda_2, \lambda_3, \ldots\ldots\ldots\ldots, \lambda_n$ as the eigenvalues of the coefficients matrix of the principal part of a second-order partial differential equation of n independent variables.

(i) If $\lambda_1, \lambda_2, \lambda_3, \ldots\ldots\ldots\ldots, \lambda_n$ are *non-zero, and possess the same sign at a point*, then the partial differential equation is said to be elliptic at that point: i.e.

$$\frac{\partial^2 u}{\partial x_1^2} + \frac{\partial^2 u}{\partial x_2^2} + \frac{\partial^2 u}{\partial x_3^2} + \ldots\ldots\ldots\ldots \frac{\partial^2 u}{\partial x_n^2} = 0 \tag{4.94}$$

(ii) If any *one* of $\lambda_1, \lambda_2, \lambda_3, \ldots\ldots\ldots\ldots, \lambda_n$ is *zero at a point*, then the equation is said to be *parabolic* at that point.

(iii) If $\lambda_1, \lambda_2, \lambda_3, \ldots\ldots\ldots\ldots, \lambda_n$ are all *non-zero and all excluding one have the same sign* at a point, then the partial differential equation is said to be *hyperbolic* at that point.

(iv) If $\lambda_1, \lambda_2, \lambda_3, \ldots\ldots\ldots\ldots, \lambda_n$ are all *non-zero and at least two of them are positive and two of them are negative* at a point, then the partial differential equation is *ultra-hyperbolic* at that point.

Example 4.5

Classify the second-order partial differential equation

$$\frac{\partial^2 u}{\partial x^2} - 3\frac{\partial^2 u}{\partial y^2} + 4\frac{\partial^2 u}{\partial x \partial z} + 4\frac{\partial^2 u}{\partial z^2} + x\frac{\partial u}{\partial x} + y\frac{\partial u}{\partial y} + z\frac{\partial u}{\partial z} = 0 \tag{4.95}$$

where $u = u(x, y, z)$ is a continuously differentiable function.

Solution

The principal part of the partial differential equation consists only of the second-order terms: i.e.

$$\text{Principal part} = \frac{\partial^2 u}{\partial x^2} - 3\frac{\partial^2 u}{\partial y^2} + 4\frac{\partial^2 u}{\partial x \partial z} + 4\frac{\partial^2 u}{\partial z^2} + \text{lower-order terms} \tag{4.96}$$

The matrix of coefficients of the "*principal part*" can be constructed by ensuring that the resulting matrix is symmetric, i.e.

$$A_{ij} = \begin{bmatrix} A_{11} & A_{12} & A_{13} \\ A_{21} & A_{22} & A_{23} \\ A_{31} & A_{32} & A_{33} \end{bmatrix} \tag{4.97}$$

where $A_{ij} = A_{ji}$ and

$$\begin{aligned} & A_{11} = 1 \quad ; A_{22} = -3 \quad ; \quad A_{33} = 4 \\ & A_{12} = A_{21} = 0 \ ; A_{13} + A_{31} = 4 \text{ or } A_{13} = A_{31} = 2 \\ & A_{23} = A_{32} = 0 \end{aligned} \tag{4.98}$$

Therefore the matrix of the "*principal part*" is given by

$$A_{ij} = \begin{bmatrix} 1 & 0 & 2 \\ 0 & -3 & 0 \\ 2 & 0 & 4 \end{bmatrix} = A_{ji} \tag{4.99}$$

The eigenvalues of A_{ij} can be obtained from the result

$$|A_{ij} - \lambda\delta_{ij}| = 0 \tag{4.100}$$

or

$$\begin{vmatrix} (1-\lambda) & 0 & 2 \\ 0 & (-3-\lambda) & 0 \\ 2 & 0 & (4-\lambda) \end{vmatrix} = 0$$

or

$$(3+\lambda)[4-(1-\lambda)(4-\lambda)] = 0 \tag{4.101}$$

or

$$(3+\lambda)\lambda(\lambda-5) = 0 \tag{4.102}$$

or

$$\lambda_1 = 0, \qquad \lambda_2 = -3 \quad ; \quad \lambda_3 = 5 \tag{4.103}$$

Since one eigen-value in the matrix constructed from the principal part is *zero*, the partial differential equation is *parabolic* for all choices of x, y, z.

4.5 PROBLEM SET 4

4.1 Define the domain ($x \in (-\infty, \infty)$), ($y \in (-\infty, \infty)$) in which the following partial differential equations can be classified as either *elliptic* or *parabolic* or *hyperbolic*. (Indicate your answers with reference to a sketch).

(i) $x\dfrac{\partial^2 u}{\partial x^2} = y\dfrac{\partial^2 u}{\partial y^2}$

(ii) $y\left(\dfrac{\partial}{\partial x} + \dfrac{\partial}{\partial y}\right)\left(\dfrac{\partial}{\partial x} - \dfrac{\partial}{\partial y}\right)u + 2\dfrac{\partial}{\partial y}\left(\dfrac{\partial}{\partial y} + x\dfrac{\partial}{\partial x}\right)u = 0$

(iii) $\sqrt{x}\nabla^2 u = y\dfrac{\partial^2 u}{\partial x \partial y}$

(iv) $\dfrac{\partial^2 u}{\partial x^2} + y\dfrac{\partial^2 u}{\partial y^2} = 0$

(v) $\left(x^2 - 1\right)\dfrac{\partial^2 u}{\partial x^2} + 2xy\dfrac{\partial^2 u}{\partial x \partial y} + \left(y^2 - 1\right)\dfrac{\partial^2 u}{\partial y^2} = 0$

(vi) $\dfrac{\partial^2 u}{\partial y^2} - x\dfrac{\partial^2 u}{\partial x \partial y} + x\dfrac{\partial u}{\partial x} + y\dfrac{\partial u}{\partial y} + u = 0$

(vii) $\dfrac{\partial^2 u}{\partial x^2} + xy\dfrac{\partial^2 u}{\partial y^2} = 0$

(viii) $\dfrac{\partial^2 u}{\partial x^2} - \dfrac{\partial^2 u}{\partial y^2} = 0$

(ix) $y^2\dfrac{\partial^2 u}{\partial x^2} + 2xy\dfrac{\partial^2 u}{\partial x \partial y} + x^2\dfrac{\partial^2 u}{\partial y^2} = 0$

(x) $\dfrac{\partial^2 u}{\partial x^2} = \dfrac{\partial^2 u}{\partial x \partial y} - 2\dfrac{\partial^2 u}{\partial y^2}$

(xi) $\dfrac{(y+b)}{b}\dfrac{\partial^2 u}{\partial x^2} + 2\dfrac{x}{a}\dfrac{\partial^2 u}{\partial x \partial y} + x^2\dfrac{\partial u}{\partial x} + y^2\dfrac{\partial u}{\partial y} = \dfrac{(y-b)}{b}\dfrac{\partial^2 u}{\partial y^2}$
(a, b are constants)

(xii) $2x\dfrac{\partial^2 u}{\partial x^2} + 2\,(x+y)\,\dfrac{\partial^2 u}{\partial x \partial y} = (1+y)\left[\dfrac{\partial u}{\partial y} - \dfrac{\partial^2 u}{\partial y^2}\right]$

4.2 Use the chain rule differentiations and product rule to derive the complete expressions (i.e. including lower-order terms) for $(\partial^2 u/\partial x^2)$, $(\partial^2 u/\partial y^2)$ and $(\partial^2 u/\partial x \partial y)$ in terms of the new coordinates ξ and η such that $\xi = \xi(x, y)$; $\eta = \eta(\xi, x)$.

4.3 Reduce the following partial differential equations to their canonical forms. In each case determine the characteristic equations and characteristic coordinates.

(i) $\dfrac{\partial^2 u}{\partial x^2} + 2\sqrt{3}\dfrac{\partial^2 u}{\partial x \partial y} + \dfrac{\partial^2 u}{\partial y^2} = 0$

(ii) $y^2\dfrac{\partial^2 u}{\partial x^2} = x^2\dfrac{\partial^2 u}{\partial y^2}$

(iii) $\dfrac{\partial^2 u}{\partial x^2} + \dfrac{\partial^2 u}{\partial x \partial y} + \dfrac{\partial^2 u}{\partial y^2} = 0$

(iv) $\dfrac{\partial^2 u}{\partial x^2} + x^2\dfrac{\partial^2 u}{\partial y^2} = 2$

4.4 Reduce the second-order partial differential for $u(x, y)$ given by

(i) $\left(y\dfrac{\partial}{\partial x} + x\dfrac{\partial}{\partial y}\right)\left(y\dfrac{\partial}{\partial x} - x\dfrac{\partial}{\partial y}\right)u\,(x, y) = 0; \quad x > 0; \quad y > 0$

(ii) $\dfrac{\partial^2 u}{\partial x^2} - x^2 y\dfrac{\partial^2 u}{\partial y^2} = 0; \quad y > 0$

(iii) $\left(\dfrac{\partial}{\partial x} + \dfrac{\partial}{\partial y}\right)\left(\dfrac{\partial}{\partial x} - 3\dfrac{\partial}{\partial y} - 1\right)u\,(x, y) = 0$

to their canonical forms and include in your solution all lower-order terms

(iv) Integrate the canonical form for (iii) to obtain an expression for $u(\xi, \eta)$ in terms of arbitrary functions.

4.5 Obtain the general forms of the solutions for the following partial differential equations.

(i) $y^2 \frac{\partial^2 u}{\partial x^2} + 2xy \frac{\partial^2 u}{\partial x \partial y} + x^2 \frac{\partial^2 u}{\partial y^2} = 0$

(ii) $2 \frac{\partial^2 u}{\partial x^2} + 4 \frac{\partial^2 u}{\partial x \partial y} + \frac{\partial^2 u}{\partial y^2} = 0$

4.6 Classify the following partial differential equations by considering the principal part of each equation

(i) $\nabla^2 u + 3 \frac{\partial^2 u}{\partial z^2} - 6 \frac{\partial^2 u}{\partial x \partial y} + 2\sqrt{2} \frac{\partial}{\partial z} \left(\frac{\partial u}{\partial x} - \frac{\partial u}{\partial y} \right) = 0$

(ii) $\frac{\partial^2 u}{\partial x \partial y} + \frac{\partial^2 u}{\partial y \partial z} + \frac{\partial^2 u}{\partial x \partial z} = 0$

(iii) $\nabla^2 u + \frac{\partial^2 u}{\partial x \partial y} + \frac{\partial^2 u}{\partial y \partial z} + \frac{\partial^2 u}{\partial x \partial z} = 0$

(iv) $\nabla^2 u + 2 \left(\frac{\partial}{\partial x} + \frac{\partial}{\partial z} \right) \left(\frac{\partial}{\partial x} + \frac{\partial}{\partial y} \right) u = \left(\frac{\partial}{\partial y} - \frac{\partial}{\partial z} \right) \left(\frac{\partial}{\partial y} - \frac{\partial}{\partial z} \right) u$

where $u = u(x, y, z)$ is a non-zero continuously differentiable function, and ∇^2 is Laplace's operator in three dimensions.

Chapter 5

Laplace's equation

The partial differential equation which is identified with the name of Pierre Simon Marquis de Laplace (1749-1827) is one of the most important equations in mathematics which has wide applications to a number of topics relevant to mathematical physics and engineering. In the area of mathematical physics, Laplace's equation can be used to formulate the general class of problems associated with the theory of gravitation, electrostatics, dielectrics and magnetostatics. In all these applications, the associated fields can be expressed as a gradient of a potential. For example, in the theory of gravitation, the force of attraction associated with matter is expressed as the gradient of a gravitational potential; in electrostatics, the electric vector is expressed as the gradient of an electrostatic potential; in magnetostatics, the magnetic vector is expressed in terms of a magnetostatic potential; in electricity, the conduction current vector is derived from the gradient of a potential function. This ability to express fundamental vector fields of interest to particular areas is an underlying theme in all applications which culminate in the development of Laplace's equation.

Similar developments follow in areas of interest to engineering applications. For example, in the area of fluid mechanics it is assumed that the velocity vector in a perfect fluid in irrotational motion can be derived as the gradient of a velocity potential; similar considerations apply to the formulation of problems dealing with ground water flow in porous media, the study of torsion of prismatic elastic solids and in the analysis of steady heat conduction in solid bodies.

The objective of this chapter is, firstly, to review the physical principles which lead to the development of Laplace's equation, by appeal to engineering applications in the areas of fluid flow, flow in porous media and heat conduction. Secondly, to outline certain basic procedures which can be used to develop solutions of interest to engineering applications. Finally the chapter contains a brief discussion of certain general principles which facilitate identification of the structure of the solutions and the theorem pertaining to the uniqueness of the solution.

5.1 Derivation of Laplace's equation

As indicated previously, the derivation of Laplace's equation can be approached by appeal to a variety of physical processes associated with science and engineering. In the ensuing, however, we shall focus on the derivation of the equation by considering applications in the general area of fluid mechanics.

5.1.1 Irrotational flow in fluid mechanics

The equation of continuity or mass conservation is central to any discussion involving problems in fluid mechanics. We focused on this aspect in section 1.7 in an illustration of the measure of divergence in vector calculus. Let us consider a stationary control volume V with unspecified shape contained within a flow region (Figure 5.1).

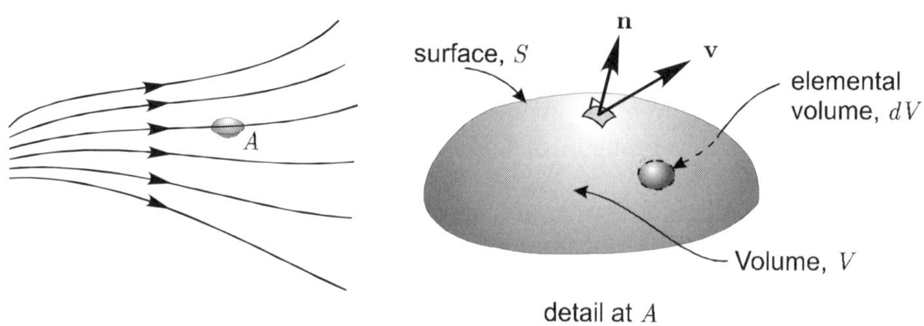

Figure 5.1: Stationary control volume in a flow domain.

The total mass M in the stationary control volume V at any given time is given by

$$M = \int\int\int_V \rho(\mathbf{x}, t)\, dV \tag{5.1}$$

where $\rho(\mathbf{x}, t)$ is the mass density of the fluid which, in general, can be a function of position and time. The rate of increase of M is

$$\frac{dM}{dt} = \frac{d}{dt}\int\int\int_V \rho(\mathbf{x}, t)\, dV = \int\int\int_V \frac{\partial\rho(\mathbf{x}, t)}{\partial t} dV \tag{5.2}$$

The rate at which mass enters the stationary control volume V through the surface S is

$$\frac{dM}{dt} = -\int\int_S \rho(\mathbf{x}, t)\, \mathbf{v}\cdot\mathbf{n}\; dS \tag{5.3}$$

where $\mathbf{v}$ is the velocity vector and the negative sign indicates that the direction of $\mathbf{n}$ corresponds to the *outward* unit normal to S. Assuming that no fluid is either created or lost, we have

$$\int\int\int \frac{\partial\rho}{\partial t}(\mathbf{x}, t)\, dV = -\int\int_S \rho(\mathbf{x}, t)\, \mathbf{v}\cdot\mathbf{n}\; dS \tag{5.4}$$

Using the divergence theorem (1.30), (5.4) can be reduced to the form

$$\int\int\int_V \left\{\frac{\partial\rho(\mathbf{x}, t)}{\partial t} dV + \operatorname{div}\left[\rho(\mathbf{x}, t)\,\mathbf{v}\right]\right\} dV = 0 \tag{5.5}$$

Since the control volume V is arbitrary, the mass conservation equation gives

$$\frac{\partial \rho}{\partial t}(\mathbf{x}, t) + \text{div}\left[\rho(\mathbf{x}, t)\,\mathbf{v}\right] = 0 \tag{5.6}$$

In arriving at (5.6) we have implicitly made use of the *Dubois-Reymond Lemma,* for the existence of a local equation.

In most situations involving fluid flow problems, we deal with fluids which are *incompressible* which implies that there is no volume change during motion; i.e. $\rho(\mathbf{x}, t)$ =const. As a result, (5.6) can be reduced to the form

$$\text{div}(\mathbf{v}) = 0 \quad or \quad \nabla \cdot \mathbf{v} = 0 \tag{5.7}$$

This statement essentially imposes a *kinematic constraint* on the velocity field within the fluid in motion. In developing (5.7) we have employed, in a weak sense, an assumption pertaining to the *constitutive behaviour* of the fluid. This, by itself, is insufficient to solve any problems of an engineering nature. In order to achieve such an objective we need to make further statements concerning the behaviour of the fluid in motion.

Firstly, we will focus on *perfect* or *non-viscous* fluids. While no such fluid exists in practice, the free stream flow of a perfect fluid outside of a narrow boundary layer is unaffected by viscosity effects. We will consider *irrotational* flow of a perfect fluid in which the vorticity Ω which is defined by

$$\Omega = \nabla \times \mathbf{v} \tag{5.8}$$

is *identically zero.* We further restrict attention to *barotropic fluids* in which the fluid mass density depends only on the fluid pressure p, i.e.:

$$\rho = \rho(p) \tag{5.9}$$

A theorem due to Lord Kelvin (1824-1907) states that if the fluid is *barotropic* and *inviscid*, then the vorticity Ω is constant in time. Also, if we invoke the *incompressibility* assumption, then ρ =const. A consequence of Kelvin's theorem is that once the flow in such a *barotropic inviscid* fluid is irrotational, it remains irrotational. Physically this makes sense since vorticity is the 'spin' imparted to a particle of fluid. The only internal physical mechanism which can achieve this through "shear stresses" is absent in an inviscid fluid.

THEOREM 5.1

Consider a vector $\mathbf{v}$ in a simply connected region V. The vorticity $\nabla \times \mathbf{v} = 0$, throughout V if, and only if, there exists a scalar field φ such that $\mathbf{v} = \nabla\varphi$. For $\mathbf{v}$ to be continuous and single-valued, the scalar field φ should have derivatives up to the second-order.

PROOF

Let us assume that there exists a scalar function φ such that $\mathbf{v} = \nabla\varphi$. Then it can be shown that $\nabla \times \mathbf{v} = \nabla \times \nabla\varphi = 0$. We now need to establish that $\nabla \times \mathbf{v} = 0$ implies the existence of φ such that $\mathbf{v} = \nabla\varphi$.

A vector field $\mathbf{v}$ which is irrotational throughout some simply connected region V can be expressed as the gradient of a scalar function

$$\varphi(\mathbf{r}) = \int_{\mathbf{r}_0}^{\mathbf{r}} \mathbf{v}(\mathbf{r}^*) \cdot d\mathbf{r}^* \tag{5.10}$$

In equation (5.10), the path of integration is any curve joining the reference point $P_0(\mathbf{r}_0)$ and $P(\mathbf{r})$ lying entirely within V (Figure 5.2).

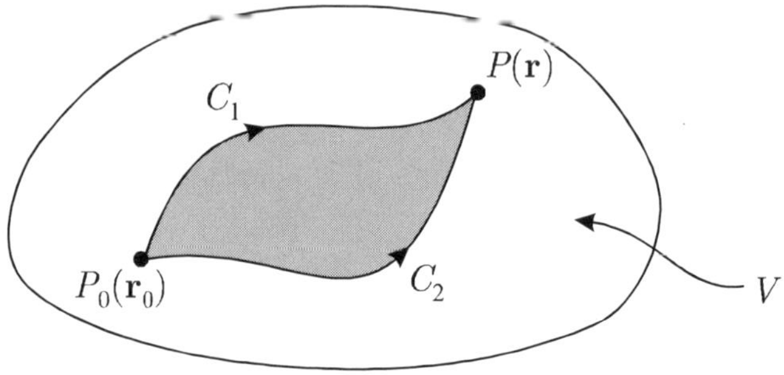

Figure 5.2: Integration paths.

If C_1 and C_2 are two distinct paths as shown in Figure 5.2, which form a closed loop, the difference in the value of the integral as obtained by evaluating (5.10) along the two separate paths is identical to the integral obtained had the entire loop $C\,(= C_1 + (-C_2))$ been considered, i.e.

$$\oint_C \mathbf{v} \cdot d\mathbf{r} = \oint_{C_1} \mathbf{v} \cdot d\mathbf{r} - \oint_{C_2} \mathbf{v} \cdot d\mathbf{r} \tag{5.11}$$

which vanishes by virtue of Stokes' theorem, provided the contour C is spanned by a surface S lying entirely in the region V, with $\nabla \times \mathbf{v} = 0$, i.e.

$$\oint_C \mathbf{v} \cdot d\mathbf{r} = \int\!\!\int_S \mathbf{n} \cdot \nabla \times \mathbf{v}\; dS = 0 \tag{5.12}$$

Hence

$$\oint_{C_1} \mathbf{v} \cdot d\mathbf{r} = \oint_{C_2} \mathbf{v} \cdot d\mathbf{r} \tag{5.13}$$

which implies that the integral is path-independent and dependent only on the values of $\mathbf{r}_0$ and $\mathbf{r}$. Thus we can express (as implied by (5.10)) the integral in the form of a function φ uniquely determined by $\mathbf{r}$.

To verify that $\nabla\varphi = \mathbf{v}$, we can take the directional derivative of φ in any direction $\mathbf{t}$ at P, i.e.

$$\mathbf{t}\nabla\varphi = \frac{d\varphi}{dS} = \mathbf{v}(\mathbf{r})\frac{d\mathbf{r}}{dS} = \mathbf{t}\mathbf{v} \tag{5.14}$$

Since the identity holds for all choices of $\mathbf{t}$ we confirm that

$$\mathbf{v} = \nabla\varphi \tag{5.15}$$

Therefore for all irrotational fluid motions, in an ideal fluid, the vorticity is zero and the velocity field can be derived as the gradient of a scalar function φ.

⊙ ⊙ ⊙

Combining the representation (5.15) with the equation of continuity given by (5.7) we obtain

$$\nabla\cdot(\nabla\varphi) = \text{ div } (\text{grad } \varphi) = 0 \tag{5.16}$$

From the results given in Chapter 1, we can write (5.16) as

$$\nabla^2\varphi = 0 \tag{5.17}$$

where ∇^2 is Laplace's operator. By applying the criteria for classification of second-order partial differential equations given in Chapter 4, it can be shown that Laplace's equation gives rise to an *elliptic* partial differential equation. In rectangular Cartesian coordinates (x, y, z),

$$\nabla^2 = \frac{\partial^2}{\partial x^2} + \frac{\partial^2}{\partial y^2} + \frac{\partial^2}{\partial z^2} \tag{5.18}$$

In cylindrical coordinates (r, θ, z) (Figure 5.3)

$$\nabla^2 = \frac{\partial^2}{\partial r} + \frac{1}{r}\frac{\partial}{\partial r} + \frac{1}{r^2}\frac{\partial^2}{\partial \theta^2} + \frac{\partial^2}{\partial z^2} \tag{5.19}$$

where

$$x = r\cos\theta \quad ; \quad y = r\sin\theta \tag{5.20}$$

In spherical coordinates (R, θ, ϕ) (Figure 5.3)

$$\nabla^2 = \frac{1}{R^2}\frac{\partial}{\partial R}\left(R^2\frac{\partial}{\partial R}\right) + \frac{1}{R^2\sin\phi}\frac{\partial}{\partial \phi}\left(\sin\phi\frac{\partial}{\partial \phi}\right) + \frac{1}{R^2\sin^2\phi}\frac{\partial^2}{\partial \theta^2} \tag{5.21}$$

where

$$x = R\sin\phi\cos\theta \quad ; \quad y = R\sin\phi\sin\theta \quad ; \quad z = R\cos\phi \tag{5.22}$$

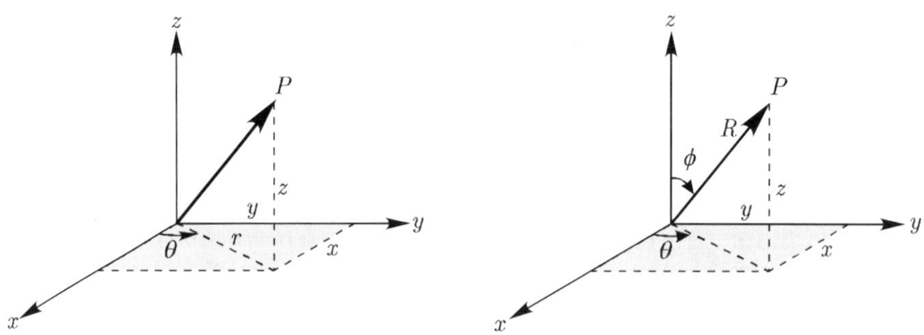

Figure 5.3: Cylindrical and spherical coordinates.

5.1.2 Flow of fluids in porous media

Laplace's equation also forms the basis for the study of flows in porous geological materials such as soils and rocks. The subject is of considerable importance to civil engineers involved in water resources engineering and geohydrology. More recently, the study of flows in porous geological media has become increasingly important to applications in the geo-environmental area. In such applications, the theory of flows in porous geological media is used to examine the transport of pollutants and other hazardous substances such as fertilizer chemicals, pesticides, radionuclides, biological contaminants, etc., as a result of groundwater movement.

Before considering the basic equation governing groundwater flow, it is important to examine the fundamental principles which define flow in a porous medium. A porous medium is characterized by the presence of a *solid matrix* and a *void phase*. The void region can be occupied by more than one phase which can include both gases and water. A further assumption is that we consider sufficiently large volumes of the porous medium so that there always exists a solid phase and a void phase. When water flows through a porous medium, the flow takes place through the void space. When all the void space is filled with the permeating fluid, the porous medium is said to be *saturated*. A scalar measure which identifies the void content is the porosity of the medium, n^*, which is defined as

$$n^* = \frac{\text{Volume of voids}}{\text{Total volume}} \tag{5.23}$$

It can also be shown that for most porous geological media with a direction-independent internal structure, on any area element within the porous medium

$$n^* = \frac{\text{Area of voids}}{\text{Total area}} \tag{5.24}$$

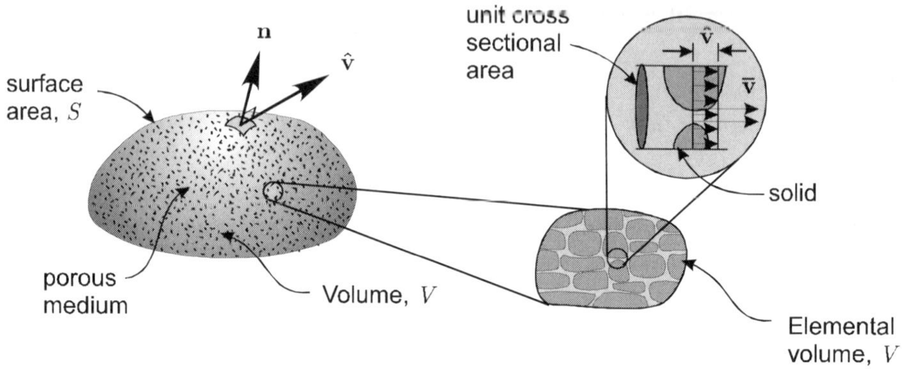

Figure 5.4: Flow in a porous medium.

Consider the flow across a unit area within the fabric of the porous medium which is fully saturated. As illustrated in Figure 5.4, over the fraction n^* of the unit cross sectional area, the average velocity of the fluid in the pores is $\overline{\mathbf{v}}$. There is no flow across the solid material region. We can define the volume flux or the fluid volume rate per unit cross sectional area, or the global flow velocity by considering the conservation equation

$$A\hat{\mathbf{v}} = A_v\overline{\mathbf{v}} \qquad \text{or} \qquad \hat{\mathbf{v}} = n^*\overline{\mathbf{v}} \tag{5.25}$$

where A_v is the area of the void phase. An interpretation of $\hat{\mathbf{v}}$ is that it corresponds to the flow velocity with which the fluid would be moving if it was to occupy the entire region and having the same volume transport and the same direction of flow as $\overline{\mathbf{v}}$. This is an extremely simple concept but has far reaching advantages in that we have now disregarded the presence of the pore structure by representing the flow averaged over the entire cross section. Since the value of n^* is numerically small, the flow velocity $\hat{\mathbf{v}}$ is very much less than $\overline{\mathbf{v}}$.

Let us now consider a fixed total control volume V of the porous medium which consists of both the solid fabric and the void space saturated with the permeating fluid.

$$\text{The mass of the interstitial fluid } = \iiint_V n^* \rho \, dV \tag{5.26}$$

where ρ is the density of the fluid in the voids and the integral is over the total volume region. In the absence of the generation of pore fluids, by chemical reactions or other means, the change in the mass of the fluid can take place only as a result of net fluid flow across the boundary surface, S.

$$\begin{aligned}&\text{The net transport of mass out of the volume}\\ &= \iint_S \rho \hat{\mathbf{v}} \cdot d\mathbf{S}\end{aligned} \tag{5.27}$$

where the integral is over the entire surface S of the region V, and the element of surface area

$$d\mathbf{S} = \mathbf{n} \, dS \tag{5.28}$$

where $\mathbf{n}$ is the outward unit normal to dS. (We note here that it is completely appropriate to perform the integration in (5.27), over the entire surface area S, which contains *both* solid and fluid areas, since the velocity measure is the transformed velocity $\hat{\mathbf{v}}$ and not the velocity in the pores $\overline{\mathbf{v}}$.) If mass is conserved, then the rate of decrease in fluid mass in the region is equal to the net transfer out of the region, i.e.

$$\begin{aligned}\frac{d}{dt} \iiint_V n^* \rho \, dV &= \iiint_V \frac{\partial}{\partial t} (n^* \rho) \; dV \\ &= - \iint_S \rho \hat{\mathbf{v}} \mathbf{n} \, dS \\ &= - \iiint_V \nabla \cdot (\rho \hat{\mathbf{v}}) \, dV\end{aligned} \tag{5.29}$$

Since the volume region V which is fixed in time is arbitrary, we have, by virtue of the *Dubois-Reymond Lemma*

$$\frac{\partial}{\partial t}(\rho n^*) + \nabla \cdot (\rho \hat{\mathbf{v}}) = 0 \tag{5.30}$$

The subtle difference between (5.6) and (5.30) is the appearance of n^* as a variable. The result (5.30) can be written in the form

$$\rho \frac{\partial n^*}{\partial t} + n^* \frac{\partial \rho}{\partial t} + \hat{\mathbf{v}} \cdot \nabla \rho + \rho \nabla \cdot \hat{\mathbf{v}} = 0 \tag{5.31}$$

We can introduce the substitution

$$D_R = \log_e \left(\frac{\rho}{\rho_0} \right) \tag{5.32}$$

where ρ_0 is a reference density (e.g. the local average) and rewrite (5.31) as

$$\nabla \cdot \hat{\mathbf{v}} = -\frac{\partial n^*}{\partial t} - \left(n^* \frac{\partial}{\partial t} + \hat{\mathbf{v}} \cdot \nabla \right) D_R \tag{5.33}$$

The first term on the right hand side involves the change in the porosity of the medium. Such porosity changes do in fact occur. An example of this would be *land subsidence* due to ground water withdrawal; the change in porosity of the medium translates to a physical consolidation of the geological medium. If, on the other hand, the fabric of the porous medium does not experience any change during fluid flow through the pores, then n^* will remain constant with time and the term $\partial n^*/\partial t$ can be neglected. The remaining terms on the right hand side of (5.33) involving rates of changes of the pore fluid density are also negligible. Invoking these assumptions, it is evident that the mass conservation equation for flow within the porous medium reduces to

$$\nabla \cdot \hat{\mathbf{v}} = 0 \tag{5.34}$$

where $\hat{\mathbf{v}}$ is identified as the velocity averaged over the entire region of flow which includes *both the solid and pore phases.*

In addition to the equation of continuity (5.34), applicable to a saturated porous medium, the formulation of the problem is complete when we can specify some law which will govern the movement of water within the porous medium. The earliest qualitative description of water flow through a porous medium is due to Henri Darcy (1803-1858) who studied the filtration of water flowing through a sand bed for improving the water supply to the town of Dijon, France. Darcy's observations were as a result of relatively rudimentary experiments, but it allowed him to put forward the postulate that the quantity of flow per unit time through the porous medium was

(i) proportional to the *cross sectional area normal to the direction of flow*, and
(ii) proportional to the *hydraulic gradient* across the element under consideration.

The *hydraulic gradient* is defined in relation to the gradient of the *hydraulic potential.* The hydraulic potential Φ as defined by D. Bernoulli (1700-1783) is composed of the following

$$\Phi = \Phi_D + \Phi_P + \Phi_v \tag{5.35}$$

where the components are defined as the datum potential (Φ_D), the pressure potential (Φ_P) and the velocity potential (Φ_v). In most problems involving flow in porous media, $\Phi_D + \Phi_P \gg \Phi_v$, consequently, the hydraulic potential takes the reduced form

$$\Phi = \Phi_D + \Phi_P \tag{5.36}$$

From Darcy's observations, the velocity of flow $\hat{\mathbf{v}}$ normal to a cross section is proportional to the gradient of the potential Φ. Therefore, we can write Darcy's Law as

$$\hat{\mathbf{v}} = -K\nabla\Phi \tag{5.37}$$

where K is the proportionality constant (with unit L/T) and is referred to as the hydraulic conductivity of the porous medium. The negative sign merely signifies that the gradient decays with increase in $\mathbf{x}$ if flow velocities are considered positive in the positive coordinate directions.

Figure 5.5: Fluid movement patterns in a hydraulically anisotropic medium.

In the representation (5.37) we have implicitly assumed that the *hydraulic conductivity* of the medium is a *scalar property* which is direction-independent. In reality the value of K depends on the pore structure geometry and pore shape. For most porous materials which do not possess a dominant directional dependence in the fabric, the assumption of *hydraulic isotropy* of the medium is a reasonable assumption. There are a variety of situations involving stratified porous media or porous media with lenticular particles (Figure 5.5) for which the fluid flow properties are direction dependent, i.e.

$$\hat{\mathbf{v}} = -\mathbf{K}\nabla\Phi \tag{5.38}$$

where, in the specific instance when the principal axes of hydraulic conductivity coincide with the coordinate axes x, y, z, the hydraulic conductivity matrix takes the form

$$\mathbf{K} = \begin{bmatrix} K_x & 0 & 0 \\ 0 & K_y & 0 \\ 0 & 0 & K_z \end{bmatrix} \tag{5.39}$$

A further aspect of the representation (5.37) is that the fluid transport characteristics of the porous medium are represented by a single parameter which does not identify the characteristics of the percolating fluid. In reality, the fluid transport behaviour is governed both by the pore geometry and the mechanical behaviour of the percolating fluid. It can be shown that the *hydraulic conductivity* K is related to the *intrinsic permeability* k of the porous medium (which has units L^2) according to

$$K = \frac{k\gamma}{\mu} \tag{5.40}$$

where γ is the unit weight of the fluid (units M/L^2T^2) and μ is the dynamic viscosity of the pore fluid (units M/LT).

Considering the discussion presented previously, it follows that the representation (5.37) also implies that the flow in the porous medium is irrotational. By considering the equation of continuity (5.34) and Darcy's Law, we can show that the irrotational flow governing the movement of a fluid in a porous medium can be reduced to the form

$$\nabla^2 \Phi\,(x, y, z) = \text{div}\;\;(\text{grad}\Phi) = 0 \tag{5.41}$$

Other applications follow suit by invoking the requisite conservation and transport equations. For example, by considering conservation of heat energy and Fourier's Law of heat conduction, it can be shown that the steady state distribution of temperature $T\,(x, y, z)$ in a heat conducting solid obeys Laplace's equation of the form

$$\nabla^2 T\,(x, y, z) = 0 \tag{5.42}$$

The derivation of the heat conduction equation for a solid will be discussed in greater detail in the chapter dealing with the diffusion equation.

5.2 Boundary conditions

As observed in Chapter 2, in order to develop a well-posed boundary value problem which is governed by Laplace's equation, it is necessary to impose certain boundary conditions which govern the various types of physical situations. We shall illustrate these boundary conditions by appeal to examples in the areas of fluid mechanics and steady state heat conduction. The boundary conditions basically fall into the categories of generalized Dirichlet-, Neumann- and other mixed (Robin)-type boundary conditions.

5.2.1 Boundary conditions for fluid flow

Consider the boundary of a flow domain which is in contact with a moving rigid surface such that the rigid surface has a velocity $\mathbf{U}$ at a typical point P on its boundary. A *kinematic boundary condition* must be satisfied if the fluid is to maintain contact without separation. We denote by $\mathbf{n}$ the unit normal vector drawn at the location P, at the surface of contact and we denote the velocity vector by $\mathbf{v}$. The velocity of the motion of the rigid region is denoted by $\mathbf{U}$ (Figure 5.6).

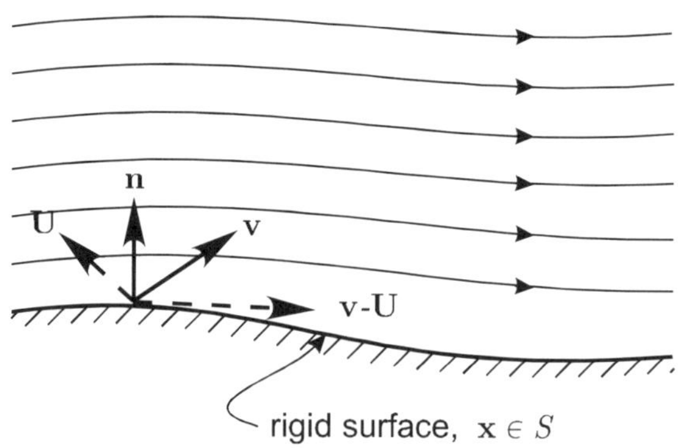

Figure 5.6: Boundary conditions for a fluid domain.

The kinematic condition at the boundary takes the form

$$\mathbf{vn} = \mathbf{Un} \quad \text{on} \quad \mathbf{x} \in S_1 \tag{5.43}$$

Using (1.43) and (5.15) in (5.43) we obtain

$$\frac{\partial \varphi}{\partial n} = -\mathbf{Un} \quad \text{on} \quad \mathbf{x} \in S_1 \tag{5.44}$$

This represents a Neumann-type boundary condition discussed in Chapter 2.

The second type of boundary condition relates to the specification of the potential itself at some prescribed boundary, i.e.

$$\varphi = \varphi_0 \quad \text{on} \quad \mathbf{x} \in S_2 \tag{5.45}$$

This represents a Dirichlet-type boundary condition. In addition to these, other types of regularity conditions can be prescribed to address specific constraints applicable to the solution domain.

5.2.2 Boundary conditions for porous media flow

The boundary conditions applicable to fluid flow in a porous medium are very similar to those applicable to the irrotational fluid flow problems described previously. For example, when the fabric of the porous medium is in contact with an impermeable boundary, there is no fluid transport across the interface; consequently

$$\hat{\mathbf{v}}\mathbf{n} = 0 \quad \text{on} \quad \mathbf{x} \in S_1 \tag{5.46}$$

where $\mathbf{n}$ is the unit normal to S_1. In terms of the hydraulic potential Φ,

(5.46) can be written as

$$\frac{\partial \Phi}{\partial n} = 0 \quad \text{on} \quad \mathbf{x} \in S_1 \tag{5.47}$$

Also, on a boundary where the hydraulic potential is prescribed we have

$$\Phi = \Phi_0 \quad \text{on} \quad \mathbf{x} \in S_2 \tag{5.48}$$

Other boundary conditions applicable to interfaces between two porous media of differing hydraulic conductivity characteristics can also be developed by considering incompressibility constraints and Darcy's Law. These can be found in the extensive literature on fluid flow in porous media.

5.2.3 Boundary conditions for heat conduction

The boundary and initial conditions applicable to the general class of transient heat conduction problems will be discussed in a subsequent Chapter dealing with the heat conduction equation which includes the transient time-dependent effects. For purposes of the discussion associated with the steady state or the time-independent problem, three types of thermal boundary conditions can be specified.

When the *heat flux* q^* at a boundary is specified (Figure 5.7), this is equiv-

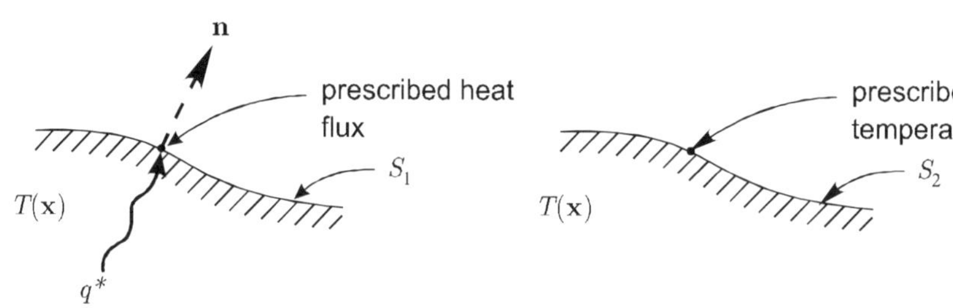

Figure 5.7: Temperature boundary conditions.

alent to a known normal derivative at the boundary, i.e.

$$-\kappa \frac{\partial T}{\partial n} = q^* \quad \text{on} \quad \mathbf{x} \in S_1 \tag{5.49}$$

where κ is the thermal conductivity of the medium. When the heat flux is specified, the boundary temperature can be computed only after the solution is found. In the special case when the boundary S_1 is *insulated*, the boundary condition (5.49) reduces to

$$\frac{\partial T}{\partial n} = 0 \quad \text{on} \quad \mathbf{x} \in S_1 \tag{5.50}$$

which implies that there is no heat flow normal to the boundary. Again in relation to the basic partial differential equation (5.42) the boundary condition (5.50) is of the Neumann-type.

If *temperature* is prescribed on a part of the boundary we require

$$T = T_0 \quad \text{on} \quad \mathbf{x} \in S_2 \tag{5.51}$$

In relation to the basic partial differential equation (5.42), the boundary condition is of the Dirichlet-type. There are a number of other boundary conditions such as *convection* and *radiation* boundary conditions that are applicable to both steady state and transient heat conduction problems. The temperature prescribed and insulated boundary conditions are those which are used quite extensively in the illustrative examples discussed in subsequent sections.

5.3 Generalized results

Prior to developing any results for specific problems governed by Laplace's equation, it is instructive to investigate the *qualitative aspects* of certain well-posed problems. This can be achieved by considering a generic volume V, on the surface of which (S) the boundary conditions will be prescribed in relation to the Dirichlet, Neumann and mixed conditions discussed in section 5.2.

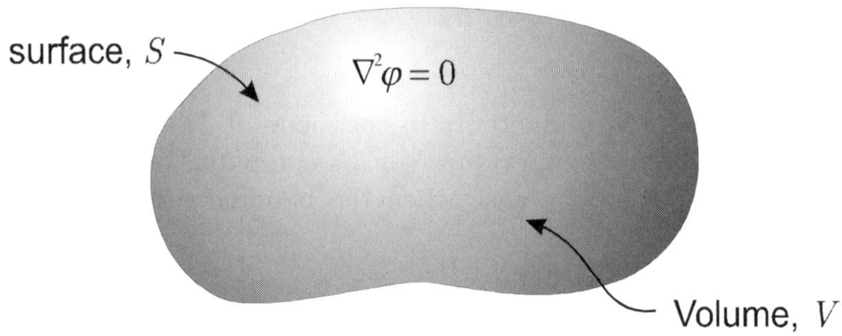

Figure 5.8: The domain V with boundary S.

We refer to a closed simply connected domain V with boundary S as shown in Figure 5.8. Considering Green's first identity (e.g. (1.45)), if φ and ψ are single-valued functions which have second derivatives at each point in V, we obtain

$$\int\int\int_V \varphi\nabla^2\psi \, dV + \int\int\int_V (\text{grad}\varphi \cdot \text{grad}\psi) \, dV = \int\int_S \varphi\frac{\partial\psi}{\partial n} \, dS \tag{5.52}$$

In the particular instance when $\varphi = \psi$ and from (5.17)

$$\nabla^2\varphi = 0 \quad \text{for} \quad \mathbf{x} \in V \tag{5.53}$$

the equation (5.52) gives

$$\int\int\int_V |\ \text{grad}\varphi\ |^2 \, dV = \int\int_S \varphi\frac{\partial\varphi}{\partial n} \, dS \tag{5.54}$$

THEOREM 5.2

If φ satisfies Laplace's equation at each point in V and if φ is zero on the boundary S, then φ is zero everywhere in V.

PROOF

The problem posed has Dirichlet-type boundary conditions everywhere on S. Physically, we can assume φ to be the potential, the gradient of which induces flow within the medium. Certainly there is no flow along the surface S. From (5.54), we note that, due to the condition $\varphi = 0$ on S, the right hand side is identically zero. Hence

$$\int\int\int_V \mid \mathrm{grad}\varphi \mid^2 dV = 0 \quad ; \quad \mathbf{x} \in V \tag{5.55}$$

The integrand of (5.55) is positive definite. Consequently, the integral can be zero only if the integrand itself is zero. This is possible only if

$$\mathrm{grad}\varphi = 0 \quad \text{or} \quad \varphi = \text{const.}, \ \mathbf{x} \in V \tag{5.56}$$

Since φ is a continuous function in a closed, simply connected region V and is zero on the boundary S, it must be zero everywhere in V

THEOREM 5.3

If φ satisfies Laplace's equation everywhere in V and if $\partial\varphi/\partial n$ is zero on S, then φ is constant everywhere.

PROOF

The problem posed has Neumann-type boundary conditions everywhere on S. Again we can assume φ to be the potential, the gradient of which induces flow within the region. There is no flow along the boundary since $\partial\varphi/\partial t$ is zero along it. Referring to (5.54), we note that since $\partial\varphi/\partial n$ is zero on S we have

$$\int\int\int_V |\,\mathrm{grad}\varphi\,|^2\, dV = 0 \quad ; \quad \mathbf{x} \in V \tag{5.57}$$

Following the arguments presented earlier in connection with Theorem 5.1, it is evident that $\mathrm{grad}\varphi = 0$ or $\varphi = \mathrm{const.}$. The value of φ is indeterminate to within an arbitrary constant since the problem does not specify φ at any location on S.

THEOREM 5.4

If φ satisfies Laplace's equation in V and if

$$\frac{\partial\varphi}{\partial n} = -\lambda\varphi \quad \text{on} \quad \mathbf{x} \in S \tag{5.58}$$

where λ is a *positive constant*, then φ is zero everywhere in V.

PROOF

The problem posed here has a Newton-type boundary condition. Considering Green's first identity in the form given by (5.54) we have

$$\int\int\int_V |\,\mathrm{grad}\varphi\,|^2\, dV = \int\int_S \varphi \frac{\partial\varphi}{\partial n} dS \tag{5.59}$$

Using the Newton boundary condition (5.58) in (5.59) we obtain

$$\int\int\int_V |\,\mathrm{grad}\varphi\,|^2\, dV + \lambda \int\int_S (\varphi)^2\, dS = 0 \tag{5.60}$$

Since the integrands in both integrals are positive definite, the equation can be satisfied only if the integrands are *separately* zero, i.e.

$$\text{grad}\varphi = 0 \quad \text{or} \quad \varphi = \text{const.} \quad \text{in} \quad \mathbf{x} \in V \tag{5.61}$$

$$\varphi = 0 \quad \text{on} \quad \mathbf{x} \in S \tag{5.62}$$

Consequently, from Theorem 5.1, if gradφ is zero in V and φ is zero in S, then φ is zero everywhere in V.

THEOREM 5.5

If φ satisfies Laplace's equation in V and if

$$\frac{\partial \varphi}{\partial n} = 0 \qquad \text{on} \quad \mathbf{x} \in S_1 \tag{5.63}$$

$$\frac{\partial \varphi}{\partial n} = f(\mathbf{x}) \quad \text{on} \quad \mathbf{x} \in S_2 \tag{5.64}$$

where $f(\mathbf{x})$ is an arbitrary function and $S_1 \cup S_2 \equiv S$, the boundary of V, then a solution to the boundary value problem will *not exist* unless

$$\iint_{S_2} f(\mathbf{x})\, dS \equiv 0 \tag{5.65}$$

PROOF

From Green's second identity, if $\varphi(\mathbf{x})$ and $\psi(\mathbf{x})$ are single-valued functions

$$\int\int\int_V [\varphi\nabla^2\psi - \psi\nabla^2\varphi]\, dV = \int\int_S \left[\varphi\frac{\partial\psi}{\partial n} - \psi\frac{\partial\varphi}{\partial n}\right] dS \tag{5.66}$$

If we set $\nabla^2\varphi = 0$ and $\psi = 1$, in (5.66), we obtain

$$\int\int_S \frac{\partial\varphi}{\partial n} dS = 0 \tag{5.67}$$

which can be written in the form

$$\int\int_S \frac{\partial\varphi}{\partial n} dS = \int\int_{S_1} \frac{\partial\varphi}{\partial n} dS = \int\int_{S_2} \frac{\partial\varphi}{\partial n} dS \tag{5.68}$$

From the boundary condition (5.63), the first integral of the right hand of (5.6) is zero. Using this reduction and the boundary condition (5.64) we obtain

$$\int\int_{S_2} \frac{\partial\varphi}{\partial n} dS = \int\int_{S_2} f(\mathbf{x})\, dS = 0 \tag{5.69}$$

Therefore for the solution to the boundary value problem to exist the function $f(\mathbf{x})$ must satisfy the *consistency condition* (5.69).

⊙ ⊙ ⊙

A physical interpretation of this consistency condition can be demonstrated by considering the following simple example involving either ideal fluid flow or flow of a fluid through a porous medium. The Figure 5.9, shows a two-dimensional region where fluid flow takes place in the x, y plane. the boundary conditions for the problem are as follows:

$$\frac{\partial\varphi}{\partial x} = 0\ ; \qquad x = 0\ ; \qquad ; \qquad y \in (0, b) \tag{5.70}$$

$$\frac{\partial\varphi}{\partial y} = 0\ ; \qquad y = 0\ ; \qquad y = b\ ; \qquad x \in (0, a) \tag{5.71}$$

$$\frac{\partial\varphi}{\partial x} = f(y); \qquad x = a\ ; \qquad\qquad ; \qquad y \in (0, b) \tag{5.72}$$

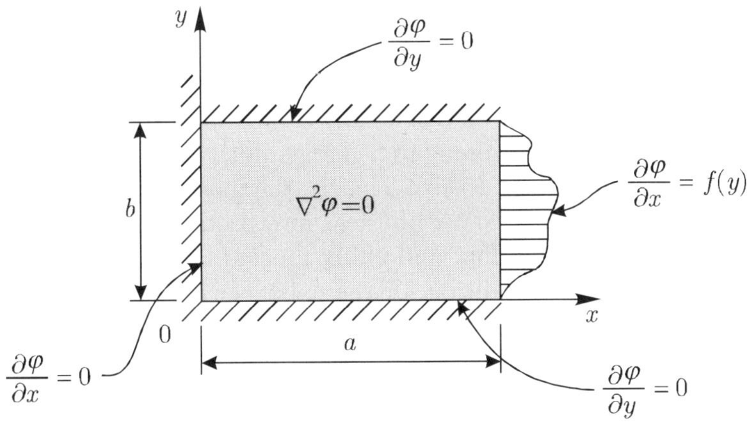

Figure 5.9: Neumann boundary conditions for a closed region.

Considering, for example, Darcy's Law governing fluid flow in porous medium, the velocity of flow $\mathbf{v}$ is directly proportional to $\nabla\varphi$. Therefore boundary conditions of the type

$$\frac{\partial\varphi}{\partial x} = 0 \quad ; \qquad \frac{\partial\varphi}{\partial y} = 0 \tag{5.73}$$

represent zero velocity or no fluid flow or impermeable boundaries through which no fluid flow is possible. The inhomogeneous boundary condition (5.72) can be interpreted as

$$v_x = -K\frac{\partial\varphi}{\partial x} = -K\ f(y) \neq 0 \tag{5.74}$$

The fluid entering a part of a boundary of an otherwise closed region will violate the divergence free condition associated with fluid mass conservation. i.e.

$$\nabla.\mathbf{v} \neq \mathbf{0} \quad ; \qquad \mathbf{x} \in V \tag{5.75}$$

Therefore a steady state solution cannot exist unless there is a mechanism to satisfy the divergence free condition. This can obviously be a *sink term* which would give rise to Poisson's equation rather than Laplace's equation. Therefore the consistency condition that will satisfy the fluid mass conservation principle without recourse to a change in the basic partial differential equation is the requirement (5.69). That is, the mass conservation should be satisfied at the boundary S_2. A physical interpretation of the requirement (5.69) is that fluid mass influx and efflux must take place on boundary S_2. This is the consistency condition for the existence of a solution to the Neumann boundary conditions being prescribed over the region S (Figure 5.10) i.e.

$$\int_0^b f^*(a,y)\, dy = 0 \tag{5.76}$$

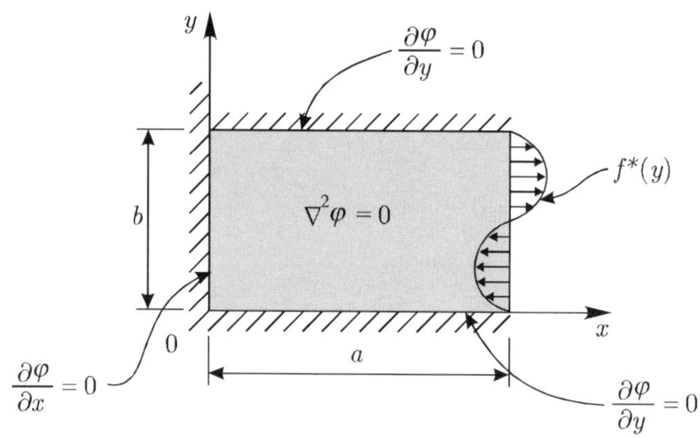

Figure 5.10: Consistent Neumann boundary conditions for Laplace's equation.

The result (5.69) can be generalized to include situations where Neumann boundary conditions on subsets of S_1, S_2 ...etc. of S.

Referring to Figure 5.11, the consistency condition for the existence of a solution to Laplace's equation is

$$\int\int_{S_1} f_1(\mathbf{x})\, dS + \int\int_{S_2} f_2(\mathbf{x})\, dS + \int\int_{S_3} f_3(\mathbf{x})\, dS = 0 \tag{5.77}$$

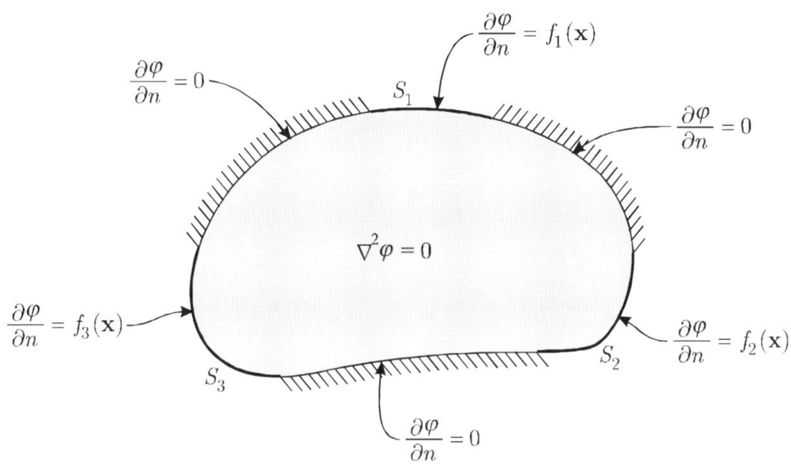

Figure 5.11: Neumann boundary conditions prescribed over sub-regions of S.

5.4 Methods of solution of Laplace's equation

The objective of this section is to outline, very briefly, some general approaches to the formulation and solution of Laplace's equation with reference to some problems of engineering interest. The presentation of the material is of necessity guided by procedures and approaches that have been developed in the literature. Also, attention is restricted to the discussion of two dimensional problems referred to either Cartesian or plane polar coordinates.

5.4.1 Direct solution procedure

In view of the presentation given in Chapter 4, it is natural to enquire whether the direct approach for the solution of the elliptic-type partial differential equation, expressed in its canonical form, can be adopted for the solution of specific boundary value problems. While this method is a valid approach, it is less versatile than other procedures which will be discussed subsequently.

Example 5.1

Consider the two-dimensional boundary value problem of steady state heat conduction in a plate square region (Figure 5.12) governed by Laplace's equation

$$\nabla^2 T = \frac{\partial^2 T}{\partial x^2} + \frac{\partial^2 T}{\partial y^2} = 0$$

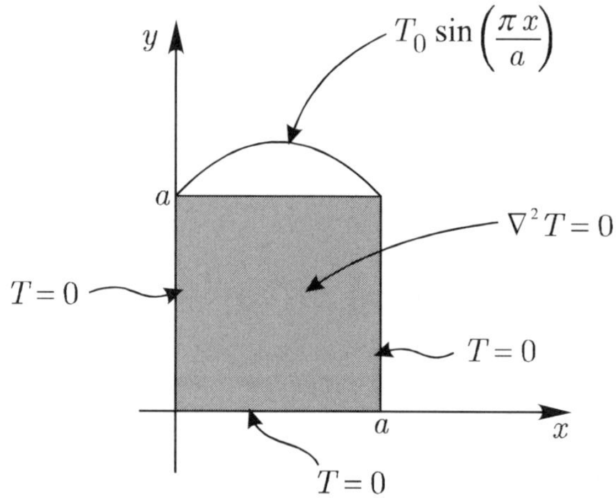

Figure 5.12: Plate subjected to boundary temperature.

and subject to the boundary conditions

$$T(0, y) = 0 \;\; ; \;\; T(a, y) = 0 \tag{5.78}$$

$$T(x, 0) = 0 \;\; ; \;\; T(x, a) = T_0 \sin\left(\frac{\pi x}{a}\right) \tag{5.79}$$

where T_0 is a constant.

Solution

Avoiding details of the analysis which follows the procedure outlined in Example 4.3 in section 4.3, the roots of the characteristic equation are

$$\alpha = +i \;\; ; \;\; \beta = -i \tag{5.80}$$

and the differential equation for determining the characteristic equations take the forms

$$\frac{dy}{dx} \pm i = 0 \tag{5.81}$$

Integrating (5.81) we have

$$x + iy = C_1 = \xi \;\; ; \;\; x - iy = C_2 = \eta \tag{5.82}$$

and the characteristics are complex. We could introduce the substitutions (see e.g. (4.64))

$$\lambda = i(\xi - \eta) = -2y \;\; ; \;\; \mu = (\xi + \eta) = 2x \tag{5.83}$$

These substitutions, however, do not result in any further simplification of the problem. We therefore utilize the canonical form (e.g. (4.41))

$$\frac{\partial^2 T}{\partial\xi\partial\eta} = 0 \tag{5.84}$$

Integrating (5.84) we obtain

$$T(\xi,\eta) = \tilde{\varphi}(\xi) + \tilde{\psi}(\eta) \tag{5.85}$$

or the general result can be written as

$$T(x,y) = \tilde{\varphi}(x+iy) + \tilde{\psi}(x-iy) \tag{5.86}$$

where $\tilde{\varphi}$ and $\tilde{\psi}$ are arbitrary functions. We can choose functions which will yield the original function when differentiated twice; i.e.

$$T(x,y) = C\left[i\cos\left\{\frac{\pi(x+iy)}{a}\right\} - i\cos\left\{\frac{\pi(x-iy)}{a}\right\}\right] \tag{5.87}$$

Since $T(x,y)$ is real, we expand the function to recover the real part of the solution (5.87) i.e.(note: $\sin(ix) = i\sinh(x); \cos(ix) = \cosh(x)$)

$$T(x,y) = C\sin\left(\frac{\pi x}{a}\right)\sinh\left(\frac{\pi y}{a}\right) \tag{5.88}$$

where C is a constant which needs to be determined by satisfying the prescribed temperature boundary condition. It is noted that the solution (5.88) already satisfies the boundary conditions on $x=0$, $x=a$ and $y=0$. The remaining boundary condition (5.79) is satisfied by setting

$$C = \frac{T_0}{\sinh \pi} \tag{5.89}$$

which gives the final solution as

$$T(x, y) = T_0 \sin\left(\frac{\pi x}{a}\right) \frac{\sinh(\pi y / a)}{\sinh(\pi)} \tag{5.90}$$

The solution of the problem is made particularly straightforward by the fact that the boundary conditions can be satisfied relatively easily. In general, this will not be the case and, for arbitrary forms of inhomogeneous boundary conditions, the problem is difficult to solve. Also, the characteristics are imaginary and yield little advantage since the solution is sought in a closed region of a real domain.

5.4.2 Separation of variables method - Cartesian coordinates

The method of separation of variables represents a most powerful technique for the solution of partial differential equations in general and more specifically for determining solutions to Laplace's equation. Although the method is generally attributed to Joseph Fourier (1763-1820), it was also extensively used by Jean Alembert (1717-1783), Daniel Bernoulli (1700-1784) and Leonhard Euler (1707-1783). In this method, we seek trial solutions either of the form

$$\varphi(x, y) = F(x) + G(y) \tag{5.91}$$

or

$$\varphi(x, y) = F(x)\, G(y) \tag{5.92}$$

Substituting (5.91) in Laplace's equation (5.17) we obtain

$$\frac{d^2F}{dx^2} = -\frac{d^2G}{dy^2} = \text{const.} \tag{5.93}$$

Integrating (5.93) we have

$$F(x) = \frac{A}{2}x^2 + Bx + C \tag{5.94}$$

$$G(y) = -\frac{A}{2}y^2 + Dy + E \tag{5.95}$$

where A, B, C, D and E are arbitrary constants. These represent a very specialized class of solutions which are applicable to situations where the boundary conditions are highly artificial or, alternatively, the solutions correspond to elementary one-dimensional states where the gradients are constant within the domain of interest.

A more general class of solutions is obtained by considering the representation (5.92). Substituting (5.92) in Laplace's equation (5.17) we obtain

$$G\frac{d^2F}{dx^2} + F\frac{d^2G}{dy^2} = 0 \tag{5.96}$$

Since $\varphi = F(x)\,G(y)$ is non-zero, (5.96) is equivalent to

$$\frac{1}{F}\frac{d^2F}{dx^2} = -\frac{1}{G}\frac{d^2G}{dy^2} = \text{const.} \tag{5.97}$$

The constant occurring in (5.97) may be *positive*, *negative* or *zero*; as such, we can consider the cases separately.

When the constant $\left(=\gamma^2\right)$ is positive, we obtain

$$\frac{d^2F}{dx^2} - \gamma^2 F = 0 \tag{5.98}$$

$$\frac{d^2G}{dy^2} + \gamma^2 G = 0 \tag{5.99}$$

and γ is an arbitrary positive constant. Solutions of (5.98) and (5.99) can be written as

$$F(x) = \begin{cases} e^{-\gamma x} \\ e^{\gamma x} \end{cases} \quad ; \quad G(y) = \begin{cases} \sin(\gamma y) \\ \cos(\gamma y) \end{cases} \tag{5.100}$$

Therefore, particular solutions of Laplace's equation are

$$\varphi(x, y) = \left\{Ae^{\gamma x} + Be^{-\gamma x}\right\} \sin(\gamma y) + \left\{Ce^{\gamma x} + De^{-\gamma x}\right\} \cos(\gamma y) \tag{5.101}$$

where A, B, C and D are arbitrary constants. Also, the multiplication of each term by any arbitrary constant, as well as summation of any number of terms, will also constitute a solution of Laplace's equation. Therefore, the general form of the solution can be written as

$$\begin{aligned} \varphi(x, y) = \sum_{n=0} &\left[\left\{A_n e^{\gamma_n x} + B_n e^{-\gamma_n x}\right\} \sin(\gamma_n y)\right. \\ &+ \left.\left\{C_n e^{\gamma_n x} + D_n e^{-\gamma_n x}\right\} \cos(\gamma_n y)\right] \end{aligned} \tag{5.102}$$

We also note that if x and y are interchanged, the resulting form of $\varphi(x, y)$ is also a solution of Laplace's equation.

When γ_n is zero, (5.98) and (5.99) imply that

$$\frac{d^2F}{dx^2} = \frac{d^2G}{dy^2} = 0 \tag{5.103}$$

which gives

$$F(x) = ax + b \quad ; \quad G(y) = cy + d \tag{5.104}$$

Therefore, from (5.92), we note that

$$\varphi(x, y) = Axy + Bx + Cy + D \tag{5.105}$$

(where A, B, C and D are arbitrary constants) is also a solution of Laplace's equation.

In the remainder of this section, the separation of variables technique will be applied to some elementary problems associated with Laplace's equation.

In the case when γ^2 is negative, the solutions will be complex valued. While solutions can be obtained in principle, all boundary conditions imposed on the problem are real. As such there are no significant advantages to considering the complex valued solutions.

Example 5.2

A half-plane can be defined as a region which occupies the domain $-\infty \leq x < \infty$ and $0 < y < \infty$. The boundary of a half-plane $y = 0$, is subjected to a potential of the form

$$\varphi(x, 0) = \varphi_0 \cos\left(\frac{\pi x}{a}\right) \tag{5.106}$$

Determine the distribution of the potential $\varphi(x, y)$ within the half-plane region if $\varphi(x, y)$ satisfies

$$\nabla^2 \varphi(x, y) = 0 \tag{5.107}$$

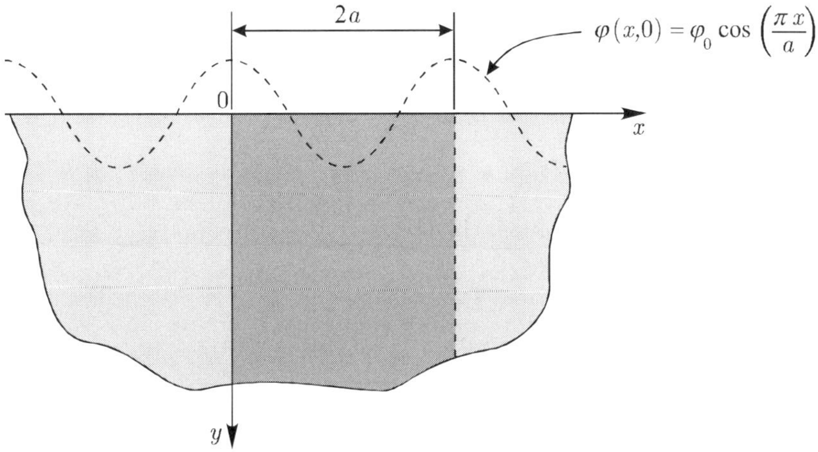

Figure 5.13: A periodic potential at the boundary of a half-plane.

Solution

The Figure 5.13 illustrates the configuration of the half-plane problem. It is evident that the problem exhibits periodic symmetry. As a result, attention can be restricted to the analysis of an infinite strip region $0 \leq x \leq 2a$; $0 \leq y < \infty$ and the boundary conditions for the reduced strip problem are

$$\varphi(x, 0) = \varphi_0 \cos\left(\frac{\pi x}{a}\right) \tag{5.108}$$

$$\left[\frac{\partial \varphi}{\partial x}\right]_{x=0} = 0 \ ; \ \left[\frac{\partial \varphi}{\partial x}\right]_{x=2a} = 0 \tag{5.109}$$

In addition to these boundary conditions we need to impose on the solution a *regularity condition*, which can be interpreted as a *boundary condition at infinity*. We shall assume that $\varphi(x, y)$ is bounded for large values of y.

Since the boundary condition is periodic with period $2a$ in x, a solution of the form (5.101) would be satisfactory with $\gamma = \pi/a$. Also, since we have imposed the regularity condition on φ, the solution should not contain positive exponential terms.

Without loss of generality we can select a solution of the form

$$\varphi(x,y) = \left[B\cos\left(\frac{\pi x}{a}\right) + D\sin\left(\frac{\pi x}{a}\right)\right] e^{-\pi y/a} \tag{5.110}$$

where B and D are constants. Also

$$\frac{\partial\varphi}{\partial x} = \left[-\frac{B\pi}{a}\sin\left(\frac{\pi x}{a}\right) + \frac{D\pi}{a}\cos\left(\frac{\pi x}{a}\right)\right] e^{-\pi y/a} \tag{5.111}$$

The boundary conditions (5.109) are both identically satisfied if we set $D = 0$. The solution now takes the reduced form

$$\varphi(x,y) = B\cos\left(\frac{\pi x}{a}\right) e^{-\pi y/a} \tag{5.112}$$

The boundary condition (5.108) can be used to evaluate the undetermined constant B. As a general treatment, however, let us assume that

$$\varphi(x,0) = \varphi^*(x) \tag{5.113}$$

where $\varphi^* = \varphi^*(x+2a)$ is a function of periodicity $2a$. The function $\varphi^*(x)$ has a Fourier series representation (see e.g. (1.197))

$$\varphi^*(x) = \frac{a_0^*}{2} + \sum_{n=1}^{\infty}\left\{a_n^*\cos\left(\frac{n\pi x}{a}\right) + b_n^*\sin\left(\frac{n\pi x}{a}\right)\right\} \tag{5.114}$$

The Fourier coefficients a_0^*, a_n^* and b_n^* are determined by

$$a_0^* = \frac{1}{a}\int_0^{2a}\varphi^*(x)\,dx \tag{5.115}$$

$$a_n^* = \frac{1}{a}\int_0^{2a} \varphi^*(x)\cos\left(\frac{n\pi x}{a}\right)dx \tag{5.116}$$

$$b_n^* = \frac{1}{a}\int_0^{2a} \varphi^*(x)\sin\left(\frac{n\pi x}{a}\right)dx \tag{5.117}$$

From (5.112) and (5.114) we have

$$\begin{aligned}\varphi^*(x,0) &= B\cos\left(\frac{\pi x}{a}\right) \\ &= \frac{a_0^*}{2} + \sum_{n=1}^{\infty}\left[a_n^*\cos\left(\frac{n\pi x}{a}\right) + b_n^*\sin\left(\frac{n\pi x}{a}\right)\right]\end{aligned} \tag{5.118}$$

This equation is satisfied when $a_0^* = 0$; $B = a_1^*$; $b_n^* = 0$ and all other $a_n^* = 0$. Therefore, from (5.116), we have

$$a_1^* = \frac{1}{a}\int_0^{2a} \varphi_0\cos^2\left(\frac{\pi x}{a}\right)dx = \varphi_0 \tag{5.119}$$

The solution of the boundary value problem takes the form

$$\varphi(x,y) = \varphi_0\cos\left(\frac{\pi x}{a}\right)e^{-\pi y/a} \tag{5.120}$$

Example 5.3

Consider the problem of steady state heat conduction in a semi-infinite strip region shown in Figure 5.14, where the boundary conditions are

$$T(0,y) = 0 \tag{5.121}$$

$$T(a, y) = 0 \tag{5.122}$$

$$T(x, 0) = f(x) \quad ; \quad 0 \leq x \leq a \tag{5.123}$$

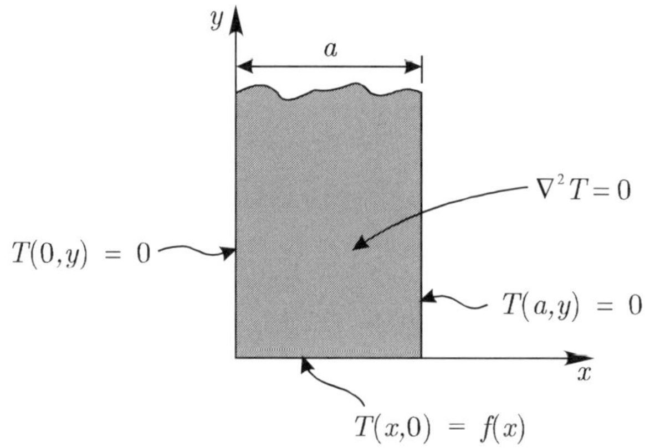

Figure 5.14: Steady state heat conduction in a semi-infinite strip.

Solution

Considering the general solution for Laplace's equation obtained via a separation of variables technique, we can write

$$\begin{aligned} T(x, y) = &\left[D \cos(\gamma x) + B \sin(\gamma x)\right] e^{-\gamma y} \\ &+ \left[C \cos(\gamma x) + A \sin(\gamma x)\right] e^{\gamma y} \end{aligned} \tag{5.124}$$

where A, B, C and D are arbitrary constants, which are to be determined by applying the boundary conditions and the regularity conditions. Applying the regularity conditions we require

$$A = C = 0 \tag{5.125}$$

Since $T = 0$ when $x = 0$, we cannot have a $\cos(\gamma x)$ in the final solution.

Therefore we set

$$D = 0 \tag{5.126}$$

Also at $x = a$, $T = 0$ for all positive values of y. Hence

$$B \sin(\gamma a) = 0 \tag{5.127}$$

For a non-trivial solution $B \neq 0$; hence

$$\sin(\gamma a) = 0 \tag{5.128}$$

which determines the possible values of γ i.e.

$$\gamma = \frac{n\pi}{a}\,, \quad \text{where} \quad n = 0, 1, 2, 3, \ldots \tag{5.129}$$

Hence for each n there exists a solution

$$T_n(x, y) = B_n e^{-n\pi y/a} \sin\left(\frac{n\pi x}{a}\right) \tag{5.130}$$

where B_n is an arbitrary constant. As with (5.102), taking the sum of all possible solutions of the type (5.130) we can construct the solution

$$T(x, y) = \sum_{n=1,2,} B_n e^{-\frac{n\pi y}{a}} \sin\left(\frac{n\pi x}{a}\right) \tag{5.131}$$

It is natural to enquire whether the solutions of the type

$$T(x,y) = A^*xy + B^*x + C^*y + D^* \tag{5.132}$$

have any relevance to the problem. Clearly, if $T = 0$ for $x = 0$ for all y, we require $C^* = D^* = 0$. Also if $T = 0$ for $x = a$ for all y, we require $A^* = B^* = 0$. Hence for this particular problem, these elementary solutions have no relevance.

The last remaining boundary condition (5.123) requires that

$$T(x,0) = f(x) \tag{5.133}$$

Considering (5.131) we have

$$T(x,0) = f(x) = \sum_{n=1,2,} B_n \sin\left(\frac{n\pi x}{a}\right) \tag{5.134}$$

This represents the half range sine series in which the coefficients B_n are given by (see e.g. (5.117))

$$B_n = \frac{2}{a}\int_0^a f(x)\sin\left(\frac{n\pi x}{a}\right)dx \tag{5.135}$$

If $f(x)$ is specified, the values of the coefficients B_n can be explicitly evaluated.

For example, if

$$f(x) = T_0\frac{x}{a} \tag{5.136}$$

we have

$$B_n = \frac{2T_0}{\pi}\left\{-\frac{\cos(n\pi)}{n}\right\} \quad ; \quad n = 1, 2, 3, \ldots \tag{5.137}$$

Hence the solution can be written as

$$T(x, y) = \frac{2T_0}{\pi}\left[\sum_{n=1,2,} \left(-\frac{\cos n\pi}{n}\right) e^{-\frac{n\pi y}{a}} \sin\left(\frac{n\pi x}{a}\right)\right] \tag{5.138}$$

Example 5.4

Consider the problem of fluid flow in a thin layer of a porous geological material as shown in Figure 5.15. The thickness of the layer $t \ll a$ or b. The layer is contained between two impervious strata. For the purposes of this problem we shall assume that flow takes place in the plane of the porous layer and the hydraulic conductivity of the layer K is isotropic and accounts for any non-uniformity in the fluid velocity distribution over its thickness. The boundary conditions governing the hydraulic potential correspond to the following:

$$\Phi(x, 0) = 0 \tag{5.139}$$

$$\Phi(x, b) = 0 \tag{5.140}$$

$$\Phi(0, y) = 0 \tag{5.141}$$

$$\Phi(a, y) = F(y) \tag{5.142}$$

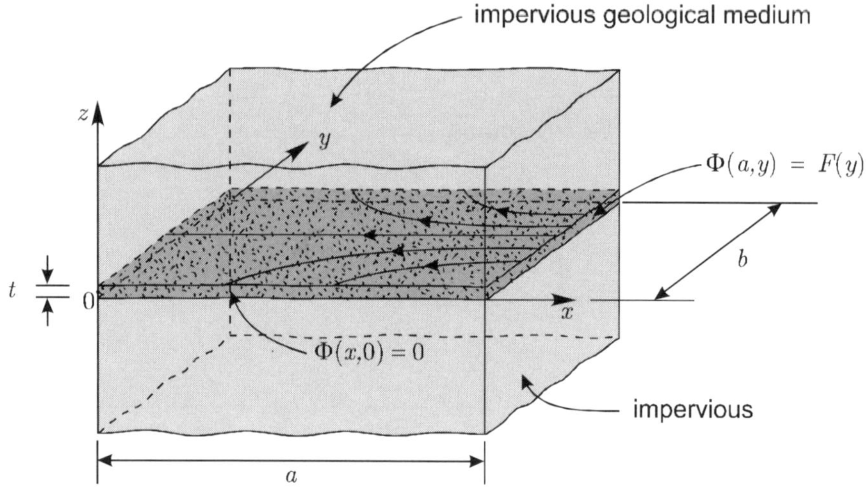

Figure 5.15: Flow in a confined porous layer.

Solution

It may be noted that the potential Φ is composed of the datum potential Φ_D and the pressure potential Φ_P. Since the porous layer is located on the plane $z = 0$ we can set the datum at $z = 0$ and Φ can be identified directly as the pressure potential or "pressure head ".

Considering a variables separation technique, we select the solution

$$\Phi(x, y) = \left\{Ae^{\gamma x} + Be^{-\gamma x}\right\} \sin(\gamma y) + \left\{Ce^{\gamma x} + De^{-\gamma x}\right\} \cos(\gamma y) \tag{5.143}$$

Since $\Phi = 0$ when $y = 0$ and $y = b$, for arbitrary x, the boundary conditions will not be satisfied in the presence of the term $\cos(\gamma y)$, in (5.143).

We have

$$\Phi(x, y) = \left\{Ae^{\gamma x} + Be^{-\gamma x}\right\} \sin(\gamma y) \tag{5.144}$$

Considering the boundary condition (5.141)

$$\Phi(0, y) = \{A + B\} \sin(\gamma y) = 0 \tag{5.145}$$

or

$$A = -B \tag{5.146}$$

and Φ can be written as

$$\Phi(x, y) = A\left\{e^{\gamma x} - e^{-\gamma x}\right\} \sin(\gamma y) \tag{5.147}$$

Invoking the boundary condition (5.140) we have

$$\Phi(x, b) = A\left\{e^{\gamma x} - e^{-\gamma x}\right\} \sin(\gamma b) = 0 \tag{5.148}$$

For a non-trivial solution of Φ for arbitrary x, we require

$$\gamma = n\pi/b \quad ; \quad n = 0, 1, 2, 3, \ldots \tag{5.149}$$

Hence, considering solutions of Φ corresponding to each γ, we can write the final result in the form

$$\Phi(x, y) = \sum_{n=1,2,} \left[A_n \left\{e^{n\pi x/b} - e^{-n\pi x/b}\right\} \sin\left(\frac{n\pi y}{b}\right)\right] \tag{5.150}$$

The coefficients A_n can be obtained by considering the final boundary condition (5.142) i.c.

$$\Phi(a,y) = F(y) = \sum_{n=1,2,} 2A_n \sinh\left(\frac{n\pi a}{b}\right) \sin\left(\frac{n\pi y}{b}\right) \tag{5.151}$$

The quantities $A_n \sinh(n\pi a/b)$ are the Fourier coefficients in the Fourier sine series of period $2b$ (see e.g. (1-200)), i.e.

$$2A_n \sinh\left(\frac{n\pi a}{b}\right) = \frac{2}{b}\int_0^b F(y) \sin\left(\frac{n\pi y}{b}\right) dy \tag{5.152}$$

The final form of the solution for the distribution of hydraulic potential Φ in the porous geological layer is given by

$$\Phi(x,y) = \sum_{n=1,2,} \tilde{A}_n^* \frac{\sinh(n\pi x/b)}{\sinh(n\pi a/b)} \sin\left(\frac{n\pi y}{b}\right) \tag{5.153}$$

where

$$\tilde{A}_n^* = \frac{2}{b}\int_0^b F(y) \sin\left(\frac{n\pi y}{b}\right) dy \tag{5.154}$$

The solution can be evaluated when explicit forms for $F(y)$ are specified. For example, when

$$F(y) = H_0 = \text{const.} \tag{5.155}$$

where H_0 is the constant pressure head imposed on $x = a$, we have

$$\widetilde{A}_n^* = \frac{2H_0}{\pi n}\left[1 + (-1)^{n+1}\right] \tag{5.156}$$

or

$$\Phi_P(x,y) = \frac{2H_0}{\pi} \sum_{n=1,2,} \left\{ \frac{1+(-1)^{n+1}}{n} \right\} \frac{\sinh(n\pi x/b)}{\sinh(n\pi a/b)} \sin\left(\frac{n\pi y}{b}\right) \quad (5.157)$$

5.4.3 Separation of variables method - plane polar coordinates

As indicated previously, the choice of the coordinate system for any problem must reflect the geometry of the region under consideration. When circular regions are encountered, it is more convenient to formulate and solve the appropriate problem by appeal to a system of plane polar coordinates. Laplace's equation in plane polar coordinates takes the form

$$\nabla^2 \varphi(r,\theta) = 0 \quad (5.158)$$

where Laplace's operator now takes the form

$$\nabla^2 = \frac{\partial^2}{\partial r^2} + \frac{1}{r}\frac{\partial}{\partial r} + \frac{1}{r^2}\frac{\partial^2}{\partial \theta^2} \quad (5.159)$$

and

$$r = \left(x^2+y^2\right)^{1/2} \quad ; \quad \theta = \tan^{-1}\left(\frac{y}{x}\right) \quad (5.160)$$

Again, we can attempt to seek solutions of (5.158) which are independent of θ; i.e. $\varphi = \mathcal{R}(r)$. In this case (5.158) reduces to

$$\frac{d^2\mathcal{R}}{dr^2} + \frac{1}{r}\frac{d\mathcal{R}}{dr} = 0 \quad ; \quad \text{or} \quad \frac{1}{r}\frac{d}{dr}\left(r\frac{d\mathcal{R}}{dr}\right) = 0 \quad (5.161)$$

Integrating (5.161), we obtain the cylindrically symmetric solution of φ as

$$\varphi(r) = A_0 \ln r + B_0 \tag{5.162}$$

where A_0 and B_0 are constants. An examination of (5.162) indicates that if the region V under consideration includes the origin where the function φ is bounded (e.g. finite pressure, finite temperature, etc.), then

$$A_0 = 0 \tag{5.163}$$

The general solution of the radially symmetric case given by (5.162) is applicable for situations which do not include the origin, e.g. an annular region where the potential is prescribed on the boundaries.

A more generalized solution can be obtained by seeking solutions of the form

$$\varphi(r, \theta) = F(r) G(\theta) \tag{5.164}$$

Using this representation, Laplace's equation (5.158) gives

$$\frac{G(\theta)}{r} \frac{d}{dr}\left(r \frac{dF}{dr}\right) + \frac{F(r)}{r^2} \frac{d^2 G}{d\theta^2} = 0 \tag{5.165}$$

Multiplying (5.165) by r^2 and dividing by FG, we obtain

$$\frac{1}{F}\left(r^2 \frac{d^2 F}{dr^2} + r \frac{dF}{dr}\right) = -\frac{1}{G} \frac{d^2 G}{d\theta^2} = \text{const.} = n^2 \tag{5.166}$$

This gives rise to two ordinary differential equations

$$r^2 \frac{d^2F}{dr^2} + r\frac{dF}{dr} - n^2F = 0 \tag{5.167}$$

and

$$\frac{d^2G}{d\theta^2} + n^2G = 0 \tag{5.168}$$

which represent the governing equations for the separated variables. The equation (5.167) is an example of *Euler's differential equation* with non-constant coefficients and equation (5.168) resembles the classical equation for *simple harmonic motion*. The solutions of (5.167) and (5.168) can be obtained in the form

$$F(r) = Ar^n + \frac{B}{r^n} \quad , \quad \text{if} \quad n \neq 0 \tag{5.169}$$

$$G(\theta) = C\cos n\theta + D\sin n\theta \quad ; \quad \text{if} \quad n \neq 0 \tag{5.170}$$

If $n = 0$, the solutions of (5.167) and (5.168) give

$$F(r) = A_0 \ln r + B_0 \tag{5.171}$$

$$G(\theta) = C_0\theta + D_0 \tag{5.172}$$

where $A, B, C, D, A_0, \ldots$ etc. are arbitrary constants. The number n is referred to as the degree of the harmonic. Solutions of Laplace's equation referred to the plane polar coordinate system (r, θ) are called circular harmonics. The general form of the solution for $\varphi(r, \theta)$ can be written as

$$\varphi_0(r, \theta) = (A_0^* \ln r + B_0^*)(C_0^*\theta + D_0^*) \quad \text{for} \quad n = 0 \tag{5.173}$$

and

$$\varphi_n(r,\theta) = \left(A_n^* r^n + \frac{B_n^*}{r^n}\right)(C_n^* \cos n\theta + D_n^* \sin n\theta) \quad \text{for} \quad n > 0 \tag{5.174}$$

In certain problems which deal with complete domains where θ is continuous, φ is a single-valued function of θ; i.e., if we increase θ from 0 to 2π we should arrive at the same magnitude for $\varphi(r,\theta)$ without a discontinuity in the function, i.e.

$$\varphi(r,\theta) = \varphi(r,\theta + 2\pi) \tag{5.175}$$

Therefore, to accommodate this class of problem, we should take only integer values of n. The complete solution for $\varphi(r,\theta)$ can be obtained by a summation of the terms of (5.173) and (5.174), i.e.

$$\begin{aligned}\varphi(r,\theta) &= A_0 \ln r + \sum_{n=1} r^n (A_n \cos n\theta + B_n \sin n\theta) \\ &\quad + \sum_{n=1} \frac{1}{r^n}(C_n \cos n\theta + D_n \sin n\theta) + C_0\end{aligned} \tag{5.176}$$

where $A_0, C_0, A_n, B_n, \ldots$ etc. are arbitrary constants.

In the ensuing, we will apply the representation (5.176) to determine solutions to certain elementary problems governed by Laplace's equation.

Example 5.5

A cylindrical rod of radius a is subjected to a boundary temperature

$$T(a,\theta) = T_0\left(1 + \cos^2\theta\right) = T^*(\theta) \tag{5.177}$$

Determine the steady state maximum temperature at the centre of a cross section of the heat conducting rod (Figure 5.16).

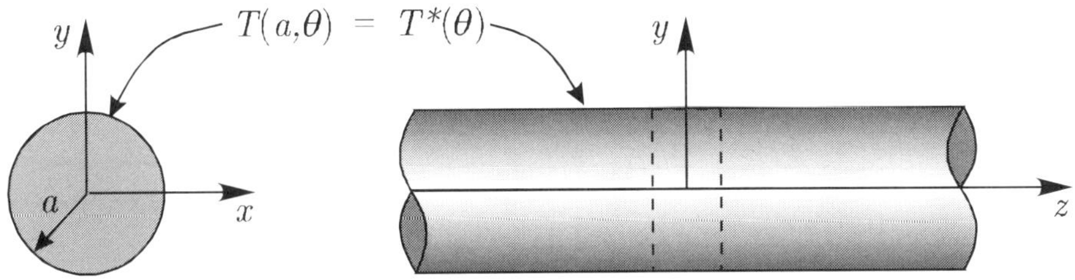

Figure 5.16: Boundary heating of a cylindrical rod.

Solution

The boundary value problem can be posed as follows: the partial differential equation governing the problem is

$$\nabla^2 T = \frac{\partial^2 T}{\partial r^2} + \frac{1}{r}\frac{\partial T}{\partial r} + \frac{1}{r^2}\frac{\partial^2 T}{\partial \theta^2} = 0 \tag{5.178}$$

The boundary condition is

$$T\,(a,\theta) = T_0 + T_0 \cos^2\theta = \frac{3T_0}{2} + \frac{T_0}{2}\cos 2\theta \tag{5.179}$$

In addition, the temperature field must be finite and single-valued in the domain of interest $0 \leq r \leq a;\quad 0 \leq \theta \leq 2\pi$.

Since the governing partial differential equation is linear, we can obtain the final solution by superposing the solutions to two problems: (i) the first problem$^{(1)}$ deals with the situation where the boundary of the rod is subjected to a constant temperature $(3T_0/2)$ and (ii) the second problem$^{(2)}$ considers the case where the boundary of the rod is subjected to a temperature distribution $(T_0 \cos 2\theta/2)$.

For the radially symmetric problem described by (i), we note that from Theorem 5.1 for the Dirichlet-type boundary condition, the steady state temperature field within the bar due to the constant boundary temperature $3T_0/2$ is

$$T^{(1)}(r,\theta) = \frac{3T_0}{2} \tag{5.180}$$

For the second problem, the boundary condition is

$$T^{(2)}(a,\theta) = \frac{T_0}{2}\cos 2\theta \tag{5.181}$$

Considering the general solution (5.176), it is evident that for the temperature field to be bounded in the region of the bar, we require

$$A_0 = 0 \;\; ; \;\; C_n = 0 \;\; ; \;\; D_n = 0 \tag{5.182}$$

and, without loss of generality, we can set $C_0 = 0$. The resulting solution is

$$T^{(2)}(r,\theta) = \sum_{n=1} r^n \left\{A_n \cos n\theta + B_n \sin n\theta\right\} \tag{5.183}$$

Applying the boundary condition (5.181) we have

$$T^{(2)}(a,\theta) = \frac{T_0}{2}\cos 2\theta = \sum_{n=1} a^n \left\{A_n \cos n\theta + B_n \sin n\theta\right\} \tag{5.184}$$

By inspection, we note that (5.184) is identically satisfied when $A_2 \neq 0$, $B_n = 0$ and all other A_n are set equal to zero, i.e.

$$A_2 = \frac{T_0}{2a^2} \tag{5.185}$$

The complete solution for the steady state distribution of temperature within the circular bar is given by

$$T(r,\theta) = T_0 \left[\frac{3}{2} + \frac{1}{2}\left(\frac{r}{a}\right)^2 \cos 2\theta\right] \tag{5.186}$$

This is a relatively simple form of a boundary condition which has a straightforward solution. In general, the boundary values of $T(a,\theta) = T^*(\theta)$ on the boundary has to be expanded in the form of a Fourier series. Considering (5.176) and the general form of the solution applicable to a solid bar we have

$$T^*(\theta) = \sum_{n=1,} a^n \left\{A_n^* \cos n\theta + B_n^* \sin n\theta\right\} + C_0^* \tag{5.187}$$

The constants A_n^*, B_n^* and C_0^* can be determined from the results of Fourier series representations which give

$$A_n^* = \frac{1}{a^n \pi} \int_0^{2\pi} T^*(\theta) \cos n\theta \; d\theta \tag{5.188}$$

$$B_n^* = \frac{1}{a^n \pi} \int_0^{2\pi} T^*(\theta) \sin n\theta \; d\theta \tag{5.189}$$

$$C_0^* = \frac{1}{2\pi} \int_0^{2\pi} T^*(\theta) \, d\theta \tag{5.190}$$

Example 5.6

A plate in the shape of a quadrant of an annulus (Figure 5.17) is subjected to boundary temperatures in the forms $T_a \cos 4\theta$ and $T_b \cos 4\theta$, respectively along the curved boundaries. The plane boundaries are kept insulated.

Formulate the boundary value problem and derive an expression for the steady state temperature distribution within the plate.

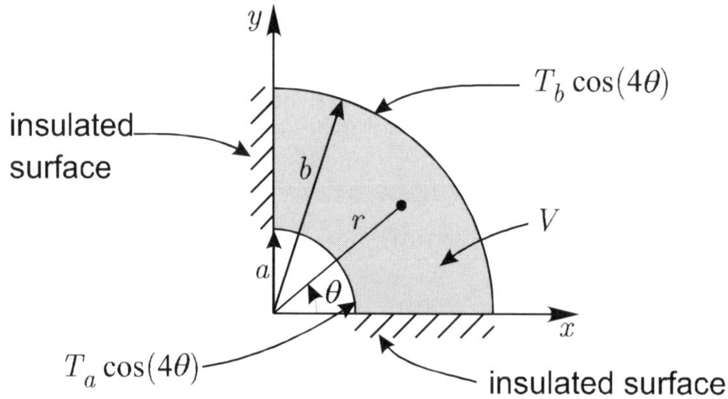

Figure 5.17: Temperature distributions in an annular sector.

Solution

The boundary value problem requires the solution of Laplace's equation

$$\nabla^2 T(r, \theta) = 0 \quad ; \qquad (r, \theta) \in V \tag{5.191}$$

subject to the boundary conditions

$$T(a, \theta) = T_a \cos 4\theta \tag{5.192}$$

$$T(b,\theta) = T_b \cos 4\theta \tag{5.193}$$

$$\frac{\partial T}{\partial \theta} = 0 \quad \text{on} \quad \theta = 0 \quad , \quad \theta = \frac{\pi}{2} \tag{5.194}$$

Considering the general solution (5.176) we note that since the origin ($r = 0$) is not included in the problem domain, all solutions are admissible. From the boundary conditions (5.194) relevant to insulated conditions on the plane ends of the plate it is clear that

$$\left(\frac{\partial T}{\partial \theta}\right) = 0 \quad \text{on} \quad \theta = 0, \quad \text{and} \quad \theta = \frac{\pi}{2} \tag{5.195}$$

for all r only when $B_n = D_n = 0$. The reduced form of the general solution, which automatically satisfies the boundary conditions on $\theta = 0, \pi/2$, takes the form

$$T(r,\theta) = A_0 \ln r + C_0 + \sum_{n=1,2} \left(A_n r^n + \frac{C_n}{r^n}\right) \cos n\theta \tag{5.196}$$

We note that the terms $A_0 \ln r$ and C_0 give finite non-zero temperatures at $\theta = \pi/8$ and $3\pi/8$; the boundary conditions, on the other hand give zero values at the boundary locations.

Furthermore, the terms $A_0 \ln r$ and C_0 account for temperature distributions which are radially symmetric (i.e. independent of θ). Therefore these terms are not applicable to temperature distributions which will depend on θ. Thus the relevant form of the solution is

$$T(r,\theta) = \sum_{n=1} \left\{A_n r^n + \frac{C_n}{r^n}\right\} \cos n\theta \tag{5.197}$$

Using the boundary conditions (5.192) and (5.193) in (5.197) we have

$$T_a \cos 4\theta = \sum_{n=1} \left(A_n a^n + \frac{C_n}{a^n} \right) \cos n\theta \tag{5.198}$$

$$T_b \cos 4\theta = \sum_{n=1} \left(A_n b^n + \frac{C_n}{b^n} \right) \cos n\theta \tag{5.199}$$

It is evident that these equations will be satisfied when A_4 and C_4 are non-zero and all other A_n and C_n are identically zero. This gives the following equations to determine A_4 and C_4, i.e.

$$A_4 a^8 + C_4 = a^4 T_a \quad ; \quad A_4 b^8 + C_4 = b^4 T_b \tag{5.200}$$

which gives

$$A_4 = \frac{b^4 T_b - a^4 T_a}{(b^8 - a^8)} \quad ; \quad C_4 = -a^4 b^4 \frac{(a^4 T_b - b^4 T_a)}{(b^8 - a^8)} \tag{5.201}$$

Hence, the final form of the temperature distribution in the plate is given by

$$T(r, \theta) = \left\{ \frac{a^4 b^4 T_b}{(b^8 - a^8)} \left[\frac{r^4}{a^4} - \frac{a^4}{r^4} \right] - \frac{a^4 b^4 T_a}{(b^8 - a^8)} \left[\frac{r^4}{b^4} - \frac{b^4}{r^4} \right] \right\} \cos 4\theta \tag{5.202}$$

It is clear that when $r = a$ and $r = b$, the boundary conditions are identically satisfied.

5.5 Integral transform solution of Laplace's equation

In this section we shall present the application of an exponential Fourier transform technique for the solution of Laplace's equation in two dimensions. We specifically consider the problem referred to a half-plane occupying the region $-\infty < x < \infty$ (Figure 5.18); $0 \le y < \infty$. The surface of the half-plane is subjected to a potential $\varphi_0(x)$. For the present,

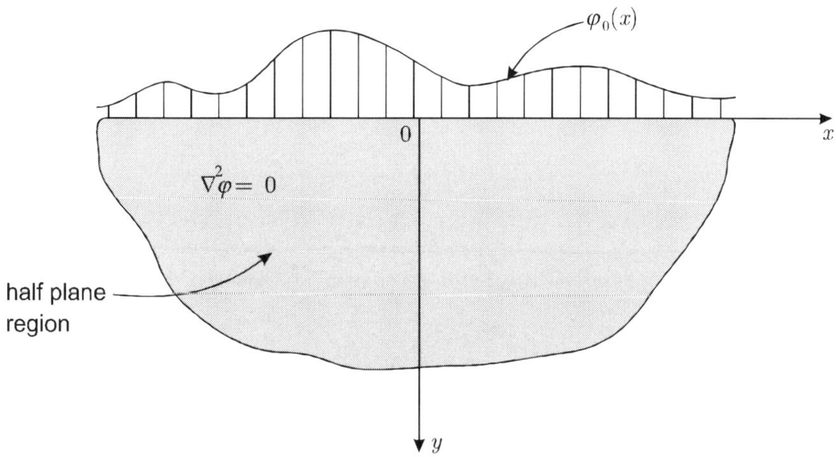

Figure 5.18: Laplace's equation for a half-plane region.

the form of the function is left unspecified except that we assume that

$$\varphi_0(x) \to 0 \qquad \text{as} \qquad |x| \to \infty \tag{5.203}$$

The boundary value problem requires the solution of

$$\nabla^2 \varphi = \frac{\partial^2 \varphi}{\partial x^2} + \frac{\partial^2 \varphi}{\partial y^2} = 0 \tag{5.204}$$

subject to the boundary condition

$$\varphi(x, 0) = \varphi_0(x) \tag{5.205}$$

and the regularity conditions

$$\varphi(x, y) \to 0 \qquad \text{as} \qquad |x| \to \infty \quad ; \quad y \to \infty \tag{5.206}$$

We define the exponential Fourier transform of $\varphi(x, y)$ with respect to the variable x as

$$\widetilde{\varphi}(\xi, y) = \mathcal{F}\{\varphi(x, y); \xi\} = \frac{1}{\sqrt{2\pi}} \int_{-\infty}^{\infty} \varphi(x, y) e^{i\xi x} dx \tag{5.207}$$

We multiply the partial differential equation (5.204) by $e^{i\xi x}$ and integrate between $-\infty < x < \infty$:

$$\int_{-\infty}^{\infty} \frac{\partial^2 \varphi}{\partial x^2} e^{i\xi x} dx + \int_{-\infty}^{\infty} \frac{\partial^2 \varphi}{\partial y^2} e^{i\xi x} dx = 0 \tag{5.208}$$

Performing the integration by parts we can show that since $\varphi(x, y) \to 0$ as $(x^2 + y^2)^{\frac{1}{2}} \to \infty$ (see also (1.173))

$$\mathcal{F}\left\{\frac{\partial^2 \varphi}{\partial x^2}; \xi\right\} = -\xi^2 \widetilde{\varphi}(\xi, y) \tag{5.209}$$

$$\mathcal{F}\left\{\frac{\partial^2 \varphi}{\partial y^2}; \xi\right\} = \frac{d^2 \tilde{\varphi}(\xi, y)}{dy^2} \tag{5.210}$$

Using (5.209) and (5.210) we obtain from (5.208) the ordinary differential equation for $\tilde{\varphi}(\xi, y)$:i.e.

$$\frac{d^2 \tilde{\varphi}}{dy^2} - \xi^2 \tilde{\varphi} = 0 \tag{5.211}$$

The solution of (5.211) gives

$$\tilde{\varphi}(\xi, y) = A(\xi) e^{\xi y} + B(\xi) e^{-\xi y} \tag{5.212}$$

We can transform the boundary condition (5.205) and the regularity condition (5.206) to give the following:

$$\tilde{\varphi}(\xi, 0) = \tilde{\varphi}_0(\xi) = \frac{1}{\sqrt{2\pi}} \int_{-\infty}^{\infty} \varphi_0(x) e^{i\xi x} dx \tag{5.213}$$

$$\tilde{\varphi}(\xi, y) \to 0 \qquad \text{as} \qquad y \to \infty \tag{5.214}$$

In order to satisfy the regularity condition (5.214) we require, for $\xi > 0$

$$A(\xi) = 0 \tag{5.215}$$

Similarly, if $\xi < 0$, we require

$$B(\xi) = 0 \tag{5.216}$$

Since ξ occupies the range $-\infty < \xi < \infty$ we can write the solution which satisfies the regularity condition (5.214) as

$$\tilde{\varphi}(\xi, y) = B^*(\xi) e^{-|\xi| y} \tag{5.217}$$

To satisfy the boundary condition (5.213) we require

$$\tilde{\varphi}(\xi, 0) = B^*(\xi) = \tilde{\varphi}_0(\xi) \tag{5.218}$$

Hence the solution to the problem, in the transformed domain, is

$$\tilde{\varphi}(\xi, y) = \tilde{\varphi}_0(\xi) e^{-|\xi| y} \tag{5.219}$$

Inverting (5.219) we obtain

$$\varphi(x,y) = \frac{1}{\sqrt{2\pi}} \int_{-\infty}^{\infty} \tilde{\varphi}_0(\xi) e^{-i\xi x} e^{-|\xi| y} \, d\xi \tag{5.220}$$

This constitutes the formal solution to the two-dimensional problem of Laplace's equation associated with a half-plane problem where $\varphi_0(x)$ is prescribed on the surface. We now consider a specific case where the potential $\varphi_0(x)$ is prescribed only on a region of the boundary $y = 0$. We specifically consider the case where

$$\varphi(x,0) = \begin{cases} \varphi_0^* \, ; & |x| < a \\ 0 \, ; & |x| > 0 \end{cases} \tag{5.221}$$

where φ_0^* is a constant and (5.221) satisfies the condition (5.203) (Figure 5.19). The problem relates to the case where Dirichlet boundary conditions are prescribed on the entire boundary $y = 0$.

We can transform the boundary potential (5.221) according to the result

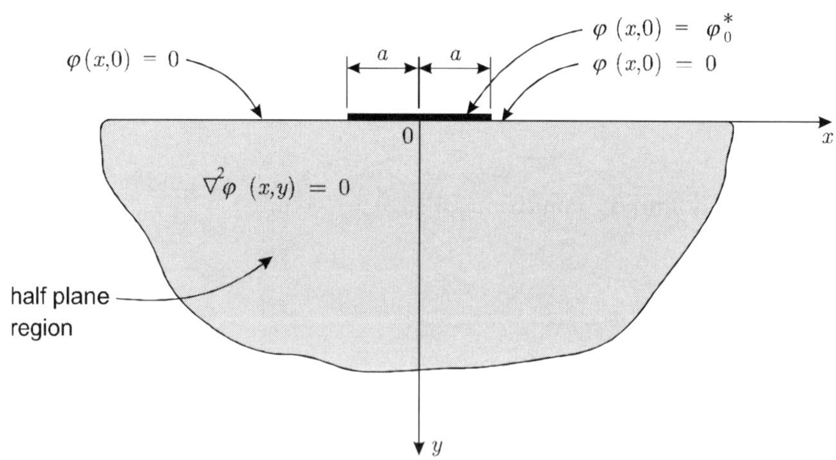

Figure 5.19: Potential prescribed on a segment of the half-plane region.

(5.213); i.e.

$$\begin{aligned}\tilde{\varphi}(\xi) &= \frac{1}{\sqrt{2\pi}} \int_{-\infty}^{\infty} \varphi(x, 0) e^{i\xi x}\, dx \\ &= \frac{\varphi_0^*}{\sqrt{2\pi}} \int_{-a}^{a} \{\cos(\xi x) + i \sin(\xi x)\} dx \end{aligned} \tag{5.222}$$

or

$$\widetilde{\varphi}_0(\xi) = \frac{2\varphi_0^*}{\sqrt{2\pi}} \frac{\sin(\xi a)}{\xi} \tag{5.223}$$

Substituting (5.223) in (5.220) we have

$$\varphi(x, y) = \frac{1}{\sqrt{2\pi}} \int_{-\infty}^{\infty} \frac{2\varphi_0^*}{\sqrt{2\pi}} \frac{\sin(\xi a)}{\xi} \{\cos(\xi x) - i \sin(\xi x)\} e^{-|\xi| y}\, d\xi \tag{5.224}$$

We can reduce (5.224) to the form

$$\varphi(x, y) = \frac{\varphi_0^*}{\pi} \int_0^{\infty} [\sin \xi(x + a) - \sin \xi(x - a)] \frac{e^{-\xi y}}{\xi}\, d\xi \tag{5.225}$$

which is an integral form of the solution to the problem. Using results of Fourier sine transforms we have

$$\int_0^{\infty} [\sin \xi(x + a)] \frac{e^{-\xi y}}{\xi}\, d\xi = \tan^{-1}\left(\frac{x + a}{y}\right) \tag{5.226}$$

$$\int_0^{\infty} [\sin \xi(x - a)] \frac{e^{-\xi y}}{\xi}\, d\xi = \tan^{-1}\left(\frac{x - a}{y}\right) \tag{5.227}$$

We can express the result (5.225) in the form

$$\varphi(x,y) = \frac{\varphi_0^*}{\pi}\left[\tan^{-1}\left(\frac{x+a}{y}\right) - \tan^{-1}\left(\frac{x-a}{y}\right)\right] \tag{5.228}$$

We note that in the limit as $y \to 0$, we have

$$\operatorname*{Lim}_{\Omega\to\infty} \tan^{-1}(\Omega) = \frac{\pi}{2} \tag{5.229}$$

$$\operatorname*{Lim}_{\Omega\to-\infty} \tan^{-1}(\Omega) = -\frac{\pi}{2} \tag{5.230}$$

and the boundary condition (5.221) is satisfied.

5.6 Line source within a half-plane region

The solution of Laplace's equation for a half-plane region in the presence of a source term in the form of Dirac delta function is examined in this section. This particular problem is of interest to the construction of a two-dimensional *Green's function* for a half-plane region occupying $-\infty < x < \infty$, $0 \le y < \infty$. By definition, a Green's function is a *particular solution* of partial differential equation applicable to a specified domain of specified dimensions. It is also possible to adopt a more formal approach for the derivation of the Green's function which will also cover the specific problem being examined. In this section, however, we shall apply the exponential Fourier transform technique to develop a solution to the internally applied line source problem (Figure 5.20)

For the present purposes we note that the development of a Green's function for Laplace's equation referred to a half-plane requires the solution of the partial differential equation

$$\nabla^2\varphi = -\,\delta(x-x^*)\delta(y-y^*) \tag{5.231}$$

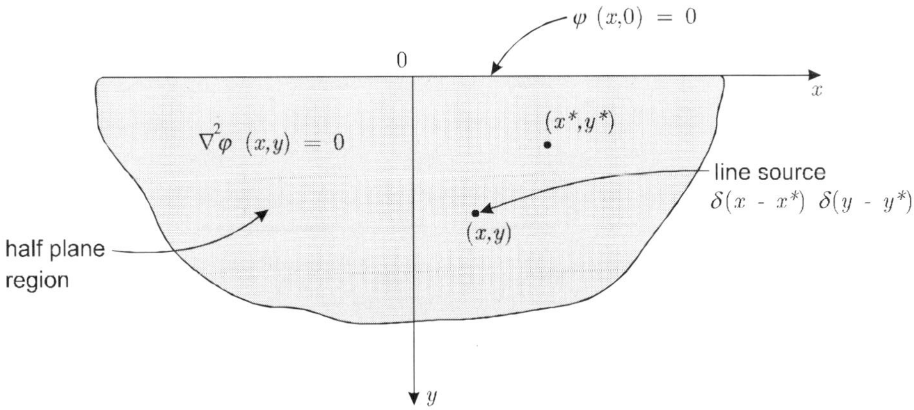

Figure 5.20: Line source applied at the interior of a half-plane.

where $\varphi(x, y; x^*, y^*)$ represents the "potential" at (x^*, y^*) due to a line source located at (x, y). In addition to the governing *non-homogeneous* partial differential equation, or Poisson's equation, we require that at distances remote from the "source", $\varphi(x, y; x^*, y^*)$ and its derivates must reduce to zero. For the solution of the problem we adopt an exponential Fourier transform technique. We multiply both sides of (5.231) by $e^{i\xi x}/\sqrt{2\pi}$ and integrate over the interval $x = -\infty$ to $x = \infty$. We note that the exponential Fourier transform of $\varphi(x, y; x^*, y^*)$ with respect to x is,

$$\overline{\varphi}(\xi, y; x^*, y^*) = \frac{1}{\sqrt{2\pi}} \int_{-\infty}^{\infty} \varphi(x, y; x^*, y^*) e^{i\xi x} \, dx \tag{5.232}$$

Assuming that

$$\operatorname*{Lim}_{|x|\to\infty} \left(\frac{\partial \varphi}{\partial x} + i\xi\varphi \right) = 0 \tag{5.233}$$

we obtain from (5.231) the following ordinary differential equation governing

$\overline{\varphi}(\xi, y; x^*, y^*)$:

$$\frac{d^2\overline{\varphi}}{dy^2} - \xi^2\overline{\varphi} = -\delta(y - y^*)\frac{e^{i\xi x^*}}{\sqrt{2\pi}} \tag{5.234}$$

The solutions

$$\overline{\varphi}(\xi, y; x^* y^*) = Ae^{-|\xi|(y-y^*)} \quad ; \quad y > y^* \tag{5.235}$$

and

$$\overline{\varphi}(\xi, y; x^*, y^*) = Be^{|\xi|(y-y^*)} \quad ; \quad y < y^* \tag{5.236}$$

where A and B are constants and both equations (5.235) and (5.236) satisfy (5.234) when $y \neq y^*$. Considering the solution (5.235) we note that as the 'source' coordinate $y \to +\infty$, the solution $\overline{\varphi}$ at y^* reduces to zero since the source point and the location have an infinite separation interval. Similarly as $y \to -\infty$, in (5.236), the solution at y^* reduces to zero. Each solution for $\overline{\varphi}$ gives the correct decay behaviour at y^* as the source term recedes to $y \to \pm\infty$. For continuity of the solution at $y = y^*$ we require that $A = B$, and the resulting solutions are

$$\overline{\varphi}(\xi, y; x^* y^*) = Ae^{-|\xi|(y-y^*)} \quad ; \quad y > y^* \tag{5.237}$$

$$\overline{\varphi}(\xi, y; x^*, y^*) = Ae^{|\xi|(y-y^*)} \quad ; \quad y < y^* \tag{5.238}$$

We now need to determine the arbitrary constant A, such that (5.237) and (5.238) satisfy not only the governing ordinary differential equation (5.234) when $y > y^*$ and $y < y^*$ but also for $y \geq y^*$ and $y \leq y^*$. We can accomplish this by integrating (5.234) about a small interval of length 2ϵ centered about $y = y^*$; i.e.

$$\int_{y=y^*-\epsilon}^{y=y^*+\epsilon} \left\{ \frac{d^2\overline{\varphi}}{dy^2} \right\} dy - \xi^2 \int_{y=y^*-\epsilon}^{y=y^*+\epsilon} \overline{\varphi}\, dy$$
$$= \frac{-e^{i\xi x^*}}{\sqrt{2\pi}} \int_{y=y^*-\epsilon}^{y=y^*+\epsilon} \delta(y-y^*) dy \tag{5.239}$$

and taking the limit as $\epsilon \to 0$. Using this approach we have

$$\operatorname*{Lim}_{\epsilon \to 0} \left[\frac{d\overline{\varphi}}{dy} \right]_{y=y^*-\epsilon}^{y=y^*+\epsilon} = \frac{-e^{i\xi x^*}}{\sqrt{2\pi}} \tag{5.240}$$

It may be noted that since $\overline{\varphi}(\xi, y; x^*, y^*)$ is continuous at $y = y^*$, the second term on the left hand side of (5.239) is zero. The result (5.240) indicates that the derivative of $\overline{\varphi}(\xi, y; x^*, y^*)$ must be discontinuous at $y = y^*$, and the discontinuity is $-e^{i\xi x^*}/\sqrt{2\pi}$. We can compute the derivative as approached from either side of $y = y^*$ and from (5.237) and (5.238) we have

$$\frac{d}{dy} \left\{ A e^{-|\xi|(y-y^*)} \right\} = -|\xi|\, A e^{-|\xi|(y-y^*)} \quad ; \quad y > y^* \tag{5.241}$$

$$\frac{d}{dy} \left\{ A e^{|\xi|(y-y^*)} \right\} = |\xi|\, A e^{|\xi|(y-y^*)} \quad ; \quad y < y^* \tag{5.242}$$

Using (5.241) and (5.242) in the condition (5.240) we have

$$\operatorname*{Lim}_{\epsilon \to 0} \left[\frac{d\overline{\varphi}}{dy} \right]_{y=y^*-\epsilon}^{y=y^*+\epsilon} = -2\,|\xi|\, A = \frac{-e^{i\xi x^*}}{\sqrt{2\pi}} \tag{5.243}$$

From (5.243) we have

$$A = \frac{e^{i\xi x^*}}{2\sqrt{2\pi}\,|\xi|} \tag{5.244}$$

We note that the general results corresponding to (5.238) and (5.239) can be written as a single result using $|y - y^*|$ in the exponential term; i.e.

$$\overline{\varphi}(\xi, y; x^*, y^*) = \frac{e^{i\xi x^* - |\xi||y-y^*|}}{2\sqrt{2\pi}\,|\xi|} \tag{5.245}$$

which is the required solution of (5.234). We can invert (5.245) to determine $\varphi(x, y; x^*, y^*)$: i.e.

$$\varphi(x, y; x^*, y^*) = \frac{1}{2\pi}\int_{-\infty}^{\infty} \frac{e^{-i\xi(x-x^*)-|\xi||y-y^*|}}{2\,|\xi|}\,d\xi \tag{5.246}$$

For example, in the special instance when the line source term acts at the surface of the half-plane (5.231) reduces to

$$\nabla^2\varphi = -\,\delta(x - x^*) \tag{5.247}$$

and the corresponding solution of $\varphi(x, y)$ takes the form

$$\varphi(x, y; x^*) = -\frac{y}{\pi\{(x - x^*)^2 + y^2\}} \tag{5.248}$$

5.7 Uniqueness theorem

In section 5.3 we discussed some general results which were useful from the point of view of identifying, *a priori*, the general character of the solutions of Laplace's equation. These general results were obtained for situations where the boundary S of the region was subjected to Dirichlet, Neumann or mixed (Robin-type) boundary conditions prescribed over some fixed surface S. The objective of this section is to utilize this general approach to establish the uniqueness of the solution to a well-posed problem or, conversely, if the solution is unique this satisfies one of the criteria required for well-posedness. The general approach for proof of uniqueness is through a *reductio ad absurdum* procedure or a *proof by contradiction*. (The concept of a proof by contradiction is generally attributed to Eudoxus (circa 408-355 B.C.) of Cnidus a great mathematician of the Greek Classical Period and second only to Archimedes (287 B.C.-212 B.C.)).

THEOREM 5.6

Let us consider the region V in which φ is a solution of $\nabla^2\varphi = 0$, and satisfies the boundary conditions

$$\varphi = f_D(\mathbf{x}) \quad , \quad \text{on} \quad \mathbf{x} \in S_D \tag{5.249}$$

$$\frac{\partial\varphi}{\partial n} = f_N(\mathbf{x}) \quad , \quad \text{on} \quad \mathbf{x} \in S_N \tag{5.250}$$

$$\frac{\partial\varphi}{\partial n} + \lambda\varphi = f_M(\mathbf{x}) \quad , \quad \text{on} \quad \mathbf{x} \in S_M \tag{5.251}$$

where $f_D(\mathbf{x})$, $f_N(\mathbf{x})$,and $f_M(\mathbf{x})$, are prescribed functions and λ is a positive constant. The equations (5.249), (5.250) and (5.251) are, respectively, Dirichlet, Neumann and Mixed (Robin-type) boundary conditions applied to the separate subregions of S (Figure 5.21), i.e.

$$S_D \cup S_N \cup S_M = S \quad ; \quad S_D \cap S_N \cap S_M = 0 \tag{5.252}$$

If the solution of φ satisfying Laplace's equation and the boundary conditions (5.249), (5.250) and (5.251) exists, it is unique.

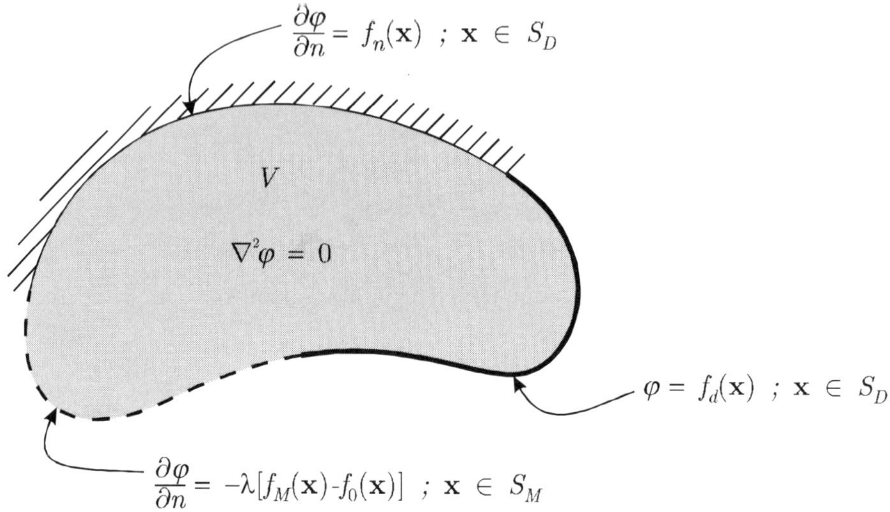

Figure 5.21: Domain for Laplace's equation

PROOF

Let us assume that the boundary value problem admits *two* solutions $\varphi_1(\mathbf{x})$ and $\varphi_2(\mathbf{x})$ each satisfying Laplace's equation and the boundary conditions on S_i $(i = D, N, M)$. Then the difference of these two solutions

$$\varphi_3(\mathbf{x}) = \varphi_1(\mathbf{x}) - \varphi_2(\mathbf{x}) \tag{5.253}$$

also constitutes a solution of Laplace's equation but satisfying the boundary conditions

$$\varphi_3 = 0 \quad , \quad \text{on} \quad \mathbf{x} \in S_D \tag{5.254}$$

$$\frac{\partial \varphi_3}{\partial n} = 0 \quad , \quad \text{on} \quad \mathbf{x} \in S_N \tag{5.255}$$

$$\frac{\partial \varphi_3}{\partial n} = -\lambda \varphi_3 \quad , \quad \text{on} \quad \mathbf{x} \in S_M \tag{5.256}$$

Now, applying Green's first identity,

$$\begin{aligned} \iiint_V |\operatorname{grad} \varphi_3|^2 \, dV &= \iint_S \varphi_3 \frac{\partial \varphi_3}{\partial n} dS \\ &= \iint_{S_D} \varphi_3 \frac{\partial \varphi_3}{\partial n} dS + \iint_{S_N} \varphi_3 \frac{\partial \varphi_3}{\partial n} dS \\ &+ \iint_{S_M} \varphi_3 \frac{\partial \varphi_3}{\partial n} dS \end{aligned} \tag{5.257}$$

Considering the boundary conditions (5.254) and (5.255), it is evident that the first two integrals on the right hand side are zero; using the third boundary condition (5.256) the equation (5.257) can now be reduced to the form

$$\iiint_V |\operatorname{grad} \varphi_3|^2 \, dV + \lambda \iint_S (\varphi_3)^2 \, dS = 0 \tag{5.258}$$

The integrands of both integrals are positive definite; consequently, the equation (5.258) is satisfied if and only if

$$\varphi_3 = 0 \quad , \quad \mathbf{x} \in \Omega \tag{5.259}$$

where $\Omega (= V \cup S)$ is the closure.

The assertion that there exists *two* solutions of Laplace's equation which satisfy the same boundary conditions is now proved to be *false*. Hence, if a solution to the boundary value problem exists, it is unique.

5.8 A maximum principle

In section 5.4 we have developed a number of solutions for the Laplace's equation which can represent steady state phenomena associated with processes such as heat conduction, flow in porous media and other physical processes in potential theory. When Laplace's equation is solved for a process such as steady state heat conduction it is intuitively clear that there can be no maximum temperature within the region of interest when boundary values of temperature are prescribed. If such a maximum temperature occurs within the region there will be heat flow from the location of maximum temperature and the process will not constitute a steady state condition. The *maximum principle* associated with Laplace's equation can be used to ensure that the maximum value of the temperature, pressure, etc., should occur at the boundary of the region. In the ensuing we shall present a proof of this principle and consider a more general problem associated with the inhomogeneous form of Laplace's equation, namely Poisson's equation.

THEOREM 5.7

Consider a function $\varphi(\mathbf{x})$ which is a solution of the equation

$$\nabla^2\varphi(\mathbf{x}) = -f(\mathbf{x}) \quad ; \quad \mathbf{x} \in \mathrm{V} \tag{5.260}$$

and assume that $f(\mathbf{x}) < 0$ in V. If $\varphi(\mathbf{x})$ is continuous in $\overline{V} \in V$, it will attain its maximum φ_M at S the boundary of V.

PROOF

From the properties applicable to local maxima we have

$$\text{grad } \varphi = 0 \quad ; \quad \overline{\mathbf{x}} \in \overline{V} \tag{5.261}$$

and

$$\nabla^2\varphi(\overline{\mathbf{x}}) \leq 0 \quad ; \quad \overline{\mathbf{x}} \in \overline{V} \tag{5.262}$$

Hence $\nabla^2\varphi \leq 0$ at $\overline{\mathbf{x}}$ contradicts the hypothesis that $f(\mathbf{x}) < 0$. Hence the maximum of φ_M of $\varphi(\mathbf{x})$ must occur on S the boundary of V.

This approach can be extended to examine the case where $f(\mathbf{x}) \leq 0$.

PROOF

Let us assume that $f \leq 0$ and $\varphi \leq \varphi_M$ on S. We now introduce an auxiliary function $\chi(\mathbf{x})$ such that

$$\chi(\mathbf{x}) = \varphi(\mathbf{x}) + \widetilde{\epsilon}(\mathbf{x} \cdot \mathbf{x}) \tag{5.263}$$

where $\widetilde{\epsilon}$ is a parameter which is non-negative. Since

$$\nabla^2\chi = -(f - 6\widetilde{\epsilon}) \quad \text{in} \quad \mathbf{x} \in V \tag{5.264}$$

and $(f - 6\widetilde{\epsilon}) < 0$, from the previous proof, χ must attain its maximum on the boundary S of the region V. Hence

$$\chi(\mathbf{x}) \leq \max_{\mathbf{x}\in S} [\varphi + \widetilde{\epsilon}(\mathbf{x} \cdot \mathbf{x})] \leq \varphi_M + \widetilde{\epsilon}R^2 \tag{5.265}$$

where R is the radius of a sphere containing V (which is assumed to be bounded). Since by definition

$$\varphi(\mathbf{x}) \leq \chi(\mathbf{x}) \tag{5.266}$$

it follows from (5.265)

$$\varphi(\mathbf{x}) \leq \varphi_M + \widetilde{\epsilon} R^2 \tag{5.267}$$

for any value of $\widetilde{\epsilon}$. If we reduce $\widetilde{\epsilon} \to 0$ we obtain

$$\varphi(\mathbf{x}) \leq \varphi_M \qquad \text{for} \qquad \mathbf{x} \in V \tag{5.268}$$

It is possible to use a corollary resulting from these theorems to prove the uniqueness of a solution associated with Laplace's equation involving Dirichlet boundary conditions. The proof given in section 5.7 is, however, more general, much simpler and easier to interpret.

5.9 PROBLEM SET 5

5.1 Verify that

$$u = \ln r \qquad ; \quad u = r^n \cos n\theta$$

$$u = r^n \sin\theta \qquad ; \quad u = \tan^{-1}\left(\frac{y}{x}\right)$$

$$u = \sin\left(x^2 - y^2\right)\cosh\left(2xy\right)$$

where $r^2 = (x^2+y^2)$ all satisfy $\nabla^2 u = 0$, where ∇^2 is the two-dimensional form of Laplace's operator.

5.2 Using results of vector calculus show that for a region which excludes the origin

$$\nabla^2\left(\frac{1}{R}\right) = 0$$

where

$$R^2 = x^2 + y^2 + z^2$$

5.3 If $\varphi(\mathbf{x})$ satisfies Laplace's equation in a region V and if the boundary conditions on S, the boundary of V, are

$$\varphi = f(\mathbf{x}), \qquad \text{on} \quad \mathbf{x} \in S_1$$

$$\frac{\partial \varphi}{\partial n} = g(\mathbf{x}), \qquad \text{on} \quad \mathbf{x} \in S_2$$

$$\frac{\partial \varphi}{\partial n} = -\lambda \left[\varphi(\mathbf{x}) - \varphi_0(\mathbf{x})\right], \quad \text{on} \quad \mathbf{x} \in S_3$$

where λ is a positive constant, $\varphi_0(\mathbf{x})$ is a reference value and

$$S_1 \cup S_2 \cup S_3 = S \quad ; \quad S_1 \cap S_2 \cap S_3 = 0$$

then prove that the solution of $\varphi(\mathbf{x})$ is unique.

5.4 Given that

$$T(r, \theta) = A \ln r + B\theta \ln r + C\theta + D$$

$$+\sum \left\{A_n r^n + \frac{B_n}{r^n}\right\} \{C_n \sin(n\theta) + D_n \cos(n\theta)\}$$

where A, B, C, D, A_n, B_n, etc., are constants, determine the steady state temperature distributions within the annular sectors shown in Figure 5.22.

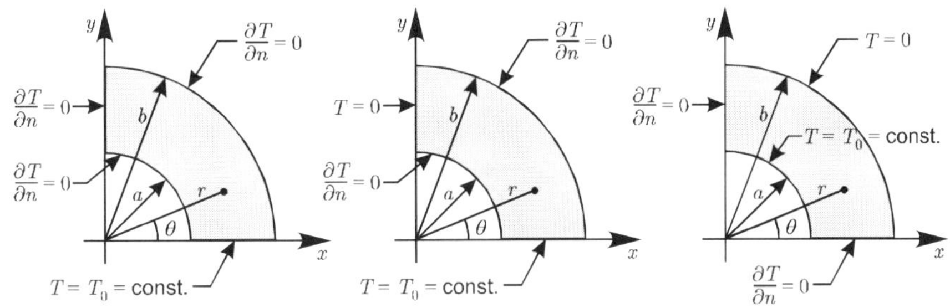

Figure 5.22: Heat conduction in an annular sector.

5.5 The Figure 5.23 shows the two-dimensional problem of a rectangular plate in which the sides are subjected to the boundary temperatures

$$T(0, y) = 0 \ ; T(a, y) - T^*(y)$$

$$T\,(x, 0) = 0 \ ; T\,(x, b) = 0$$

Show that the generalized result for the temperature distribution within the plate can be expressed in the form of a series

$$T\,(x, y) = \sum_{n=1,2,} C_n^* \sinh\left(\frac{n\pi x}{b}\right) \sin\left(\frac{n\pi y}{b}\right)$$

Derive expressions for the constants C_n^* ($n = 1, 2, 3$ only) when

(i) $T^*\,(y) = T_0 y$

(ii) $T^*\,(y) = T_0 y\,(b - y)$

where T_0 is a constant.

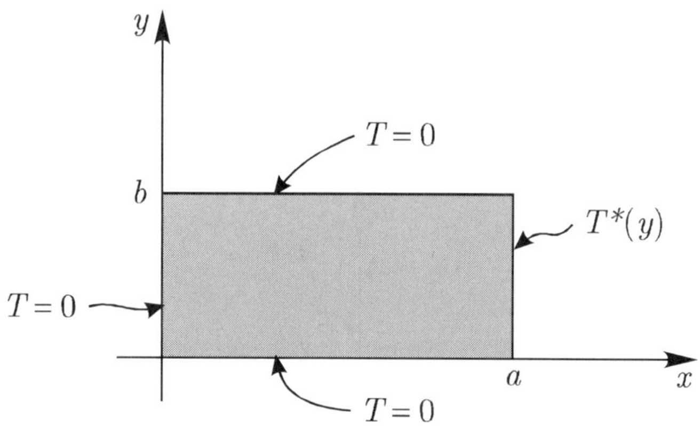

Figure 5.23: Boundary temperatures for a plate region.

5.6 The Figure 5.24 shows the configuration of a heat conducting rectangular plate where two sides are perfectly insulated. The side $y = 0$ is maintained at zero temperature and the side $y - b$ is subjected to a temperature distribution

$$T(x,b) = T_0\left[1+\cos\left(\frac{2\pi x}{a}\right)\right]$$

Formulate the boundary value problem for steady state temperature distribution in the plate and derive the expression for the temperature distribution.

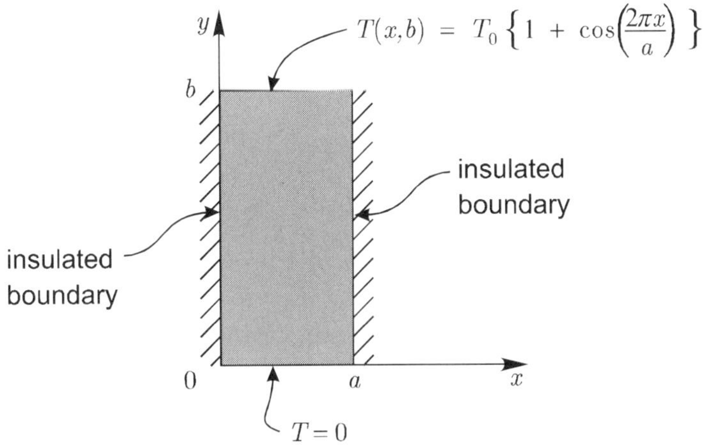

Figure 5.24: The insulated plate problem.

5.7 (i) The Figure 5.25 shows the cross section through a heat conducting medium of infinite extent (in the radial (r) and axial (z) directions) containing a long cylindrical cavity of radius a. Heating elements located on the surface of the cavity maintain a temperature of

$$T(a,\theta) = T_0 \sin\theta\ (1 + 2\cos\theta)$$

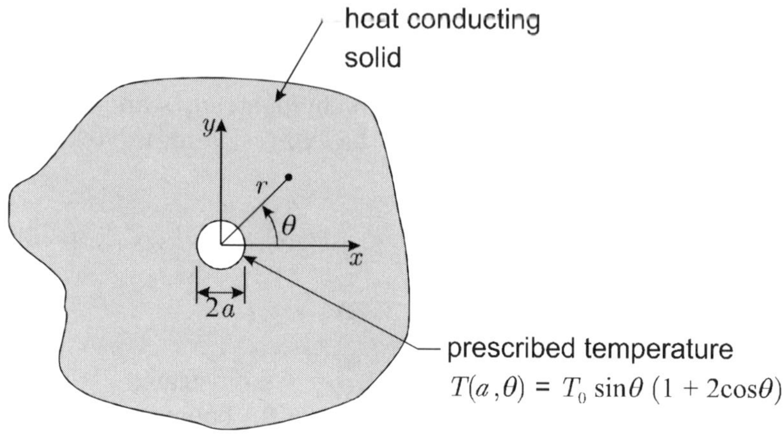

Figure 5.25: An exterior Dirichlet problem for Laplace's equation.

where T_0 is a constant. The steady state temperature distribution for this *exterior Dirichlet problem* is governed by

$$\nabla^2 T(r, \theta) = 0$$

Obtain an expression for the temperature distribution in the solid.
(ii) Verify the accuracy of your solution by using *Green's First Identity.*

5.8 The function $\varphi(x, y)$ satisfies Laplace's equation. Consider the quarter-plane region occupying $0 \leq x < \infty$; $0 \leq y < \infty$ (Figure 5.26). the entire boundary $x = 0$ of this region is subjected to a *homogeneous Neumann* boundary condition and the boundary $y = 0$ is subjected to *Dirichlet* boundary conditions as follows:

$$\begin{aligned} &\varphi(x,0) = 0 \;\; ; 0 < x < a \\ &\varphi(x,0) = \varphi_0 \; ; a < x < b \\ &\varphi(x,0) = 0 \;\; ; b < x < \infty \end{aligned}$$

where φ_0 is a constant. Use *Fourier cosine transform technique* to obtain the solution to

$$\nabla^2 \varphi(x, y) = 0 \;\; ; \;\; 0 \leq x < \infty \;\; ; \;\; 0 \leq y < \infty$$

(You may give the solution in integral form.)

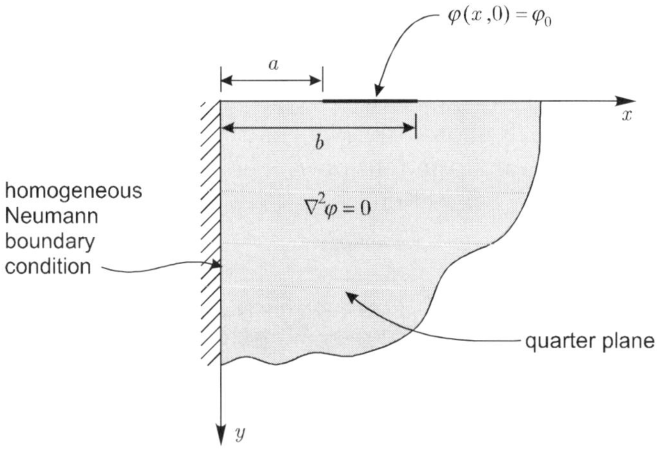

Figure 5.26: Laplace's problem for a quarter-plane.

5.9 Consider the fluid flow in a region where the velocity potential φ is constant over the boundary of a simply connected region V. If the fluid is incompressible and the fluid motion is irrotational, prove that φ is constant over the entire region.

5.10 Consider the problem of two dimensional, steady inviscid flow around a rigid stationary cylinder of radius a as shown in Figure 5.27.

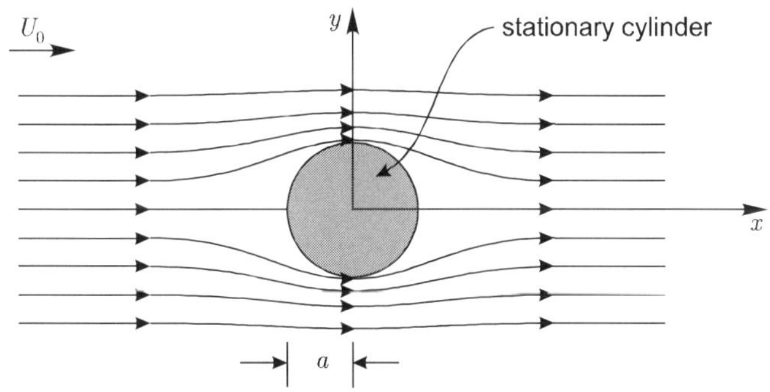

Figure 5.27: Cylinder in a uniform flow region.

The uniform far field velocity of the flow region is a velocity U_0 in the x-direction.

(i) Pose the boundary value problem described by the above.

(ii) Obtain an expression for the velocity field in the flow region.

5.11 The Figure 5.28 shows the plan and sectional views of a membrane which is under a uniform tension T_0 and subjected to a transverse loading $p(x, y)$ in the z-direction.

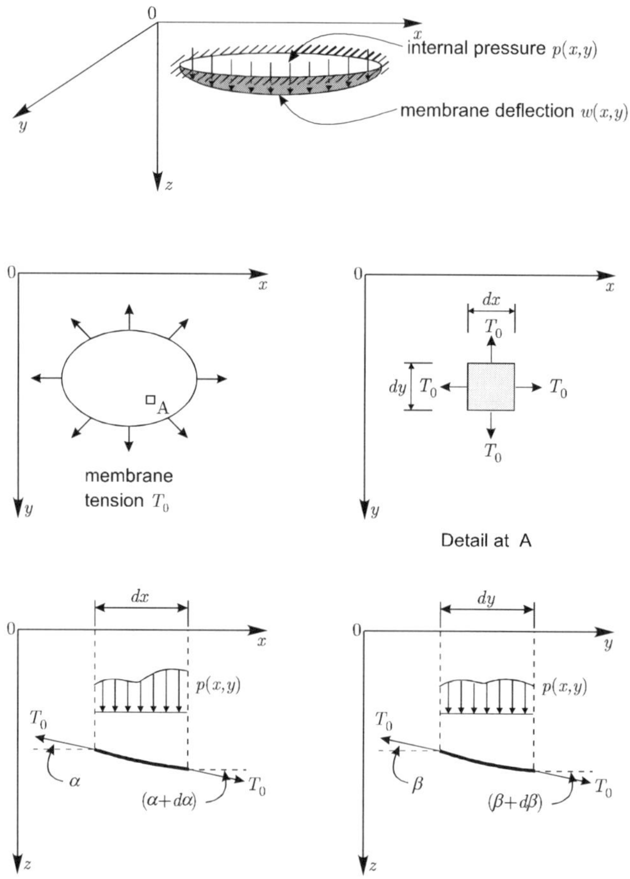

Figure 5.28: Equilibrium of a membrane element.

The small transverse deflection of the membrane is $w(x, y)$. Considering the equation of equilibrium of an infinitesimal element show that the partial differential equation governing $w(x, y)$ is

$$T_0 \nabla^2 w(x, y) + p(x, y) = 0$$

5.12 The unloaded, weightless membrane roof of an enclosure for a stadium consists of an annular region where the outer circular boundary $r = b$, is held fixed and the inner disk shaped region is subjected to a displacement (Figure 5.29)

$$w(a, \theta) = \Delta_0 + \Delta_1 \sin\theta$$

where Δ_0 and Δ_1 are constants. The deflection $w(r, \theta)$ of the unloaded membrane satisfies Laplace's equation

$$\nabla^2 w(r, \theta) = 0$$

where ∇^2 is Laplace's operator, defined by

$$\nabla^2 = \frac{\partial^2}{\partial r^2} + \frac{1}{r}\frac{\partial}{\partial r} + \frac{1}{r^2}\frac{\partial^2}{\partial \theta^2}.$$

(i) Formulate the boundary value problem.

(ii) Develop an expression for the deflection of the membrane, given that

$$w = A \ln r + B\theta \ln r + C\theta + D$$
$$+ \sum_{n=1,2,..} \left(A_n r^n + \frac{B_n}{r^n} \right) (C_n \sin n\theta + D_n \cos n\theta)$$

where $A, B, C, D, A_n, B_n, C_n$ and D_n are constants.

(iii) Calculate the resultant force and moment necessary to maintain the inner rigid disk-shaped region in the displaced position.

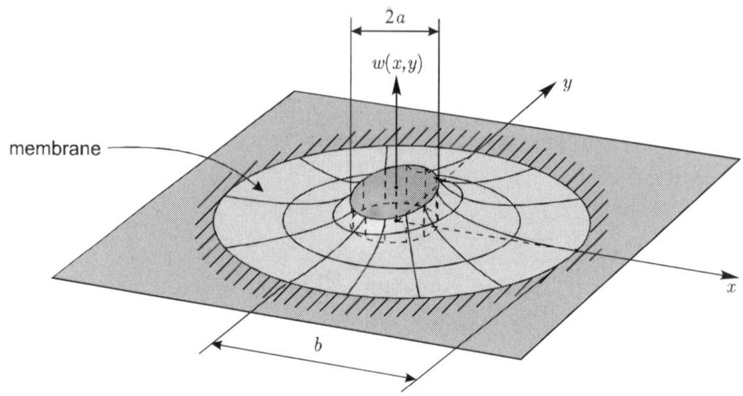

Figure 5.29: Deflection of a stretched annular membrane.

5.13 Figure 5.30(i) shows a flat, square metal frame. A membrane with constant tension T is attached to this frame. The edges OA and OC are now given displacements in the form of *Half-Sine functions* as shown in Figure 5.30(ii).
Derive an expression for the deflected shape of the membrane.
[Given: $w(x, y) = \left\{A^* e^{\lambda x} + B^* e^{-\lambda x}\right\} \sin(\lambda y) + \left\{C^* e^{\lambda x} + D^* e^{-\lambda x}\right\} \cos(\lambda y)$, is one of the possible solutions of Laplace's equation.]

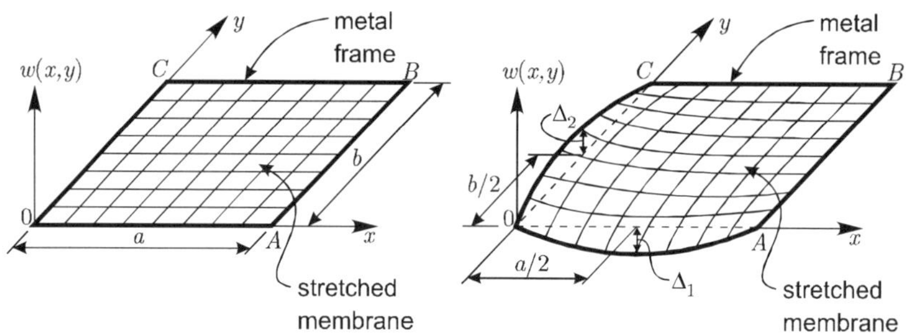

(i) (ii)

Figure 5.30: Boundary deflections of a stretched membrane.

5.14 The static deflection $w(x,y)$ of an unloaded, weightless stretched membrane satisfies Laplace's equation

$$\nabla^2 w(x,y) = 0$$

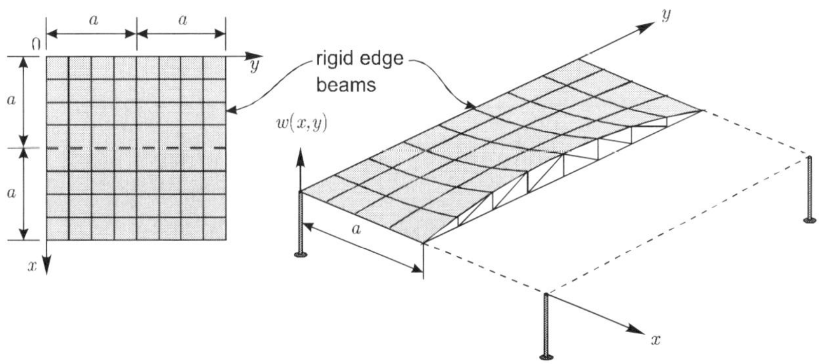

Figure 5.31: The membrane roof.

The Figure 5.31 shows the boundary displacements of the half segment of a membrane where the boundary is restrained and the deflections along the centre line are prescribed by the geometry of a roof truss as shown in Figure 5.31 (i.e. $w(a,y) = w_0 \left(\frac{y}{a}\right)\left(\frac{2a-y}{a}\right)$.

(i) Derive a series form solution for the deflection of the segment of the roof shown in Figure 5.31.

(ii) Obtain an expression for the resultant of vertical forces acting along the length of the roof truss if the tension in the membrane is T_0

5.15 The Figure 5.32 shows the cross-section through a concrete gravity dam which is founded on a deposit of hydraulically non-homogeneous porous alluvium. The arrangement of the cross-section can be idealized according to the sketch shown in Figure 5.32. If the flow in both alluvium regions is governed by

$$\nabla^2\varphi = 0$$

develop an expression for the seepage rate through the supporting alluvium foundation.

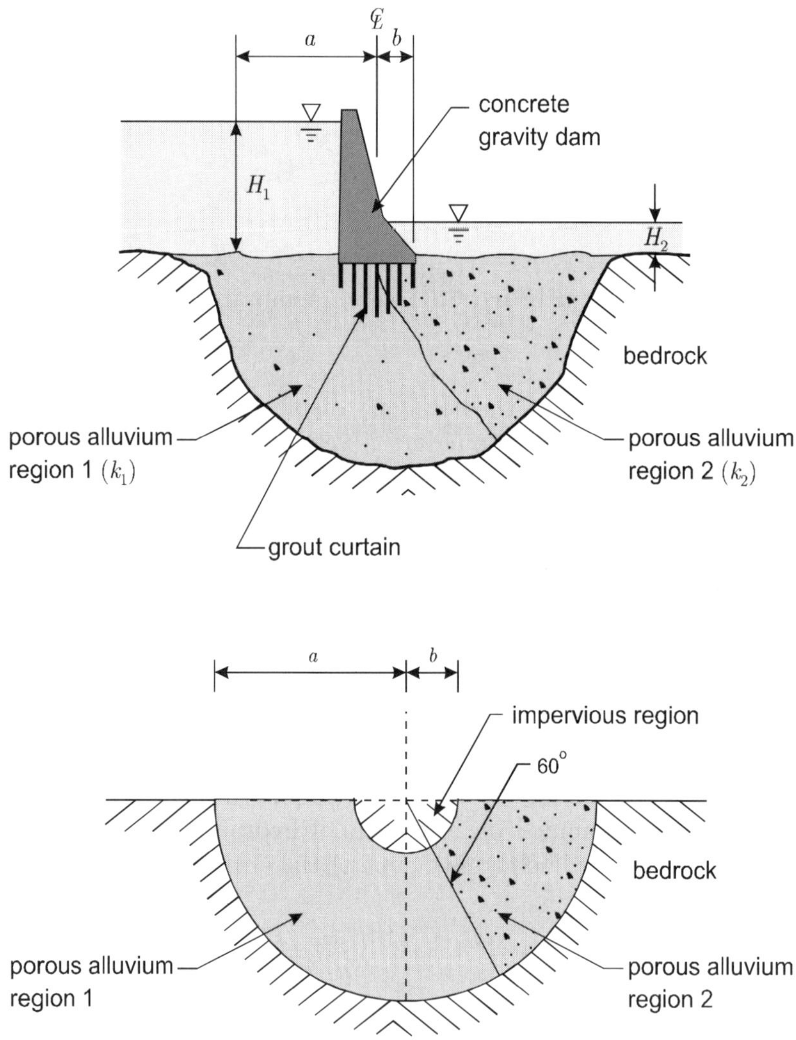

Figure 5.32: Seepage beneath gravity dam.

5.16 The regional ground water flow patterns can be estimated from a very elementary model of a saturated porous aquifer region which is subjected to a spatially varying but *constant* water level at the surface of the porous aquifer region (Figure 5.33).
The hydraulic potential causing ground water flow is composed of the datum and pressure components, i.e.

$$\Phi = \Phi_D + \Phi_P$$

Assuming that the flow in the porous medium can be described by Laplace's equation,

(i) formulate the boundary value problem, and

(ii) derive an expression for the variation of Φ in the porous aquifer region if the steady spatial variation in the water level can be represented in the form

$$H(x) = H_0 \left[1 - \sin\left(\frac{\pi x}{L}\right)\right]$$

(L can be about 20 km; D can be about 1 km and H_0 can be about 100 m)

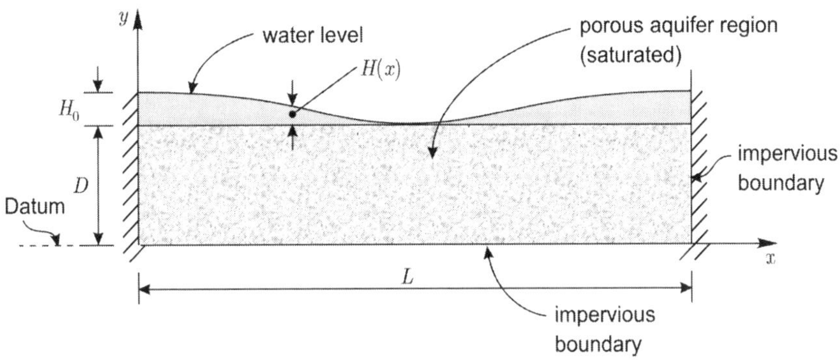

Figure 5.33: Idealized regional aquifer model.

5.17 The bedrock region between a river and a lake is connected by a thin horizontally oriented porous seam as shown in Figure 5.34. The flow pattern of the river is such that, at the location A, the depth of water is H_1 metres and decreases linearly to a depth of H_2 metres at the location B. If the hydraulic conductivity of the porous seam is k metres/sec and the flow in the *Confined Aquifer* is governed by Laplace's equation, derive an expression for the seepage from the river to the lake.

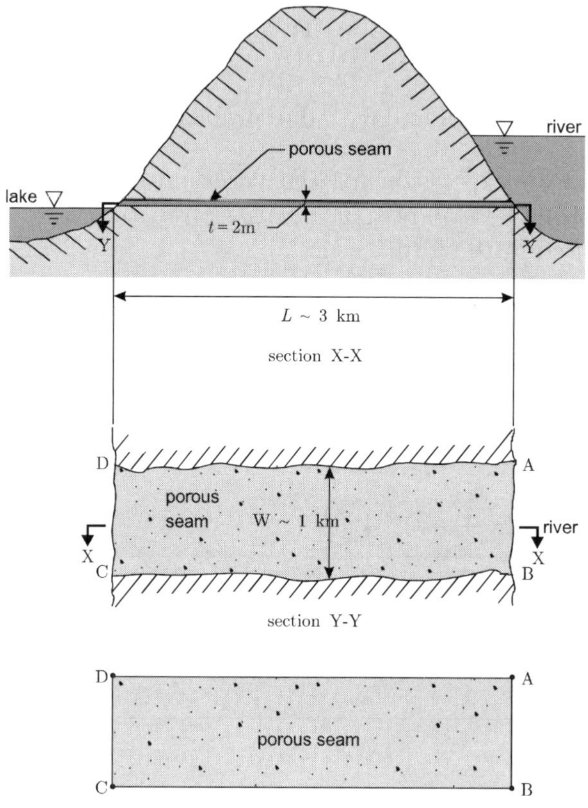

Figure 5.34: Seepage in a porous seam.

5.18 The Figure 5.35 shows the section and plan view of a hydraulic structure which consists of a pier-sluice gate structure system. Due to severe alkali-aggregate reactions, the base of the concrete has deteriorated and transformed to a porous granular medium of thickness t and effective hydraulic conductivity k.

(i) Develop an expression for the distribution of total head in the deteriorated base.

(ii) Derive a formal expression for the seepage loss through the deteriorated concrete.

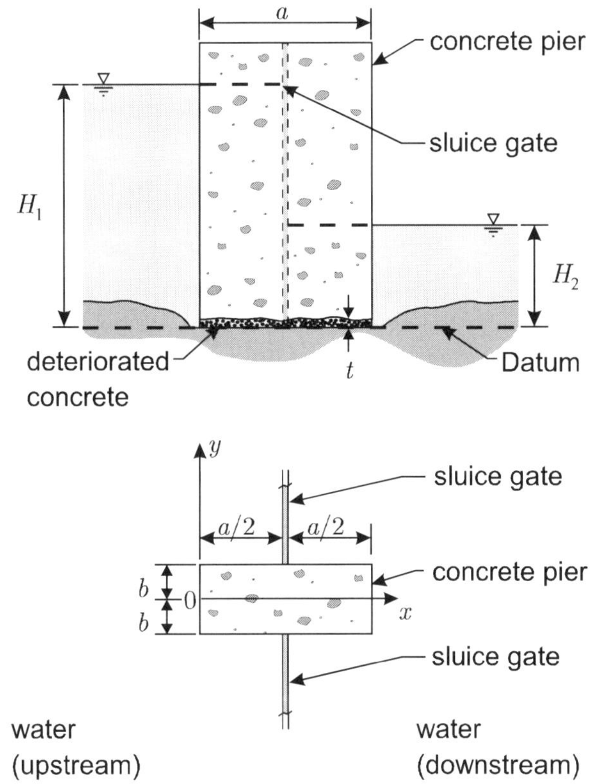

Figure 5.35: Seepage beneath a pier structure.

5.19 The Figure 5.36 shows a homogeneous porous layer with a rectangular plan shape, which is embedded at the edge of an impermeable rock formation. Groundwater flow takes place in the x, y plane with uniform velocity over the thickness of the porous layer. The total head in the layer is denoted by $\varphi(x, y)$, and the datum level is taken as the x, y plane. The total head along the boundaries of the region vary according to the following: (i) the total head has a constant value of φ_0 from O to A and zero from A to B along the x-axis, (ii) the total head is zero from O to D along the y-axis.

(i) Formulate the boundary value problem, which needs to be solved to determine $\varphi(x, y)$.

(ii) Use a Fourier series representation of the solution to determine the variation of $\varphi(x, y)$.

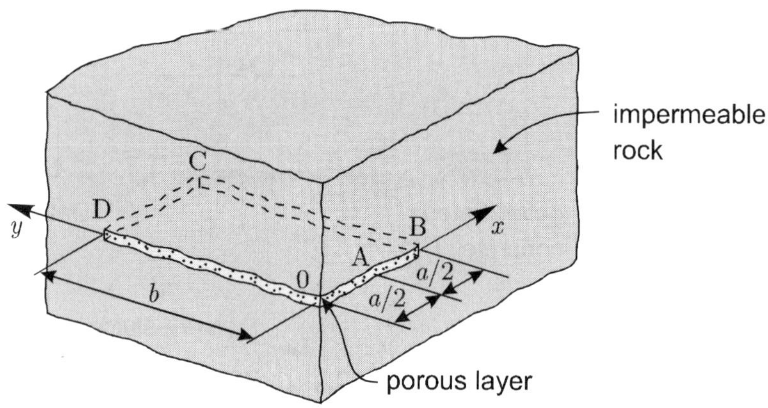

Figure 5.36: Seepage in a porous outcrop.

Chapter 6

The diffusion equation

There are a variety of situations in science and engineering where processes exhibit both spatial and temporal variations as a result of phenomena which are referred to as *diffusive*. These include thermal processes such as heat conduction in materials, fluid pressure transients in porous media, drying of solids due to moisture depletion, transport of chemicals and pollutants in the environment by gradual reduction in their concentrations, migration of chemicals within concrete and other structural materials, sedimentation and consolidation of geomaterials and in the study of transmission lines. In all these areas of application, "*diffusive processes*" such as thermal gradients, moisture gradients, pressure gradients, chemical concentration gradients, etc., will induce the transport processes. The diffusion equation acquires its importance by virtue of its ability to model the complete time-dependent processes related to the phenomena under consideration. The transient phenomena that are observed during the analysis of a diffusion-type equation ultimately reach their *steady state* or long term equilibrium values. In appropriate cases the steady state distributions of temperature, moisture content, chemical concentration, etc., are determined by solving Laplace's equation.

6.1 Derivation of the diffusion equation

In this chapter we shall present the derivation of the equations which are characterized by diffusion-type phenomena with appropriate transport characteristics. Three areas of application of particular interest relate to the topics of heat conduction in solids, pressure transients in fluid saturated media and the transport of chemicals in the subsurface environment.

6.1.1 Heat conduction in solids

It is known from experience that heat will flow from points of higher temperature to points of lower temperature. Heat is therefore a form of energy which is released by warmer regions of a body to the cooler regions. Conduction refers to the mode of heat transfer where this energy transport takes place in solids and fluids which are at rest. This is in contrast to *convective heat transfer* which can involve, for example, motion of a conducting fluid. A further mode of heat transfer, namely *radiative heat transfer*, is attributed to electro-magnetic wave propagation which can occur in a total vacuum as well as within a medium. Although heat flow cannot be measured directly, its presence manifests in terms of a measurable scalar quantity which is the *temperature*. From a knowledge of the temperature distribution $T(\mathbf{x}, t)$ within a region, it is possible to compute the heat flow within that region. As a consequence, the major part of the study of heat conduction in a solid or a fluid at rest focuses on the determination of the temperature distributions within the medium which is subject to appropriate boundary conditions and initial conditions applicable to $T(\mathbf{x}, t)$.

The basic law that determines the relationship between heat flow and the temperature gradient within a conducting body was first postulated by the mathematical physicist Joseph Fourier (1768 - 1830) in his classic treatise on the "*Analytical Theory of Heat* ". We consider a conducting medium which possesses *homogeneity* and *isotropy* with respect to its heat conduction characteristics. In this sense, all heat parameters which govern the heat conduction process are assumed to be independent of either the direction or spatial position of the point under consideration. It is assumed that the amount of heat ΔQ crossing an element of surface area ΔA in time Δt is proportional to the greatest rate of decrease in the temperature T, i.e.

$$\left\{ \frac{(\Delta Q)}{(\Delta A)(\Delta t)} \right\} \alpha \left\{ \frac{dT}{dn} \right\} \tag{6.1}$$

Generalizing the above concept, we can define a vector of heat flow $\mathbf{q}$ by the result

$$\mathbf{q} = -k\nabla T = -k \text{ grad } T \tag{6.2}$$

where k is defined as the thermal conductivity of the solid (with units: W/m °C).

Let us consider an arbitrary region of a heat conducting solid of volume V and surface area S. The total heat flux across S or the quantity of heat leaving the surface per unit time is

$$\int\int_S \mathbf{q} \cdot d\mathbf{S} = \int\int_S (-k\nabla T) \cdot \mathbf{n} \, dS \tag{6.3}$$

Hence, the quantity of heat *entering* the surface S per unit time is

$$\begin{aligned}\int\int_S (k\nabla T) \cdot \mathbf{n} \, dS &= \int\int\int_V \nabla \cdot (k\nabla T) \, dV \\ &= \int\int\int_V \text{div } (k \text{ grad } T) \, dV\end{aligned} \tag{6.4}$$

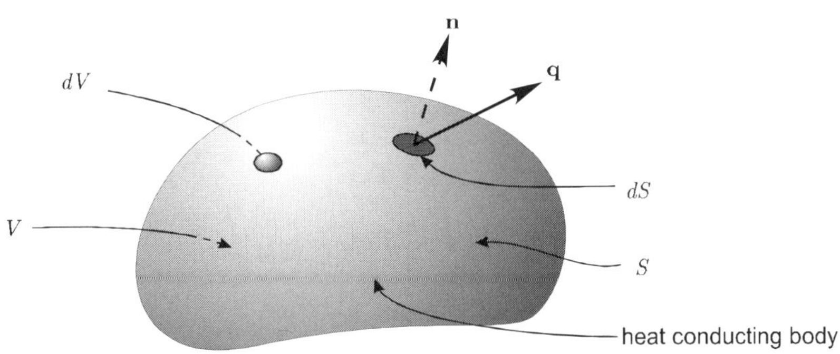

Figure 6.1: Heat conduction in a solid

Now consider an element of volume dV within the conducting body V. The heat dQ contained in the volume dV is

$$dQ = \rho c \, T dV \tag{6.5}$$

where ρ is the mass density of the conducting solid and c is its specific heat. The rate of increase of heat within the conducting body is

$$\frac{dQ}{dt} = \frac{d}{dt} \int\int\int_V \rho c \, T dV = \int\int\int_V \rho c \, \frac{\partial T}{\partial t} dV \tag{6.6}$$

In some instances heat can be generated from within the body; if $g(\mathbf{x}, t)$ is the specific rate at which heat is generated (i.e. rate of heat generation per unit volume) then the rate of total heat generated (G) within the control volume V is

$$G = \int\int\int_V g \, dV \tag{6.7}$$

If we consider the energy balance for the control volume, the rate at which the heat enters through the boundary S (i.e. (6.4)) together with the rate at which heat is generated from within the body V (i.e. (6.7)) must balance the rate at which heat is being stored in the volume V (i.e. (6.6)); thus, for a heat conducting body in which the thermal conductivity k is constant in space (i.e. thermally homogeneous)

$$\int\int\int_V k \, \nabla . (\nabla T) \, dV + \int\int\int_V g \, dV = \int\int\int_V \rho c \frac{\partial T}{\partial t} dV \tag{6.8}$$

which gives

$$\int\int\int_V \left[k \, \nabla . (\nabla T) + g - \rho c \frac{\partial T}{\partial t} \right] dV = 0 \tag{6.9}$$

Since the volume dV considered is arbitrary, the integrand of (6.9) is a continuous function and the thermal parameters are non-degenerate, (6.9) gives by virtue of the *Dubois-Reymond Lemma*,

$$k\nabla^2 T + g = \rho c \frac{\partial T}{\partial t} \tag{6.10}$$

In classical heat conduction problems, the heat generation g is absent (which is of course not the case in radioactive materials which can produce radiogenic heat or common materials, such as freshly mixed concrete, which give out heat due to exothermic chemical reactions between the water and the cement particle during the hydration process). In the absence of internal heat production, (6.10) reduces to the classical heat conduction equation developed by Joseph Fourier in 1822; i.e.

$$\alpha \nabla^2 T = \frac{\partial T}{\partial t} \tag{6.11}$$

where α is defined as the *thermal diffusivity* of the conducting medium defined by

$$\alpha = \frac{k}{\rho c} \tag{6.12}$$

and has dimensions L^2/T. From a physical point of view the thermal diffusivity of a medium is indicative of the speed of propagation of heat into the solid during temperature changes. The higher the value of α, the faster will be the propagation of heat within the medium. From (6.11) it is evident that as the heat conduction process reaches a steady state, the temperature field $T(\mathbf{x}, t)$ becomes only a function of the spatial variable $\mathbf{x}$ and (6.11) reduces to Laplace's equation

$$\nabla^2 T = 0 \tag{6.13}$$

The heat conduction equation is therefore intimately linked to Laplace's equation through the steady state limiting condition. As the thermal diffusivity of the material α increases, transient processes acquire diminishing importance and much of the focus is on the steady state heat conduction process which is governed by *spatial gradients* rather than *temporal variations* of temperature within the medium.

6.1.2 Pressure transients in porous media

Flow processes in porous media generally tend to achieve a steady state with progress of time. The steady state flow patterns in the porous medium are, of course, governed by Laplace's equation. The manner in which this steady state is achieved will depend on the permeability or hydraulic conductivity characteristics of the porous medium. It is reasonable to expect that, in porous media with relatively high permeability, there are no observable transient or time-dependent phenomena as the fluid potential in the flow domain reaches steady state conditions. For porous media with relatively low permeability, however, the fluid potential can exhibit a prolonged delay before achieving steady state conditions. In this latter instance, the study of the transient behaviour of the flow potential, which precedes the steady state is equally important. In this section, we shall develop a mathematical relationship which can be used to describe the development of transient effects in a porous medium especially in relation to pressure distributions within it. The analysis of transient phenomena in fluid saturated porous media can be of significant importance to geophysical methods used in resource exploration, land subsidence analysis and in the study of techniques used to determine the permeability characteristics of porous media.

In section 5.1.2 we discussed the description of flow within the porous structure of a medium which is saturated with a fluid. From (5.29) we have

$$\int\int\int_V \frac{\partial}{\partial t}(\rho n^*)dV = -\int\int_S \rho\hat{\mathbf{v}}\mathbf{n}\, dS \tag{6.14}$$

Before proceeding to examine the influence of *compressibility* of the *fluid* and that of the porous medium on the result (6.14), it is necessary to introduce certain relationships which can describe how volume changes can occur in these two phases as a result of changes in the fluid pressure. The fluid can undergo a volume change as a result of changes in pressure. This volume change manifests as a change in the density of the fluid. For example, experimental evidence suggests that such changes in density can be represented in the form

$$\rho = \rho_0 \exp[C_f(p - p_0)] \tag{6.15}$$

where ρ_0 is a reference density, p_0 is a reference pressure and C_f is defined as the "compressibility" of the fluid. If the fluid is "incompressible" then $C_f = 0$ and $\rho = \rho_0$. The result (6.15), however, gives a variation of density with pressure as

$$\frac{d\rho}{dp} = \rho C_f \tag{6.16}$$

With regard to the volume changes in the fabric of the porous medium, these can be interpreted as changes in the pore volume n^*. This is a very simplified model of a very complex medium; however, as a first approximation, this is justifiable. As an approximate expression for this porosity change with pressure we can assume that

$$\frac{dn^*}{dp} = C_s \tag{6.17}$$

where C_s is defined as the compressibility of the *porous fabric* and *not* that of the *solid material* composing the fabric. From (6.14) we obtain

$$\frac{\partial}{\partial t}(\rho n^*) = -\operatorname{div}(\rho \hat{\mathbf{v}}) \tag{6.18}$$

Considering the left hand side we can write

$$\frac{\partial}{\partial t}(\rho n^*) = \rho \frac{\partial n^*}{\partial t} + n^* \frac{\partial \rho}{\partial t} = \rho \left(\frac{dn^*}{dp}\right)_0 \frac{\partial p}{\partial t} + n^* \left(\frac{d\rho}{dp}\right)_0 \frac{\partial p}{\partial t} \tag{6.19}$$

where $(d\rho/dp)_0$ and $(dn^*/dp)_0$ are derivatives at the initial state, which are given by (6.16) and (6.17) respectively. We can write (6.19) in the form

$$\frac{\partial}{\partial t}(\rho n^*) = \rho C_f \left(n^* + \frac{C_s}{C_f}\right)\frac{\partial p}{\partial t} = \rho C_f \left(n^* + \frac{C_s}{C_f}\right)\left(\frac{dp}{d\rho}\right)_0 \frac{\partial \rho}{\partial t}$$
$$= \left(n^* + \frac{C_s}{C_f}\right)\frac{\partial \rho}{\partial t} \tag{6.20}$$

From Darcy's Law we have

$$\hat{\mathbf{v}} = -K \text{ grad } \Phi \tag{6.21}$$

where $\Phi(= \Phi_D + \Phi_P)$ is the total hydraulic potential and Φ_D and Φ_P are datum and pressure potentials. Also

$$\Phi_P = \frac{p}{\gamma} \tag{6.22}$$

where p is the pressure at a location and γ is the unit weight of the fluid. Also we can write

$$\text{grad } p = \left(\frac{dp}{d\rho}\right)_0 \text{grad } \rho = \frac{1}{\rho C_f} \text{ grad } \rho. \tag{6.23}$$

Using (6.22) and (6.23) in (6.21) we have

$$\hat{\mathbf{v}} = -\frac{K}{\gamma \rho C_f} \text{ grad}\rho - K \text{ grad } \Phi_D \tag{6.24}$$

Using (6.20) and (6.24) in (6.18) we obtain

$$\left(n^* + \frac{C_s}{C_f}\right)\frac{\partial \rho}{\partial t} = - \operatorname{div}\left\{\frac{K \text{ grad } \rho}{\gamma C_f} - \rho K \text{ grad } \Phi_D\right\} \tag{6.25}$$

Since Φ_D varies linearly with $\mathbf{x}$, the second term on the right hand side reduces the zero, and (6.25) can be written as

$$(n^* C_f + C_s)\frac{\partial \rho}{\partial t} = \frac{K}{\gamma}\nabla^2 \rho \tag{6.26}$$

For most porous media flow problems, we can neglect powers of C_f higher than the first in a Taylor series expansion (named after Brook Taylor (1685-1735)) of the fluid compressibility relationship (6.15);
i.e.

$$\rho = \rho_0 + C_f \rho_0 (p - p_0) \tag{6.27}$$

Using (6.27), we can reduce (6.26) to the form

$$a^2 \nabla^2 p = \frac{\partial p}{\partial t} \tag{6.28}$$

where

$$a^2 = \frac{K}{\gamma(n^* C_f + C_s)} \tag{6.29}$$

This diffusion-type partial differential equation describes the time-dependent variation of a pressure transient in the pore fluid in a porous medium. Most importantly, the formulation accounts for the "*compressibility*" of both the *pore fluid* and the *porous fabric*. The equation (6.28) is also referred to as the *Elastic Drive Equation* in groundwater mechanics. It is important to note that the transient or time-dependent nature of the problem results directly from the existence of compressible effects in the saturated porous medium. If both C_f and C_s are zero, the entire system is incompressible and the distribution of pressure within the system is governed by Laplace's equation. In terms of the physical parameters of the system it is important to note that all constants K, γ, n^*, C_f and C_s are all positive and hence $a^2 > 0$.

6.1.3 Chemical mass transport in porous media

In recent years the role of mathematical modelling of *geo-environmental problems* has acquired increasing importance. The migration of contaminants and other hazardous materials in the geosphere forms an important part of this activity. Although the study of these processes involves the combined influences of *geochemistry* and other *absorptive transport* processes, an important understanding of the movement of contaminants and other hazardous materials can be gained by examining the dominant processes of *advection* and *diffusion*. Advection is the process whereby a chemical species or a solute which is dissolved in groundwater can be transported within a porous medium due to flow processes in the medium. Diffusion is the process whereby a solute in groundwater will move from a region of greater concentration to a region where it is less concentrated. This molecular diffusion-type process will occur even in a fluid which is at a state of rest as long as a concentration gradient exits (Figure 6.2). The mass of the chemical diffusing through the pore structure of a geological medium is described by the law of diffusion postulated by physiologist Adolph Fick (1829-1901).

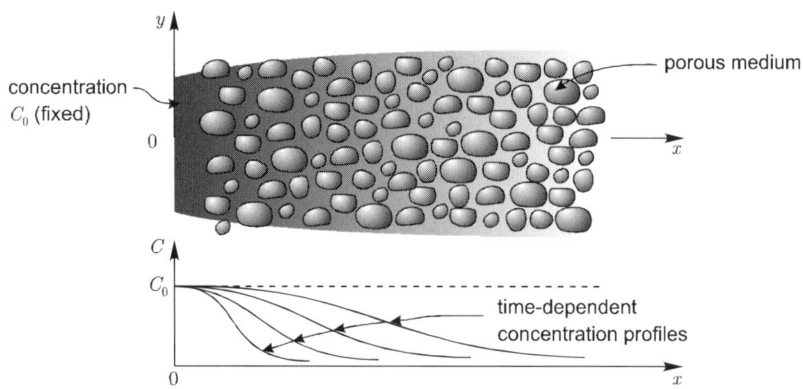

Figure 6.2: Diffusive mass transport in a porous media

We consider a fully saturated porous medium where advection and diffusion of a dissolved solute takes place in the presence of a flow velocity $\hat{\mathbf{v}}(\mathbf{x}, t)$ and the gradient of a concentration $C(\mathbf{x}, t)$. Note that the concentration, just like temperature, pressure, density, etc., is a *scalar*. Consider a control volume V of the medium which has a total surface area S (Figure 6.3). Let the porosity of the porous medium be n^*; this is a constant for a homogeneous region, i.e.

$$n^* = \frac{V_v}{V} = \frac{S_v}{S} \tag{6.30}$$

where the suffix v refers to the void phase which contains the fluid.

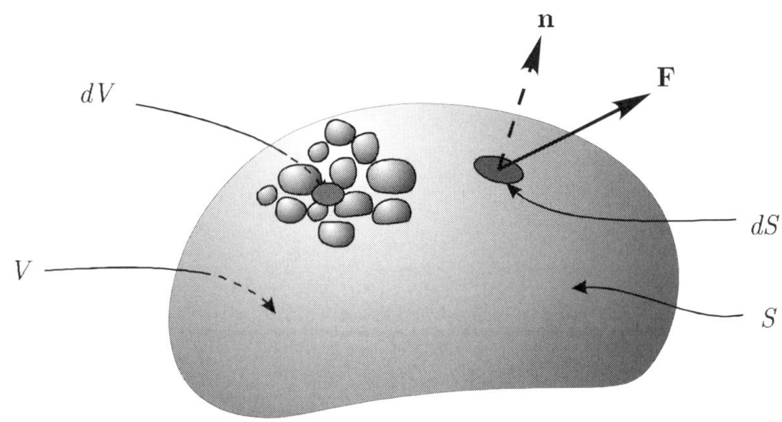

Figure 6.3: Mass transport in a porous medium

We shall assume that during the *advective* and *diffusive* transport processes of the solute the fabric of the porous medium remains intact. This is a reasonable assumption to a first approximation; in reality, however, if chemical interactions take place between the solute and the porous medium itself; this can lead to either a "*consolidation*" of the medium or "*swelling*" of the medium. A further example is "*electro-osmosis*" where electro-chemical phenomena can induce reductions in the pore space thereby inducing artificial consolidation of materials such as soils.

Let the solute concentration, which is the dissolved mass per unit volume of the solution or the pore fluid region in which the chemical migrates, be $C(\mathbf{x}, t)$. The solute concentration per unit volume of the entire porous medium $\widehat{C}(\mathbf{x}, t)$ is given by

$$\widehat{C}(\mathbf{x}, t) = \frac{V_v}{V} C(\mathbf{x}, t) = n^* C(\mathbf{x}, t) \tag{6.31}$$

Here again, by introducing the concentration $\widehat{C}(\mathbf{x}, t)$, averaged over the entire region of the porous medium, we essentially disregard the internal structure of the porous medium and represent it with an equivalent medium. This is quite similar to the approximation we introduced in Section 5.1.2, when defining the average flow velocity $\hat{\mathbf{v}}$ in a porous medium. Let us now consider an element dS of the region S. The 'flux' or the mass per unit area per unit time consists of the *advective* and *diffusive* flow rates; i.e.

$$\mathbf{F} = \mathbf{F}_a + \mathbf{F}_d \tag{6.32}$$

The mass of solute transported per unit area per unit time by *advection* is

$$\mathbf{F}_a = \hat{\mathbf{v}} \widehat{C} \tag{6.33}$$

Similarly, the mass of solute transported per unit area, per unit time, by *diffusion* is

$$\mathbf{F}_d = -D \text{ grad } \widehat{C} = -D \nabla \widehat{C} \tag{6.34}$$

where D is called the *coefficient of diffusion* which has units L^2/T. The negative sign indicates that the movement is from areas of greater concentration to lesser concentration. The mass flow rate m_i^* into the region through S is given by

$$m_i^* = -\int\int_S \mathbf{Fn}\, dS \tag{6.35}$$

Using (6.32), (6.33) and (6.34) in (6.35) and applying the divergence theorem we have

$$m_i^* = \int\int\int_V n^* \nabla \cdot \{C\hat{\mathbf{v}} - D\nabla C\} dV \tag{6.36}$$

The rate of mass accumulation m_a^* within the control element is

$$m_a^* = \frac{d}{dt}\int\int\int_V n^* C dV = \int\int\int_V n^* \frac{\partial C}{\partial t} dV \tag{6.37}$$

If no solute is either *lost* or *produced* within the control volume, the rate of accumulation of the solute within V must be identical to the rate of mass influx into the region through S, i.e.

$$-\int\int\int_V n^* \nabla \cdot \{C\hat{\mathbf{v}} - D\nabla C\} dV = \int\int\int_V n^* \frac{\partial C}{\partial t} dV \tag{6.38}$$

Since V is arbitrary, C and $\hat{\mathbf{v}}$ are continuous within V and D and n^* are non-zero constants, (6.38) gives, by virtue of the *Dubois-Reymond Lemma*,

$$D\nabla^2 C - \nabla \cdot (C\hat{\mathbf{v}}) = \frac{\partial C}{\partial t} \tag{6.39}$$

This is referred to as the "*Advection-Diffusion*" equation governing the transport of a *chemical* or a *contaminant* within a porous medium. In the special case where the transport is solely due to diffusion, (6.39) reduces to

$$D\nabla^2 C = \frac{\partial C}{\partial t} \tag{6.40}$$

Note that (6.39) and (6.40) also apply to $\widehat{C}$, and we shall assume that the partial differential equation, and the associated boundary conditions and initial condition governing either $C(\mathbf{x}, t)$ or $\widehat{C}(\mathbf{x}, t)$ are interpreted consistently and $(\widehat{\ })$ is omitted for convenience.

6.1.4 Drying of porous solids

The mathematical modelling of processes associated with drying of porous and particulate media has important applications in both science and engineering. These include chemical processing, food processing, kiln drying of ceramics and building materials, agricultural engineering and geo-environmental applications in the isolation of heat emitting nuclear wastes with engineered geological barriers. In these applications, the drying process is strongly dependent on a number of factors including temperature, relative humidity and the physical and chemical characteristics of the porous medium. As a first attempt towards the study of moisture movement in porous media, the modelling can be attempted by appeal to a *diffusive-type process* where the spatial gradients in the moisture content can induce moisture transport. The moist porous medium is considered to be *unsaturated* and the degree of saturation S is defined by

$$S(\mathbf{x}, t) = \frac{V_w(\mathbf{x}, t)}{V_v} \tag{6.41}$$

where V_w is the volume of water in the pore space. We now define a *volumetric water content* $\Omega(\mathbf{x}, t)$ as

$$\Omega(\mathbf{x}, t) = \frac{V_w}{V} = \frac{V_w}{V_v} \cdot \frac{V_v}{V} = n^* S(\mathbf{x}, t) \tag{6.42}$$

where V_v is the voids content and n^* is the porosity. Since $\Omega(\mathbf{x}, t)$ represents the value of the moisture content averaged over the entire region V of the porous medium we can effectively disregard the internal structure of the porous medium. In the modelling of moisture transport it is assumed that the moisture transport rate $\mathbf{F}_m$ is governed by a diffusion-type process which depends on the spatial gradient of the volumetric moisture content, i.e.

$$\mathbf{F}_m = -D_m \text{ grad } \Omega = -D_m \nabla \Omega \tag{6.43}$$

where D_m is the coefficient of isothermal moisture diffusivity with units of L^2/T.

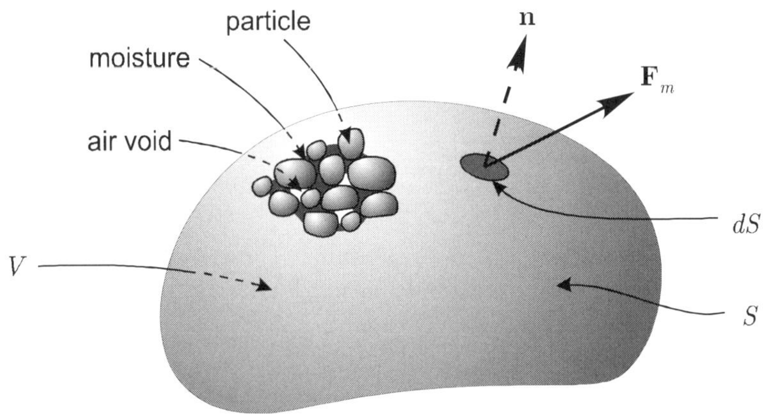

Figure 6.4: Moisture transport in a porous medium

Referring to Figure 6.4, the rate of moisture influx w_i into the region V is given by

$$w_i = -\int\int_S \mathbf{Fn}\, dS \tag{6.44}$$

Using (6.43) and the divergence theorem, (6.44) gives

$$w_i = \int\int\int_V D_m \nabla \cdot (\nabla \Omega) dV \tag{6.45}$$

The rate of moisture accumulation w_a within the region V is given by

$$w_a = \int\int\int_V \frac{\partial \Omega}{\partial t} dV \tag{6.46}$$

Again, if no moisture is either lost or produced during the moisture transport process and if Ω is a continuous function in V, it can be shown that the moisture transport process is governed by the diffusion-type partial differential equation

$$D_m \nabla^2 \Omega = \frac{\partial \Omega}{\partial t} \tag{6.47}$$

As the moisture transport processes stabilize, the steady state moisture distribution is governed by Laplace's equation for $\Omega(\mathbf{x})$.

6.1.5 Thermal oxidation of silicon

Silicon dioxide (Si 0_2) is formed by the oxidation of Silicon (Si). The formation of Si 0_2 on silicon is a problem of great importance to the manufacture of semiconductor devices. These *overlayers* of silicon dioxide form a protective barrier against the diffusion of impurities and provide the electrical insulation and dielectric properties required for the manufacture and operation of semiconductor devices. In its pure form silicon is a highly active element and will form a stable amorphous oxide on a surface exposed to an oxygen bearing medium (e.g. a mixture of an inert gas such as nitrogen or hydrogen with either oxygen or moisture). A thin layer of the order of 10 Angstroms (or 10^{-9} m) of the oxide can form even at room temperature. (The so-called "native layer" of Si 0_2 on a surface which is assumed to be clean Si). This oxidation process can be activated by pressure and temperature, particularly temperatures in excess of 700°C. The thickness of the Si 0_2 layer can thus be controlled and these surface layers of Si 0_2 are contiguous with the unoxidized Si. This can also conform to the configuration of the substrate.

It is assumed that the oxidation process is primarily due to *diffusive migration* of 0_2 molecules through the Si 0_2 to the Si 0_2/Si interface, where they react with the Si to form new Si 0_2.

The Figure 6.5 shows the configuration of the Si 0_2 and Si phases at any arbitrary time t. If $C(x, t)$ is the oxygen concentration in excess of what is contained in the silicon dioxide network, its evolution can be described by a diffusion equation of the type

$$D_0 \frac{\partial^2 C}{\partial x^2} = \frac{\partial C}{\partial t} \tag{6.48}$$

where D_0 is a diffusion constant. The oxygen concentration at $x = 0$ is C_0 and the concentration at $x = X(t)$ will depend on the chemical reaction rate which transforms $\text{Si} + 0_2 \rightarrow \text{Si}\ 0_2$. In general, this is a *first-order* reaction rate r which can be assumed to be dependent on the oxygen concentration according to

$$r = \lambda_D C_X \tag{6.49}$$

where λ_D is a temperature dependent constant and C_X is the concentration at $X(t)$ which varies with the position of the interface. In reality the problem of migration of a Si 0_2/Si interface is a *moving boundary problem* which is considerably more difficult to solve. It is, however, important to note that

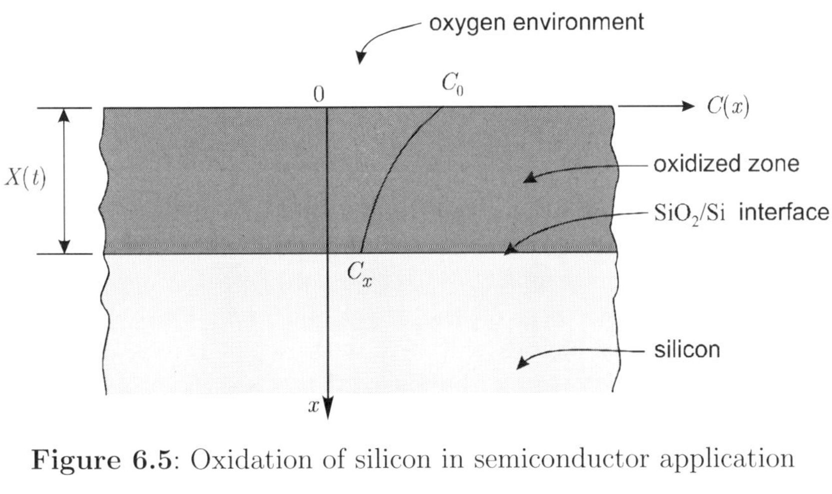

Figure 6.5: Oxidation of silicon in semiconductor application

the assumption of a diffusive process for the oxygen is an important to the correct formulation and analysis of the problem.

6.1.6 Motion of a plate on a viscous fluid

A diffusive-type equation also results in the analysis of the following problem in fluid mechanics. Consider the problem of a viscous fluid of semi-infinite extent occupying the region $0 \le x < \infty$; $-\infty < y < \infty$; $-\infty < z < \infty$. The surface of the viscous fluid contains a rigid plate which is subjected to a steady velocity v_0 (Figure 6.6).

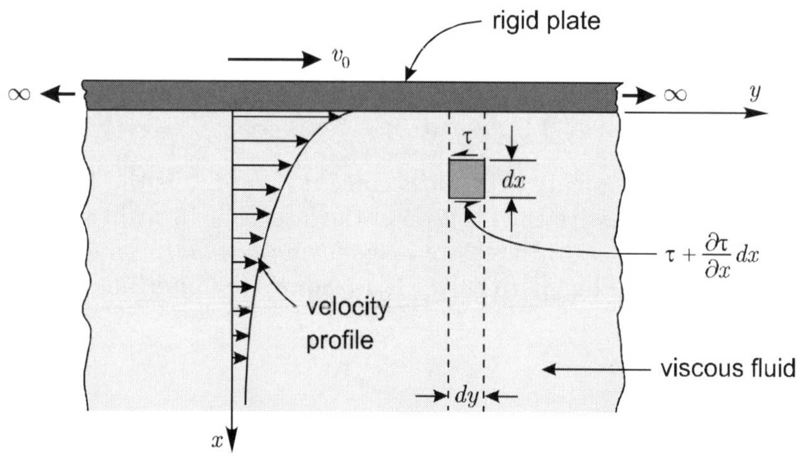

Figure 6.6: Fluid motion induced by moving plate

This problem in its highly idealized form is attributed to Sir G.G. Stokes (1819-1903) and is referred to as Stokes' first problem. Considering the infinitesimal element shown in Figure 6.6 we note that the movement of the plate induces "shear stresses" on the upper and lower faces of the element as shown. Since the velocities will vary with depth x it is essential that these shear stresses also vary with depth. Also, since the problem is purely one-dimensional only with respect to x, the vertical planes y = const., and z=const., are planes of symmetry and as such there will be no shear stresses on these planes.

We can consider the dynamic equilibrium of the element $dx\,dy$ by applying Newton's Law, i.e. (Force = (mass)(acceleration)):

$$\tau dy - (\tau + \frac{\partial \tau}{\partial x} dx)dy = \rho\, dx\, dy \left(\frac{\partial v}{\partial t} \right) \tag{6.50}$$

where the element length in the x-direction is assumed to be unity, ρ is the mass density and v is the velocity of the fluid. Simplifying (6.50) we have

$$-\frac{\partial \tau}{\partial x} = \rho \frac{\partial v}{\partial t} \tag{6.51}$$

We now need a relationship between the shear stress τ and the velocity v, which is the constitutive relationship for the viscous fluid. For a Newtonian viscous fluid the shear stress τ is directly proportional to the spatial gradient of the velocity, i.e.

$$\tau = -\eta_D \frac{\partial v}{\partial x} \tag{6.52}$$

where η_D is called the dynamic viscosity of the fluid. The negative sign indicates that v decreases with increasing x. Substituting (6.52) in (6.51) we obtain

$$\eta_K \frac{\partial^2 v}{\partial x^2} = \frac{\partial v}{\partial t} \tag{6.53}$$

where $\eta_K = \eta_D/\rho$ is referred to as the *kinematic viscosity* of the fluid. Therefore the variation of the velocity within the viscous fluid is governed by a diffusion-type equation and the process is controlled by the kinematic viscosity of the fluid. This problem is highly idealized and precludes any extension to two dimensions. Since the plate at the surface is of infinite extent, the direction y can be chosen to coincide with the direction along which the plate is set in motion, thereby reducing the problem always to the one-dimensional diffusion-type equation. A more detailed account of the study of diffusive-type phenomena which occur during slow motions of Newtonian viscous fluids is presented in Section 8.12.13.

6.2 Initial conditions and boundary conditions

In order to develop solutions for the partial differential equations of the diffusion-type for processes associated with heat conduction in solids, pressure transients in porous media, contaminant transport in geomaterials, fluid dynamics, etc., described previously, it is necessary to specify appropriate boundary conditions and initial conditions applicable to the specific problem.

The general class of boundary conditions and initial conditions can be described according to a general diffusive-type process which is described by the partial differential equation

$$a^2\nabla^2\Psi = \frac{\partial \Psi}{\partial t} \quad ; \quad \mathbf{x} \in V \tag{6.54}$$

where Ψ can refer to temperature, pressure, concentration, etc.

6.2.1 Dirichlet-type boundary condition

Considering the surface S of the region V, the Dirichlet-type boundary conditions specify the function Ψ on either S or parts of S, i.e.

$$\Psi = \begin{cases} F_1(\mathbf{x},t) & \text{on} \quad \mathbf{x} \in S_1 \\ F_2(\mathbf{x},t) & \text{on} \quad \mathbf{x} \in S_2 \end{cases} \tag{6.55}$$

where $S = S_1 \cup S_2$. These Dirichlet-type boundary conditions are applicable to situations where the functions depend on either spatial coordinates (i.e. $F_i(\mathbf{x})$) or depend on time (i.e. $F_i(t)$), or are constant. The special case when

$$\Psi = 0 \qquad \text{on} \qquad \mathbf{x} \in S \tag{6.56}$$

represents a homogeneous Dirichlet boundary condition. For example, a thermal boundary condition which maintains the temperature at zero degrees satisfies the homogeneous equation. If a boundary surface is maintained at a constant temperature T_0^*, we can treat the boundary condition as a homogeneous boundary condition provided the temperature $T(\mathbf{x}, t)$ within the body is interpreted as the value in relation to T_0^*.

6.2.2 Neumann-type boundary conditions

For certain types of problems, the normal derivative of the function can be prescribed as a function of both position and time. In this case we have

$$\frac{\partial \Psi}{\partial n} = F(\mathbf{x}, t) \qquad \text{on} \qquad \mathbf{x} \in S \tag{6.57}$$

where $\partial/\partial n$ denotes the differential along the outward unit normal to S. For example, in heat conduction problems (i.e. $\Psi = T$), this boundary condition is equivalent to that of prescribing the magnitude of heat flux along the boundary surface. This is evident if both sides of the equation (6.57) are multiplied by of the thermal conductivity of the medium. Also, in pressure transient problems, this effectively prescribes the velocity of flow of the fluid normal to the boundary S. In the special case when the normal derivative at the boundary vanishes, we have

$$\frac{\partial \Psi}{\partial n} = 0 \qquad \text{on} \qquad \mathbf{x} \in S \tag{6.58}$$

In heat conduction problems, this corresponds to a perfectly insulated boundary; in fluid pressure transient problems, this corresponds to an impermeable boundary and with other diffusive-type processes, (6.58) implies the absence of mass transport normal to the surface S.

6.2.3 Combined boundary conditions

These boundary conditions involve a linear combination of Dirichlet and Neumann-Type boundary conditions, which can take the generalized form

$$c_1 \frac{\partial \Psi}{\partial n} + c_2 \Psi = F^*(\mathbf{x}, t) \quad ; \quad \text{on} \quad \mathbf{x} \in S \tag{6.59}$$

where c_1 and c_2 are constants. The boundary condition (6.59) effectively represents a generalization of the boundary conditions described by (6.55) and (6.57). The boundary condition (6.59) acquires a physical significance only upon identification of the application under consideration. For example, in heat conduction, (6.59) describes the dissipation of heat at the surface S by convection, which is governed by Newton's Law; this implies that the heat transfer is proportional to the temperature difference between the surface (T_s) and the ambient environment (T_a). For the heat conduction problem we can identify c_1 in (6.59) with the thermal conductivity k and c_2 as a transfer coefficient h.

For energy balance along the boundary surface S (Figure 6.7), we have

$$-k \frac{\partial T}{\partial n} = h(T_s - T_a) \tag{6.60}$$

which is equivalent to the boundary condition (6.59).

The combined boundary condition can also be applied to the pressure transient problem and the contaminant migration problem where processes such as controlled leakage from the pressurized region and chemically induced mass transfer rates at boundaries can lead to controlled mass transfer boundary conditions of the type (6.59).

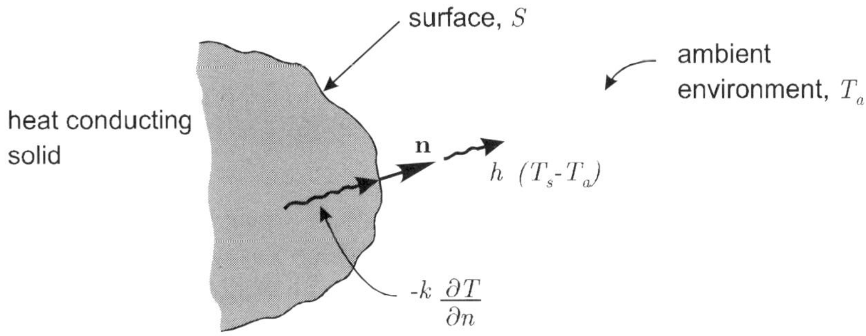

Figure 6.7: Heat transfer by convection at surface of heat conducting solid

6.2.4 Mixed boundary conditions

In certain instances, both Dirichlet- and Neumann-type boundary conditions can be prescribed on separate regions of the boundary S, e.g.

$$\Psi = F_d(\mathbf{x}, t) \quad \text{on} \quad \mathbf{x} \in S_1 \tag{6.61}$$

$$\frac{\partial \Psi}{\partial n} = F_n(\mathbf{x}, t) \quad \text{on} \quad \mathbf{x} \in S_2 \tag{6.62}$$

where $F_d(\mathbf{x}, t)$ and $F_n(\mathbf{x}, t)$ are prescribed functions and $S = S_1 \cup S_2$. These mixed boundary conditions are frequently encountered in situations involving heat conduction in solids, fluid pressure transients in porous media and diffusion of chemicals in porous media, with specific reference to the surfaces of regions of semi-infinite extent. Here the mixed boundary conditions can be prescribed along regions of the plane surface of the half-space.

6.2.5 Initial conditions

In addition to the boundary conditions discussed in sections 6.2.1. to 6.2.5., in the transient problem involving diffusive-type processes it is necessary to prescribe the values of the function Ψ within the region V at the start of diffusion-type process. Such an initial condition can imply a spatial variation of Ψ which is consistent with pre-existing boundary conditions. The initial conditions can take the form

$$\Psi(\mathbf{x}, 0) = \Psi_0(\mathbf{x}) \qquad \text{on} \qquad \mathbf{x} \in V \tag{6.63}$$

where $\Psi_0(\mathbf{x})$ is a known function. The partial differential equation (6.54) together with any combination of the boundary conditions defined in Sections (6.2.1) to (6.2.4) and the initial condition (6.63) will form a *well-posed initial boundary value problem.* Alternatively the initial boundary value problem is also referred to as the *Cauchy problem.* The following important attributes of initial boundary value problems should be noted.

- If the partial differential equation is of nth-order in the time variable t, then values of the independent variable and its $(n-1)^{th}$ time derivatives need to be specified for $t = 0$ for all other values of the independent variables $\mathbf{x}$.
- In general, it is not possible to solve an initial boundary value problem by prescribing the behaviour of the dependent variable at some finite time t^*. The term *initial boundary value problem* implies that conditions are prescribed at $t = 0$.
- In initial boundary value problems, in addition to the above initial conditions, Dirichlet and/or Neumann boundary conditions can be prescribed on the physical boundary of a region. In other words, the initial boundary value problem can have mixed boundary conditions.
- Initial boundary value problems arise mainly in connection with *parabolic-* and *hyperbolic*-type partial differential equations.
- The distinction between an initial boundary value problem (Cauchy problem) and a boundary value problem (Dirichlet, Neumann and/or Robin-Type problem) is that in the *former*, the solution can be advanced at successive intervals of time starting from the initial conditions whereas in the *latter* the solution requires the evaluation of the dependent variable for the entire domain of interest.
- The principle of superposition can be applied to any linear and homogeneous boundary condition which can be written in the form of a linear

homogeneous operator equation of the form $\mathcal{D}\Psi = 0$, where $\mathcal{D}$ is implied in (6.55) to (6.63) provided the function $\mathcal{D}\Psi$ is continuous at the boundary.

- The initial boundary value problem for the diffusion equation posed by the partial differential equation (6.54), the associated boundary conditions and the initial condition is well-posed in the sense that the uniqueness of the solution can be rigorously proved. This proof is presented in Section 6.6.

6.2.6 Change in dependent variable for homogeneous initial conditions

In the particular case when the initial conditions governing $\varphi(\mathbf{x}, t)$ are homogeneous, the dependent variable needs to be changed to make the solution procedure non-trivial. For the purposes of illustration, consider the initial boundary value problem defined by the diffusion equation

$$a^2\nabla^2\varphi = \frac{\partial\varphi}{\partial t} \quad ; \quad \mathbf{x} \in V \tag{6.64}$$

with boundary conditions

$$\varphi = \varphi_D(\mathbf{x}, t) \quad ; \quad \mathbf{x} \in S_D \tag{6.65}$$

$$\frac{\partial\varphi}{\partial n} = \varphi_N(\mathbf{x}, t) \quad ; \quad \mathbf{x} \in S_N \tag{6.66}$$

$$\frac{\partial\varphi}{\partial n} + \lambda\varphi = \varphi_M(\mathbf{x}, t) \quad ; \quad \mathbf{x} \in S_M \tag{6.67}$$

where $S_D \cup S_N \cup S_M = S$, the boundary of V and the initial condition

$$\varphi(\mathbf{x}, 0) = 0 \quad ; \quad \mathbf{x} \in V \tag{6.68}$$

The change in the dependent variable is achieved by assuming that $\varphi(\mathbf{x}, t)$ admits a solution of the form

$$\varphi(\mathbf{x}, t) = \varphi^*(\mathbf{x}) + \widetilde{\varphi}(\mathbf{x}, t) \quad ; \quad \mathbf{x} \in V \tag{6.69}$$

The initial boundary value problem defined by (6.64) to (6.68) now can be rewritten as

$$a^2 \nabla^2 \widetilde{\varphi} = \frac{\partial \widetilde{\varphi}}{\partial t} - a^2 \nabla^2 \varphi^* \quad ; \quad \mathbf{x} \in V \tag{6.70}$$

with boundary conditions for $\widetilde{\varphi}(\mathbf{x}, t)$ of the forms

$$\widetilde{\varphi} = \varphi_D(\mathbf{x}, t) - \varphi^*(\mathbf{x}) \quad ; \quad \mathbf{x} \in S_D \tag{6.71}$$

$$\frac{\partial \widetilde{\varphi}}{\partial n} = \varphi_N(\mathbf{x}, t) - \frac{\partial \varphi^*}{\partial n} \quad ; \quad \mathbf{x} \in S_N \tag{6.72}$$

$$\frac{\partial \widetilde{\varphi}}{\partial n} + \lambda \widetilde{\varphi} = \varphi_M(\mathbf{x}, t) - \left(\frac{\partial \varphi^*}{\partial n} + \lambda \varphi^* \right) \quad ; \quad \mathbf{x} \in S_M \tag{6.73}$$

and the initial condition governing $\widetilde{\varphi}(\mathbf{x}, t)$ now takes the form

$$\widetilde{\varphi}(\mathbf{x}, 0) = -\varphi^*(\mathbf{x}) \quad ; \quad \mathbf{x} \in V \tag{6.74}$$

Since $\varphi^*(\mathbf{x})$ is time-independent we select a solution such that it satisfies the following boundary value problem associated with Laplace's equation:

$$\nabla^2 \varphi^* = 0 \quad ; \quad \mathbf{x} \in V \tag{6.75}$$

with boundary conditions

$$\varphi^* = \varphi_D(\mathbf{x}) \quad ; \quad \mathbf{x} \in S_D \tag{6.76}$$

$$\frac{\partial \varphi^*}{\partial n} = \varphi_N(\mathbf{x}) \quad ; \quad \mathbf{x} \in S_N \tag{6.77}$$

$$\frac{\partial \varphi^*}{\partial n} + \lambda \varphi^* = \varphi_M(\mathbf{x}) \quad ; \quad \mathbf{x} \in S_M \tag{6.78}$$

From Theorems 5.1 to 5.5, a solution to the boundary value problem posed by (6.75) to (6.78) exists, and the initial boundary value problem governing $\widetilde{\varphi}(\mathbf{x}, t)$ is defined by the partial differential equation

$$a^2 \nabla^2 \widetilde{\varphi} = \frac{\partial \widetilde{\varphi}}{\partial t} \quad ; \tag{6.79}$$

with boundary conditions

$$\widetilde{\varphi} = 0 \quad ; \quad \mathbf{x} \in S_D \tag{6.80}$$

$$\frac{\partial \widetilde{\varphi}}{\partial n} = 0 \quad ; \quad \mathbf{x} \in S_N \tag{6.81}$$

$$\frac{\partial \widetilde{\varphi}}{\partial n} + \lambda \widetilde{\varphi} = 0 \quad ; \quad \mathbf{x} \in S_M \tag{6.82}$$

and the initial condition

$$\widetilde{\varphi}(\mathbf{x}, 0) = -\varphi^*(\mathbf{x}) \tag{6.83}$$

The solution of the original initial boundary value problem posed by (6.64) to (6.68) is now reduced to the summation of solutions $\varphi^*(\mathbf{x})$ which is governed by the boundary value problem (6.75) to (6.78) and an initial boundary value problem governed by (6.79) to (6.83). Both problems are well-posed and solutions can be obtained by the methods that will be discussed in the ensuing sections.

6.3 Methods of solution of the diffusion equation

In this section we shall briefly review some of the general procedures which can be used to develop solutions to problems associated with the diffusion equation. The presentation of the material is largely guided by the techniques that have been proposed in the literature. These techniques have also been developed to accommodate special features of the physical domain in which the problems are formulated. (e.g. finite regions, semi-infinite regions, circular regions, etc.). In the ensuing, attention is largely restricted to spatially one-dimensional problems in rectangular Cartesian and plane polar coordinates and spatially two-dimensional problems which are formulated in Cartesian coordinates.

6.3.1 Direct solution procedure

It is instructive to examine how effectively the direct solution procedures for partial differential equations of the parabolic-type, given in Chapter 4, can be applied to the solution of the spatially one-dimensional diffusion equation. The problem can be posed as follows. The partial differential equation governing a one dimensional diffusion process in a region of finite thickness L and infinite extent (Figure 6.8) is given by

$$D_0 \frac{\partial^2 C}{\partial x^2} = \frac{\partial C}{\partial t} \tag{6.84}$$

where $C(x,t)$ is the concentration and D_0 is a constant diffusion parameter.

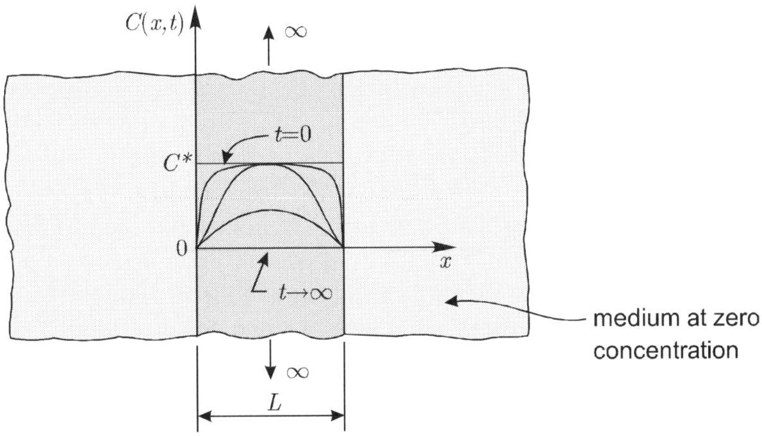

Figure 6.8: Diffusion from a finite source

The boundary conditions governing the problem are

$$C(0,t) = 0 \tag{6.85}$$

$$C(L,t) = 0 \tag{6.86}$$

and the initial conditions governing the problem are

$$C(x,0) = C^* \tag{6.87}$$

where C^* is a constant. Applying the criteria described in Chapter 4, we note that the differential equation (6.84) is already in a canonical form for parabolic equations. It is however not possible to integrate this equation directly.

6.3.2 Trial function approach

The selection of a trial function solution for any partial differential equation requires considerable experience and an appreciation of the physical principles governing the problem. With regard to the diffusion problem, an inspection of the processes indicates that the slab, which is at uniform concentration at time $t = 0$, will diffuse into the medium which is maintained at zero concentration for all $t > 0$. It is also clear that if this diffusion process continues, the concentration within the slab will reduce to zero with progress of time (i.e. $t \to \infty$). Therefore, we can select a trial function for the concentration which has a form

$$C(x,t) = Be^{-\lambda_1 t} F(x) \tag{6.88}$$

where B and λ_1 are constants and $F(x)$ is an arbitrary function. We can select specific forms for $F(x)$ by inspecting the boundary conditions (6.85) and (6.86) and the symmetry conditions that should be satisfied by $C(x,t)$. An example of such a function is

$$F(x) = \sin\left(\frac{\lambda_2 x}{L}\right) \tag{6.89}$$

where λ_2 is a constant.

The function satisfies the boundary conditions at $x = 0, L$ and gives continuous derivatives up to the second-order and higher-order in x. The function also gives a maximum value at $x = L/2$, the centre of the slab. For the purposes of the discussion, let us consider (6.89) as a part of the trial function. In order to ensure that the boundary conditions (6.85) and (6.86) are satisfied we require λ_2 to be an integer; i.e. $\lambda_2 = n\pi$ where n is an integer. Hence, (6.88) now takes the form

$$C(x,t) = Be^{-\lambda_1 t} \sin\left(\frac{n\pi x}{L}\right). \tag{6.90}$$

Substituting (6.90) in (6.84) we can show that λ_1 is given by

$$\lambda_1 = \frac{n^2\pi^2 D_0}{L^2} \tag{6.91}$$

With the determination of λ_1, the trial solution is now indeterminate to within the arbitrary constant B. We can choose values of B_n corresponding to $n = 1, 2, 3, \ldots$ etc. and construct a solution of the form

$$C(x,t) = \sum_{n=1,2,}^{\infty} B_n e^{-\left(\frac{n^2\pi^2 D_0 t}{L^2}\right)} \sin\left(\frac{n\pi x}{L}\right) \tag{6.92}$$

The arbitrary constants B_n can be evaluated by considering the Fourier integral representations similar to that used in connection with the development of solutions for Laplace's equation, i.e.

$$C(x,0) = C^* = \sum_{n=1,2,}^{\infty} B_n \sin\left(\frac{n\pi x}{L}\right) = \text{const.} \tag{6.93}$$

Using the result (5.135) we can show that

$$B_n = \frac{2C^*}{L}\int_0^L \sin\left(\frac{n\pi x}{L}\right) dx \tag{6.94}$$

Hence

$$C(x,t) = C^* \sum_{n=1,2,}^{\infty} 2e^{-\left(\frac{n^2\pi^2 D_0 t}{L^2}\right)} \sin\left(\frac{n\pi x}{L}\right) \int_0^1 \sin(n\pi\xi) d\xi \tag{6.95}$$

In this case we are particularly fortunate that the trial function technique gives a solution which satisfies *both* the boundary conditions and initial conditions accurately and exhibits the time-dependence characteristics of a diffusive-type process. The procedure is *not* expected to be so straightforward for other choices of $F(x)$ where either the governing equation or the boundary condition or the initial condition may have to be satisfied in an *approximate* or *average* sense.

Other heuristic methods based on similarity transformations exist; such techniques will attempt to transform the partial differential equation to an ordinary differential equation. The applicability of such methods will be based on experience concerning the use of such transformations for analogous problems where diffusive-type phenomena are accounted for in the formulation of the problem.

6.3.3 Separation of variables method - Cartesian coordinates

The solution of initial boundary value problems which are governed by the diffusion equation can also be attempted by employing a method of separation of variables. The application of the method also recognizes the physical domain of interest to a particular problem.

In the ensuing, the method of separation of variables will be applied to determine specific solutions of the diffusion equation which are applicable to *semi-infinite* and *finite* domains, which involve only a single spatial variable.

Example 6.1

A heat conducting region of large extent is initially maintained at a constant temperature T_0. A section of this body is now exposed to the ambient environment which has a zero temperature. Considering the exposed surface to conform to the boundary of a semi-infinite region (Figure 6.9), formulate the initial boundary value problem and develop an expression for the time-dependent variation of temperature in the vicinity of the exposed surface.

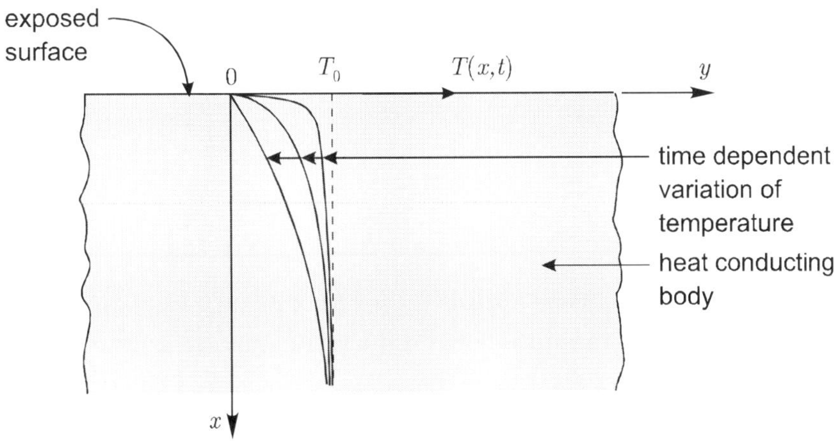

Figure 6.9: Heat conduction in a semi-infinite region

Solution

It is important to recognize at the outset that the notion of a semi-infinite medium is only a mathematical idealization which is used to facilitate the analysis of the problem. In reality, all domains of practical interest are finite in extent. The semi-infinite region in which the transient heat conduction process takes place is defined by $0 \leq x < \infty$; $-\infty < y < \infty$ and $-\infty < z < \infty$. Since the entire surface $x = 0$ is subjected to the temperature change, we can reduce the problem to a "*one dimensional*" heat conduction problem as shown in Figure 6.9. The partial differential equation governing the heat conduction problem is given by (6.11); the one dimensional form is

$$\alpha \frac{\partial^2 T}{\partial x^2} = \frac{\partial T}{\partial t} \tag{6.96}$$

where α is the thermal diffusivity of the heat conducting solid. The boundary condition governing $T(x,t)$ is

$$T(0,t) = 0 \tag{6.97}$$

The initial condition of the temperature field is

$$T(x,0) = T_0 \tag{6.98}$$

and it should also be noted that the temperature field should be "bounded" for physical reasons.

In keeping with the variables separation technique, we seek solutions of $T(x,t)$ of the form

$$T(x,t) = F(x)G(t) \tag{6.99}$$

where $F(x)$ and $G(t)$ are arbitrary functions. Substituting (6.99) in (6.96) we obtain the following:

$$\frac{d^2F/dx^2}{F} = \frac{1}{\alpha}\frac{dG/dt}{G} = -\xi^2 \tag{6.100}$$

where $\xi > 0$ and the negative sign is chosen for ξ^2 will ensure that the time-dependent effects will reduce to zero as $t \to \infty$, since α and t are both positive quantities. We obtain from (6.100)

$$\frac{d^2F}{dx^2} + \xi^2 F = 0 \tag{6.101}$$

$$\frac{dG}{dt} + \xi^2 \alpha G = 0 \tag{6.102}$$

The solutions of these ordinary differential equations take the forms

$$F(x) = A^* \cos(\xi x) + B^* \sin(\xi x) \tag{6.103}$$

and

$$G(t) = C^* e^{-\xi^2 \alpha t} \tag{6.104}$$

where A^*, B^* and C^* are arbitrary constants. The general solution of the partial differential equation (6.96) can be written as

$$T(x,t) = e^{-\xi^2 \alpha t}\{A\cos(\xi x) + B\sin(\xi x)\} \tag{6.105}$$

where A and B are arbitrary constants. Using the boundary condition (6.97) we observe that

$$A = 0 \tag{6.106}$$

and the solution for the temperature field reduces to

$$T(x,t) = Be^{-\xi^2 \alpha t}\sin(\xi x) \tag{6.107}$$

There are no restrictions on the parameter ξ; therefore we can generalize the solution (6.107) by replacing B by $B(\xi)$. Since the partial differential equation is linear, we can superpose solutions for discrete values of ξ and still obtain a valid solution. This is identical to the superposition theorem that is used in connection with Fourier series (e.g. (1.189)). We can therefore generalize the solution (6.107) to the form

$$T(x,t) = \int_0^\infty B(\xi)e^{-\xi^2 \alpha t}\sin(\xi x)d\xi \tag{6.108}$$

From the initial condition we have

$$T(x,0) = T_0 = \int_0^\infty B(\xi)\sin(\xi x)d\xi \tag{6.109}$$

This is an integral equation for the determination of the function $B(\xi)$ which is an unknown function. From the results of Fourier integrals for an odd function $\phi(x)$, (1.189) if

$$\phi(x) = \sqrt{\frac{2}{\pi}}\int_0^\infty \chi(\xi)\sin(\xi x)d\xi \tag{6.110}$$

then

$$\chi(\xi) = \sqrt{\frac{2}{\pi}}\int_0^\infty \phi(\xi)\sin(\xi x)dx \tag{6.111}$$

Using (6.109), (6.110) and (6.111) we have

$$B(\xi) = \frac{2T_0}{\pi}\int_0^\infty \sin(\xi x)dx = \frac{2T_0}{\pi}\int_0^\infty \sin(\xi\eta)d\eta \tag{6.112}$$

Using (6.112) in (6.108) we obtain the general integral expression for the temperature distribution in the semi-infinite medium as

$$T(x,t) = \frac{2T_0}{\pi}\int_0^\infty\int_0^\infty e^{-\alpha\xi^2 t}\sin(\xi\eta)\sin(\xi x)d\xi d\eta \tag{6.113}$$

It is possible to simplify this expression further by using certain trigonometric identities and results for infinite integrals. We can write

$$\sin(\xi\eta)\sin(\xi x) = \frac{1}{2}[\cos\xi(\eta - x) - \cos\xi(\eta + x)] \tag{6.114}$$

and represent the solution (6.113) in the form

$$T(x,t) = \frac{T_0}{\pi}\int_0^\infty \left[\int_0^\infty e^{-\alpha\xi^2 t}\cos\xi(\eta - x)d\xi \right.$$
$$\left. - \int_0^\infty e^{-\alpha\xi^2 t}\cos\xi(\eta + x)d\xi\right] d\eta \tag{6.115}$$

Using the integral results

$$\int_0^\infty e^{-\alpha\xi^2 t}\cos\xi(\eta - x)d\xi = \sqrt{\frac{\pi}{4\alpha t}}\exp\left\{-\frac{(x-\eta)^2}{4\alpha t}\right\} \tag{6.116}$$

$$\int_0^\infty e^{-\alpha\xi^2 t}\cos\xi(\eta + x)d\xi = \sqrt{\frac{\pi}{4\alpha t}}\exp\left\{-\frac{(x+\eta)^2}{4\alpha t}\right\} \tag{6.117}$$

we can express (6.115) as

$$T(x,t) = \frac{T_0}{(4\pi\alpha t)^{\frac{1}{2}}}\left[\int_0^\infty \exp\left\{-\frac{(x-\eta)^2}{4\alpha t}\right\} d\eta \right.$$
$$\left. - \int_0^\infty \exp\left\{-\frac{(x+\eta)^2}{4\alpha t}\right\} d\eta\right] \tag{6.118}$$

Introducing the new variables

$$-\zeta = \frac{(x-\eta)}{\sqrt{4\alpha t}} \quad ; \quad d\eta = \sqrt{4\alpha t}\ \ d\zeta \tag{6.119}$$

$$\zeta = \frac{(x+\eta)}{\sqrt{4\alpha t}} \quad ; \quad d\eta = \sqrt{4\alpha t}\ \ d\zeta \tag{6.120}$$

respectively, for the first and second integrals of (6.118) we can write

$$\frac{T(x,t)}{T_0} = \frac{1}{\sqrt{\pi}}\left[\int_{-x/\sqrt{4\alpha t}}^{\infty} e^{-\zeta^2}\,d\zeta - \int_{x/\sqrt{4\alpha t}}^{\infty} e^{-\zeta^2}\,d\zeta\right] \qquad (6.121)$$

Since $e^{-\zeta^2}$ is symmetric about $\zeta = 0$, the above result can be written as

$$\frac{T(x,t)}{T_0} = \frac{2}{\sqrt{\pi}}\int_{0}^{x/\sqrt{4\alpha t}} e^{-\zeta^2}\,d\zeta \qquad (6.122)$$

The integral on the right hand side of the equation (6.122) is the "*error function*" which is similar to the "*probability integral*". The error function can be computed for given values of x, t and α. Tabulated values of the integral are given in the literature. Hence we can write

$$\frac{T(x,t)}{T_0} = \operatorname{erf}\left(\frac{x}{\sqrt{4\alpha t}}\right) \qquad (6.123)$$

It is useful to record here an approximate series expression for $\operatorname{erf}(\Delta)$ which has the following form

$$\operatorname{erf}(\Delta) = \frac{2\Delta}{\sqrt{\pi}}\left[1 - \frac{\Delta^2}{1!3} + \frac{\Delta^4}{2!5} - \frac{\Delta^6}{3!7} + \cdots\right] \qquad (6.124)$$

provided $\Delta^2 < \infty$. We can also define a "*complementary error function*" or $\operatorname{erfc}(\Delta)$ which is given by

$$\operatorname{erfc}(\Delta) = 1 - \operatorname{erf}(\Delta). \qquad (6.125)$$

Example 6.2

We consider the spatially one-dimensional heat conduction problem for an insulated bar of uniform cross section which is subjected to an initial temperature of $T_0(x)$ along the length of the bar (Figure 6.10). The ends of the bar $x = 0$ and $x = L$ are subjected to the following boundary conditions

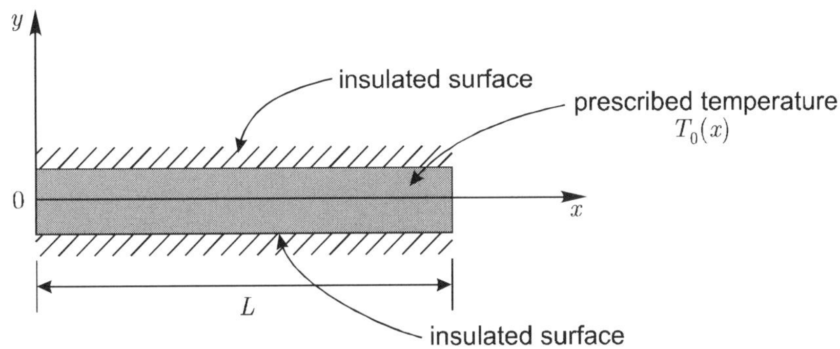

Figure 6.10: One dimensional heat conduction in a bar

$$T(0, t) = 0 \tag{6.126}$$

$$T(L, t) = 0 \tag{6.127}$$

Formulate the initial boundary value problem and determine an expression for the time-dependent variation of temperature along the length of the bar

Solution

The heat conduction problem is considered to be a one-dimensional problem which is governed by the partial differential equation (6.96) and subject to the boundary conditions (6.126) and (6.127) and the initial condition

$$T(x, 0) = T_0(x) \tag{6.128}$$

The spatially one-dimensional nature of the problem is ensured by the insulation provided along the surface of the bar. The temperatures are therefore constant across any surface $x =$ const.

The problem described is a well-posed initial boundary value problem. We can also consider the problem as a boundary value problem in the x, t plane, where the solution $T(x,t)$ is sought for a semi-infinite strip, $0 \leq x \leq L$, $0 \leq t \leq \infty$ (Figure 6.11).

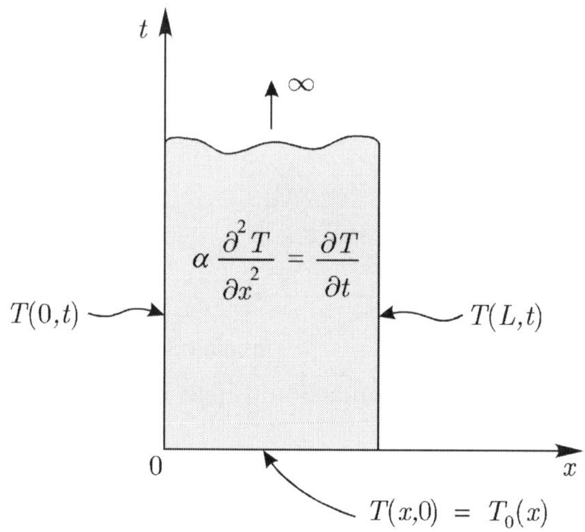

Figure 6.11: Representation of the one-dimensional heat conduction problem as a boundary value problem

Following the conventional procedure for developing a variables separable solution, we seek solutions of the form (6.99) (i.e. $T(x,t) = M(x)N(t)$). We obtain the two ordinary differential equations (6.101) and (6.102) for $M(x)$ and $N(t)$ respectively; i.e.

$$\frac{d^2M}{dx^2} + \zeta M = 0 \tag{6.129}$$

$$\frac{dN}{dt} + \zeta \alpha N = 0 \tag{6.130}$$

We seek solutions of (6.96) which satisfy the boundary conditions (6.126) and (6.127) which will place restrictions on ζ.

From the boundary condition (6.126) we have

$$T(0,t) = M(0)N(t) = 0 \tag{6.131}$$

If we satisfied (6.131) by setting $N(t) = 0$ for all t, then $T(x,t)$ is zero for all x and t. This would then violate the initial condition (6.128). Hence, in order to satisfy (6.131) we require

$$M(0) = 0 \tag{6.132}$$

Using a similar argument we can show that

$$M(L) = 0 \tag{6.133}$$

The equation (6.129) can now be solved, by imposing the boundary conditions (6.132) and (6.133). For a two-point boundary value problem, in which the values of the ends are prescribed $M(x) = 0$ is a solution which is valid for all ζ. This again gives a trivial solution $T(x,t) = 0$. For non-trivial solutions we can show that ζ must be real and the character of the solution will be governed by whether $\zeta = 0$, $\zeta < 0$ or $\zeta > 0$.

Case (i): $\zeta = 0$

Substituting $\zeta = 0$ in (6.129) and integrating the result we have

$$M(x) = c_1 x + c_2 \tag{6.134}$$

To satisfy (6.132) and (6.133) we require $c_1 = c_2 = 0$; which is a trivial result.

Case (ii): $\zeta < 0$

If we replace ζ by $-\xi^2$ where $\xi > 0$ is a new parameter, (6.129) gives

$$\frac{d^2M}{dx^2} - \xi^2 M = 0 \tag{6.135}$$

which has a general solution

$$M(x) = d_1 \cosh(\xi x) + d_2 \sinh(\xi x) \tag{6.136}$$

From boundary condition (6.132) and the result (6.136) we have

$$M(0) = d_1 = 0 \tag{6.137}$$

Applying the boundary condition (6.133) to the remaining solution we have

$$M(L) = d_2 \sinh(\xi L) = 0 \tag{6.138}$$

Since ξ and L are positive, $\sinh(\xi L) > 0$; consequently (6.138) is satisfied, if and only if, $d_2 = 0$. Again there are no non-trivial solutions if $\xi < 0$.

Case (iii): $\zeta > 0$

If we replace ζ by ξ^2 where $\xi > 0$, (6.129) gives

$$\frac{d^2M}{dx^2} + \xi^2 M = 0 \tag{6.139}$$

which has a general solution

$$M(x) = f_1 \cos(\xi x) + f_2 \sin(\xi x) \tag{6.140}$$

Again, we note that the boundary condition (6.132) requires $f_1 = 0$ and the boundary condition (6.133) gives the equation

$$f_2 \sin(\xi L) = 0 \tag{6.141}$$

Obviously, $f_2 = 0$ yields a trivial result. Alternatively, we can take $\sin(\xi L) = 0$ which gives $\xi L = n\pi$ for $n = 1, 2, 3, ...$ and we have $\xi = n\pi/L$ and

$$\zeta = -\frac{n^2\pi^2}{L^2} \quad ; \quad n = 1, 2, 3... \tag{6.142}$$

Solving the ordinary differential equations (6.129) and (6.130) and using the result (6.142) and summing up solutions applicable to $n = 1, 2, 3, ...$, the series solution for $T(x, t)$ can be evaluated in the form

$$T(x, t) = \sum_{n=1,2,3}^{\infty} A_n e^{-n^2\pi^2\alpha t/L^2} \sin\left(\frac{n\pi x}{L}\right) \tag{6.143}$$

To evaluate the arbitrary constants A_n we use the initial condition (6.128), i.e.

$$T_0(x) = \sum_{n=1,2,3}^{\infty} A_n \sin\left(\frac{n\pi x}{L}\right) \tag{6.144}$$

If we expand $T_0(x)$ as a half range sine series we have

$$A_n = \frac{2}{L} \int_0^L T_0(x) \sin\left(\frac{n\pi x}{L}\right) dx \tag{6.145}$$

This constitutes the formal solution of the heat conduction problem. When the initial temperature along the rod $T_0(x)$ is specified A_n can be evaluated explicitly.

Example 6.3

Consider the problem of a fluid saturated porous medium of infinite extent. Physically, this could be visualized as a sheathed long circular bar of the material where the length is considerably longer than the diameter (Figure 6.12).

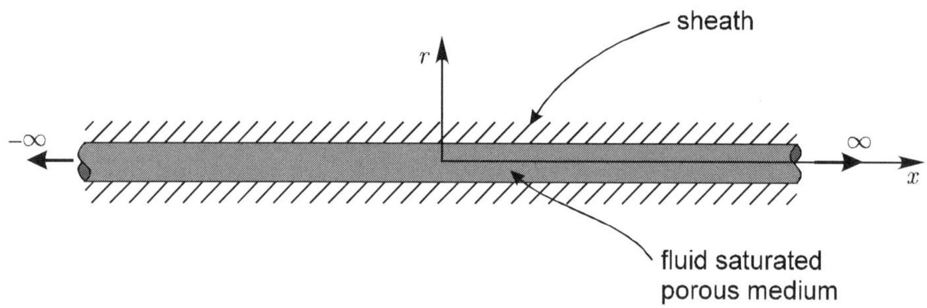

Figure 6.12: One dimensional porous medium of infinite extent

The diffusion of pressure in the porous medium is governed by the partial differential equation

$$a^2 \frac{\partial^2 p}{\partial x^2} = \frac{\partial p}{\partial t} \quad ; \quad x \in (-\infty, \infty) \ ; \ t > 0 \tag{6.146}$$

where a^2 is the diffusion parameter defined by (6.29). The initial distribution of pressure is given by

$$p(x, 0) = p^*(x) \quad ; \quad x \in (-\infty, \infty) \tag{6.147}$$

Determine the time-dependent variation of the pressure distribution in the one-dimensional infinite region if $p^*(x)$ corresponds to

$$p^*(x) = \begin{cases} p_0 & ; \quad -\ell < x < \ell \\ 0 & ; \quad |\, x \,| > \ell \end{cases} \tag{6.148}$$

Solution

The solution procedure closely follows the technique utilized in Problem 6.1 in connection with the one-dimensional region of semi-infinite extent. Unlike in the *semi-infinite region*, no boundary conditions can be prescribed since there are no "boundaries" in the problem domain. We will, however, implicitly assume that the pressures are "bounded" in the domain under consideration.

The separation of variables technique gives rise to ordinary differential equations which govern temporal and spatial components of the solution. The solution of these equations give the following form for $p(x,t)$;

$$p(x,t) = e^{-a^2\xi^2 t}[A(\xi)\cos(\xi x) + B(\xi)\sin(\xi x)] \tag{6.149}$$

where $A(\xi)$ and $B(\xi)$ are arbitrary functions. Again, since there are no restrictions on the parameter ξ, we can generalize the result (6.149) to the integral form

$$p(x,t) = \int_0^\infty e^{-a^2\xi^2 t}[A(\xi)\cos(\xi x) + B(\xi)\sin(\xi x)]d\xi \tag{6.150}$$

Considering the initial condition (6.147), we have

$$p^*(x) = \int_0^\infty [A(\xi)\cos(\xi x) + B(\xi)\sin(\xi x)]d\xi \tag{6.151}$$

We can identify (6.151) as the generalization of the Fourier formula for the integral representation of the function $p^*(x)$ defined in the interval $-\infty < x < \infty$. The coefficients $A(\xi)$ and $B(\xi)$ are given by

$$A(\xi) = \frac{1}{\pi}\int_{-\infty}^{\infty} p^*(\tau)\cos(\xi\tau)d\tau \tag{6.152}$$

$$B(\xi) = \frac{1}{\pi}\int_{-\infty}^{\infty} p^*(\tau)\sin(\xi\tau)d\tau \tag{6.153}$$

If we substitute (6.152) and (6.153) into (6.151), combine the trigonometric terms and change the order of integration, we obtain

$$p^*(x) = \int_0^{\infty}\left[\frac{1}{\pi}\int_{-\infty}^{\infty} p^*(\tau)\cos\xi(x-\tau)d\tau\right]d\xi \tag{6.154}$$

This representation is valid only if $p^*(x)$ and dp^*/dx are sectionally continuous on every finite interval on the x-axis ; $p^*(x)$ is defined as its mean value at each point of discontinuity and the integral

$$\int_{-\infty}^{\infty} \mid p(x) \mid dx$$

exists. Comparing (6.151) and (6.154) we have

$$A(\xi)\cos(\xi x) + B(\xi)\sin(\xi x) = \frac{1}{\pi}\int_{-\infty}^{\infty} p^*(\tau)\cos\xi(x-\tau)d\tau \tag{6.155}$$

Hence, the integral result for the variation in fluid pressure in the porous medium is given by

$$p(x,t) = \frac{1}{\pi}\int_0^{\infty} e^{-a^2\xi^2 t}\left[\int_{-\infty}^{\infty} p^*(\tau)\cos\xi(x-\tau)d\tau\right]d\xi \tag{6.156}$$

Considering the integral

$$\int_0^{\infty} e^{-a^2\xi^2 t}\cos\xi(x-\tau)d\xi = \sqrt{\frac{\pi}{4a^2 t}}\exp\left[-\frac{(x-\tau)^2}{4a^2 t}\right] \tag{6.157}$$

the solution (6.156) can be rewritten as

$$p(x,t) = \frac{1}{[4\pi a^2 t]^{\frac{1}{2}}}\int_{-\infty}^{\infty} p^*(\tau)\exp\left[-\frac{(x-\tau)^2}{4a^2 t}\right]d\tau \tag{6.158}$$

In the special case when

$$p^*(x) = \begin{cases} p_0 & \text{for} \quad -\ell < x < \ell \\ 0 & \text{for} \quad \mid x \mid > 0 \end{cases} \tag{6.159}$$

equation (6.158) gives

$$p(x,t) = \frac{p_0}{[4\pi a^2 t]^{\frac{1}{2}}}\int_{-\ell}^{\ell}\exp\left\{-\frac{(x-\tau)^2}{4a^2 t}\right\}d\tau \tag{6.160}$$

If we define a new variable

$$\zeta = \frac{(x-\tau)}{\sqrt{4a^2 t}} \qquad ; \qquad d\tau = -\sqrt{4a^2 t}\, d\zeta \tag{6.161}$$

the result (6.160) gives

$$p(x,t) = \frac{p_0}{2}\left[\frac{2}{\sqrt{\pi}}\int_0^{(\ell+x)/\sqrt{4a^2 t}} e^{-\zeta^2}d\zeta + \frac{2}{\sqrt{\pi}}\int_0^{(\ell-x)/\sqrt{4a^2 t}} e^{-\zeta^2}d\zeta\right] \tag{6.162}$$

This result can be expressed in the form

$$p(x,t) = \frac{p_0}{2}\left[\mathrm{erf}\left(\frac{\ell + x}{\sqrt{4a^2t}}\right) + \mathrm{erf}\left(\frac{\ell - x}{\sqrt{4a^2t}}\right)\right] \tag{6.163}$$

Example 6.4

An infinitely long heat conducting plate of width $2b$ is subjected to a uniform temperature T_0. The boundaries of the plate (Figure 6.13) are subjected to Newton-type cooling on the surfaces $y = \pm b$. Formulate the well-posed initial boundary value problem associated with the heat conduction problem and develop an expression for the time-dependent distribution of temperatures within the plate

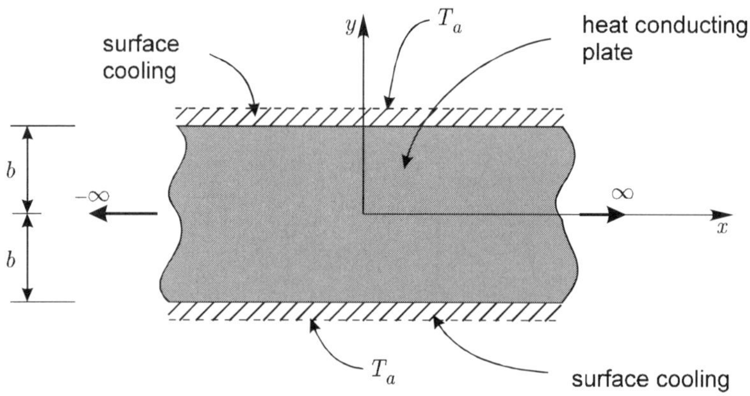

Figure 6.13: Heat conducting plate with surface cooling

Solution

The partial differential equation governing the heat conduction problem is

$$\alpha\frac{\partial^2 T}{\partial y^2} = \frac{\partial T}{\partial t} \quad ; \quad -b < y < b \quad ; \quad t > 0 \tag{6.164}$$

The initial condition is

$$T(y,0) = T_0 \quad ; \quad -b < y < b \tag{6.165}$$

and the boundary conditions are

$$k\frac{\partial T}{\partial y} = -h(T - T_a) \quad ; \quad y = \pm b \tag{6.166}$$

where k is the thermal conductivity of the medium, h is a heat transfer coefficient and T_a is the temperature of the ambient environment at a significant distance from the cooling boundary.

In view of the boundary condition (6.166) it is convenient to introduce a new dependent variable

$$T^*(y,t) = T(y,t) - T_a \tag{6.167}$$

The expressions corresponding to (6.164), (6.165) and (6.166) now reduce to

$$\alpha\frac{\partial^2 T^*}{\partial y^2} = \frac{\partial T^*}{\partial t} \quad ; \quad -b < y < b \ ; \ t > 0 \tag{6.168}$$

$$T^*(y,0) = T_0 - T_a \ ; \tag{6.169}$$

$$\frac{\partial T^*}{\partial y} = -\mathcal{B}\frac{T^*}{b} \quad ; \quad y = \pm b, \ \ t > 0 \tag{6.170}$$

where $\mathcal{B} = hb/k$ is referred to as the "Biot-modulus". If we consider variables separable solutions of the type

$$T^*(y,t) = Y(y)G(t) \tag{6.171}$$

we can show that

$$T^*(y,t) = e^{-\xi^2\alpha t}[A\cos(\xi y) + B\sin(\xi y)] \tag{6.172}$$

Considering the symmetry of the problem about $y = 0$ we can choose the solution

$$T^*(y,t) = e^{-\xi^2\alpha t}[A\cos(\xi y)] \tag{6.173}$$

(Alternately, we can include the terms which are asymmetric about $y = 0$ and show that they do indeed vanish if the constraint

$$\frac{\partial T^*}{\partial y} = 0 \quad ; \quad y = 0 \tag{6.174}$$

is invoked.)

Using (6.173), the boundary condition (6.170) can be expressed in the form

$$\frac{(\xi b)}{\mathcal{B}} = \cot(\xi b) \tag{6.175}$$

for the parameter ξ. The transcendental equation (6.175) has an infinite number of positive roots, ξ_1, ξ_2, ... and we can express the general solution for $T^*(y,t)$ as a series solution in terms of these eigenvalues, in the form

$$T^*(y,t) = \sum_{n=1}^{\alpha} A_n e^{-\xi_n^2\alpha t}\cos(\xi_n y) \tag{6.176}$$

A method of obtaining these eigenvalues is through computer evaluation of the roots of (6.175) or by means of an accurate graphical plot of the equation similar to that shown on Figure 6.14.

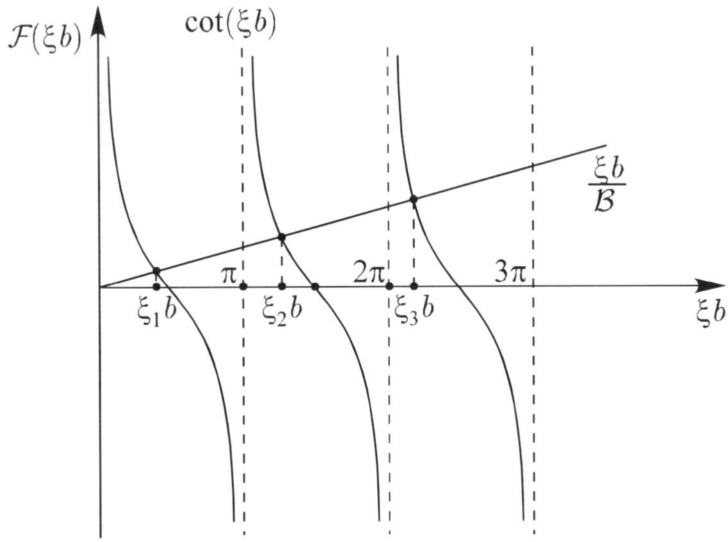

Figure 6.14: Graphical evaluation of the roots of $\cot \xi b = (\xi b/\mathcal{B})$.

Example 6.5

The entire surface of a heat conducting semi-infinite medium (Figure 6.15) is subjected to a periodic temperature variation of the form

$$T(0,t) = T_0 \sin(wt) \tag{6.177}$$

Determine the temperature of distribution $T(x, y)$ throughout the medium when a "periodic state" is established within it.

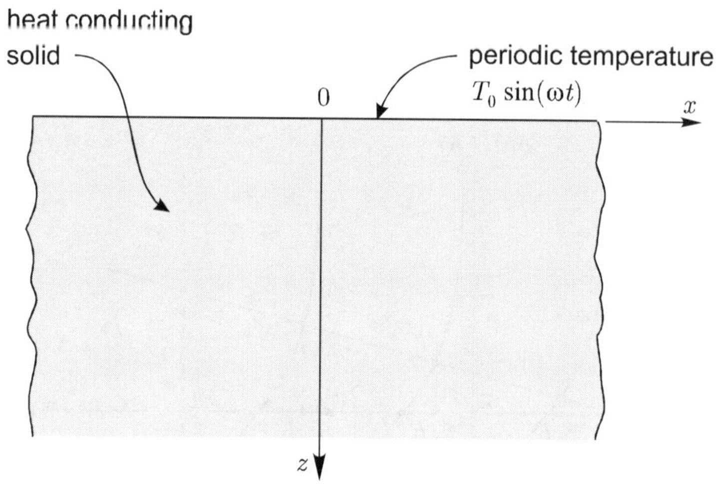

Figure 6.15: Periodic temperature variation at the surface of a semi-infinite medium

Solution

Since the temperature boundary condition is prescribed over the entire surface of the semi- infinite medium, the heat conduction problem is one-dimensional, governed by the partial differential equation

$$\alpha \frac{\partial^2 T}{\partial z^2} = \frac{\partial T}{\partial t} \quad ; \quad 0 \leq z < \infty \quad ; \quad t > 0 \tag{6.178}$$

and subject to the boundary condition (6.177). In addition, it is necessary to impose a regularity condition such that

$$T(\infty, t) \to 0 \tag{6.179}$$

In this problem we are only interested in the *periodic* state of temperature within the heat conducting solid, i.e., the temperature distribution is time-dependent on account of the periodic nature of the surface temperature. When such a temperature is first imposed on the surface, the temperature distribution within the medium will respond with a combination of *transient* and *periodic* components. As the periodic temperature is maintained at the surface for a sufficient time, the transient effects decay and the temperature variation is governed only by the periodic component with evidence of a *phase shift* to account for the thermal diffusivity of the material.

We assume a solution of the form

$$T(z,t) = \mathrm{Im}\{\tilde{T}(z)e^{i\omega t}\} \tag{6.180}$$

where Im refers to the *imaginary* part of the solution. For the time being, let us suppress the Im symbol and substitute (6.180) into (6.178). We can show that

$$\frac{d^2\tilde{T}}{dz^2} - \frac{i\omega}{\alpha}\tilde{T} = 0 \tag{6.181}$$

If we assume that

$$\lambda^2 = \frac{i\omega}{\alpha} = \frac{\omega}{\alpha}e^{i(\frac{\pi}{2}+2k\pi)} \quad ; \quad k = 0,1,2 \tag{6.182}$$

then the solution of (6.181) can be written as

$$\tilde{T}(z) = Ae^{\lambda z} + Be^{-\lambda z} \tag{6.183}$$

where A and B are arbitrary constants. From (6.182) we have

$$\lambda = \left[\frac{\omega}{\alpha}\right]^{\frac{1}{2}} e^{i(\pi/4)} \tag{6.184}$$

and the positive root can be written as

$$\lambda = \left[\frac{\omega}{2\alpha}\right]^{\frac{1}{2}}(1+i) \tag{6.185}$$

If we use the positive root of λ in (6.183) and apply the regularity condition (6.179)

$$A \equiv 0 \tag{6.186}$$

Applying the boundary condition to the remaining solution of (6.183) we have

$$B = T_0 \tag{6.187}$$

The temperature distribution corresponding to the periodic state can be written as

$$T(x,t) = \text{Im}\{T_0 e^{-\lambda z} e^{i\omega t}\} \tag{6.188}$$

or

$$T(x,t) = T_0 \exp\left\{-\left(\frac{\omega}{2\alpha}\right)^{\frac{1}{2}} z\right\} \sin\left[\omega t - \left(\frac{\omega}{2\alpha}\right)^{\frac{1}{2}} z\right] \tag{6.189}$$

As an exercise prove that even if the negative root of (6.182) is chosen, the final result will be the same as (6.189).

Example 6.6

A heat conducting slab occupies the region $x \in (-h, h)$; $y \in (-\infty, \infty)$ and $z \in (-\infty, \infty)$. the initial temperature of the slab is at zero degrees Celsius. The surfaces $z = -h$ and $z = h$ are subjected to temperature changes of $-T_0$ and T_0 respectively, and these temperatures are maintained constant with time (Figure 6.16). Assuming that the transient temperature

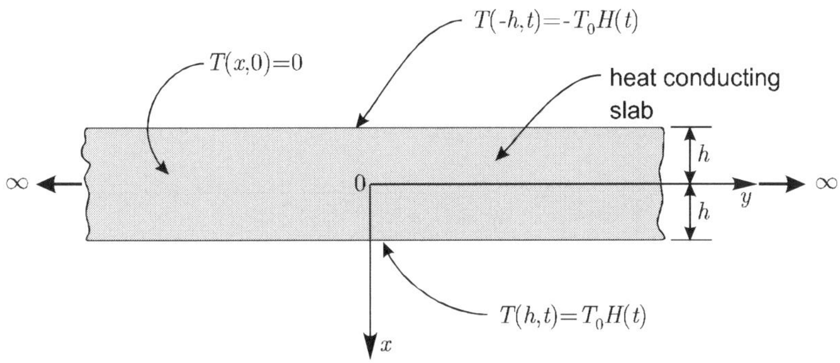

Figure 6.16: Boundary heating of an infinite slab.

distribution in the plate is governed by the one-dimensional diffusion equation derive an expression for the time-dependent distribution of temperature within the slab.

Solution

Considering the homogeneous nature of the initial temperature distribution within the heat conducting infinite slab, we adopt the solution procedure which involves a change in the dependent variable (see e.g. Section 6.2.6). Since the heat conduction process is one-dimensional, it is assumed that the solution $T(x, t)$ can be represented in the form

$$T(x, t) = T^*(x) + \widetilde{T}(x, t) \tag{6.190}$$

where $T^*(x)$ is a solution of the boundary value problem

$$\frac{d^2T}{dx^2} = 0 \tag{6.191}$$

with boundary conditions

$$T^*(h) = T_0 \quad ; \quad T^*(-h) = -T_0 \tag{6.192}$$

and $\widetilde{T}(x,t)$ is a solution of the initial boundary value problem defined by

$$\alpha \frac{\partial^2 \widetilde{T}}{\partial x^2} = \frac{\partial \widetilde{T}}{\partial t} \tag{6.193}$$

with boundary conditions

$$\widetilde{T}(h,t) = 0 \quad ; \quad \widetilde{T}(-h,t) = 0 \tag{6.194}$$

and initial condition

$$\widetilde{T}(x,t) = -\overset{*}{T}(x) \tag{6.195}$$

Integrating (6.191) and making use of the boundary conditions (6.192) it can be shown that

$$\overset{*}{T}(x) = T_0 \frac{x}{h} \tag{6.196}$$

since the initial condition associated with $\widetilde{T}(x,t)$ is asymmetric about $x = 0$, we need to select a variables separable solution which displays asymmetry about $x = 0$. Considering the general form of the variables separable solution (see e.g. (6.172)) we can write

$$\widetilde{T}(x,t) = Ae^{-\xi^2 \alpha t} \sin(\xi x) \tag{6.197}$$

where A is either a constant or a function of ξ. Using the boundary conditions (6.194) we obtain

$$\xi = \frac{n\pi}{h} \quad ; \quad n = 1, 2, ... \tag{6.198}$$

and the series solution of (6.197) takes the form

$$\widetilde{T}(x,t) = \sum_{n=1,2,} A_n e^{-\frac{n^2 \pi^2 \alpha t}{h^2}} \sin\left(\frac{n\pi x}{h}\right) \tag{6.199}$$

Applying the initial condition (6.195) to (6.199) we have

$$-T_0 \frac{x}{h} = \sum_{n=1,2,} A_n \sin\left(\frac{n\pi x}{h}\right) \tag{6.200}$$

The constants A_n can be obtained by using the approach where both sides of (6.200) are multiplied by $\sin(n\pi x/h)$ and integrated over the interval $x \in (-h, h)$. This gives

$$A_n = 2T_0 \frac{\cos(n\pi)}{(n\pi)} \tag{6.201}$$

The final result for the time-dependent distribution of temperature within the heat conducting slab is given by

$$T(x,t) = T_0 \frac{x}{h} + \frac{2T_0}{\pi} \sum_{n=1,2,} \frac{\cos(n\pi)}{n} e^{-\frac{n^2\pi^2\alpha t}{h^2}} \sin\left(\frac{n\pi x}{h}\right) \tag{6.202}$$

It is evident that as $t \to 0$, the *steady state* temperature distribution corresponds to the result (6.196). In the limit $t \to 0$, (6.202) gives

$$\begin{aligned} T(x,0) = T_0 \frac{x}{h} - \frac{2T_0}{\pi} \Big\{ &- \sin\left(\frac{\pi x}{h}\right) \\ &+ \frac{1}{2} \sin\left(\frac{2\pi x}{h}\right) - \frac{1}{3} \sin\left(\frac{3\pi x}{h}\right) + \ldots \Big\} \end{aligned} \tag{6.203}$$

From the results (6.200) and (6.201) it is evident that $T(x,0) \to 0$ for $-h < x < h$. At the boundaries of the slab, $x = h$ and $x = -h$, the temperatures are equal to T_0 and -T_0 respectively for $t \geq 0$.

6.3.4 Separation of variables method - plane polar coordinates

Up to now we have focused attention on the class of spatially one-dimensional diffusion-type problems which are best formulated in relation to a Cartesian frame of reference, primarily due to the geometry of the problem. The plane polar coordinate equivalent of the diffusion equation is the most efficient way of analyzing problems which involve cylindrical boundaries. The method of separation of variables can also be applied to the analysis of spatially one-dimensional problems referred to the plane polar coordinate system (r, θ) where the dependent variable will only be a function of r. In this case the problem is said to be *radially symmetric*. Examples of problems belonging to this category include, heat conduction over the cross section of a bar, chemical mass transport into a cylindrical cavity region, drying of solids, etc. In the ensuing we shall apply the separation of variables technique to solve a typical mass transport problem which exhibits radial symmetry.

Example 6.7

A long cylindrical specimen of a porous medium is saturated with a chemical at a uniform concentration C_0. The cylindrical specimen is placed in a fluid region which is constantly replenished (Figure 6.17). If this action maintains the concentration at the surface of the specimen at a zero value, formulate the initial boundary value problem and derive an expression for the variation of chemical concentration within the cylindrical sample

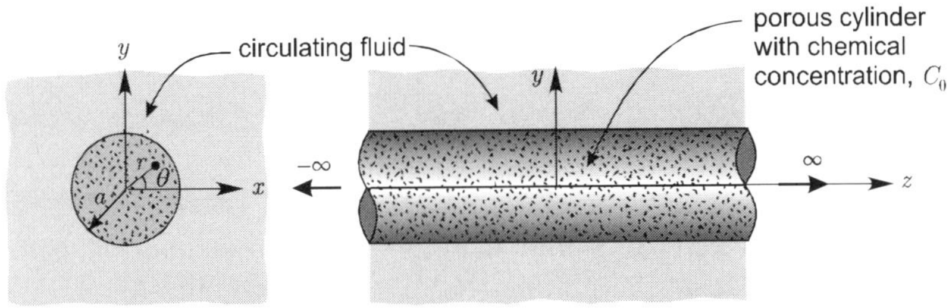

Figure 6.17: Radially symmetric diffusion from cylinder

Solution

Since the problem is assumed to be radially symmetric, the chemical concentration is $C(r,t)$. The partial differential equation (6.40) governing the diffusion problem now reduces to

$$D\left(\frac{\partial^2 C}{\partial r^2} + \frac{1}{r}\frac{\partial C}{\partial r}\right) = \frac{\partial C}{\partial t} \tag{6.204}$$

The initial condition governing the problem is

$$C(r,0) = C_0 \tag{6.205}$$

and the boundary condition applicable to the outer boundary of the cylindrical sample is

$$C(a,t) = 0 \tag{6.206}$$

where a is the radius of the cylinder. In addition to these conditions, it is necessary to ensure that the concentration is finite within the cylindrical region. Since the diffusive process is expected to decay as time increases it is reasonable to assume a variables separable solution of the form

$$C(r,t) = e^{-\lambda t} R(r) \tag{6.207}$$

where λ is a positive constant and $R(r)$ is an arbitrary function. Substituting (6.207) in (6.204) and assuming that $e^{-\lambda t} \neq 0$, we obtain

$$D\left(\frac{d^2R}{dr^2} + \frac{1}{r}\frac{dR}{R}\right) = -\lambda R \tag{6.208}$$

Since D is non-zero, (6.208) can be written as

$$r^2\frac{d^2R}{dr^2} + r\frac{dR}{dr} + \frac{\lambda r^2}{D}R = 0 \tag{6.209}$$

If we set

$$\frac{\lambda}{D} = k^2 \tag{6.210}$$

we can identify the resulting equation as *Bessel's equation of order zero*, the solutions of which take the forms

$$R(r) = AJ_0(kr) + BY_0(kr) \tag{6.211}$$

where $J_0(kr)$ is the *Bessel function of the first kind of order zero* and $Y_0(kr)$ is the *Bessel function of the second kind of zero-order*. The variation of $J_0(kr)$ and $Y_0(kr)$, are shown in Figure 6.18. Observing the variations of $J_0(kr)$ and $Y_0(kr)$ it is evident that since the concentration is finite at $r = 0$, we require

$$B \equiv 0 \tag{6.212}$$

and the solution for $C(r,t)$ can be written as

$$C(r,t) = e^{-k^2Dt}[AJ_0(kr)] \tag{6.213}$$

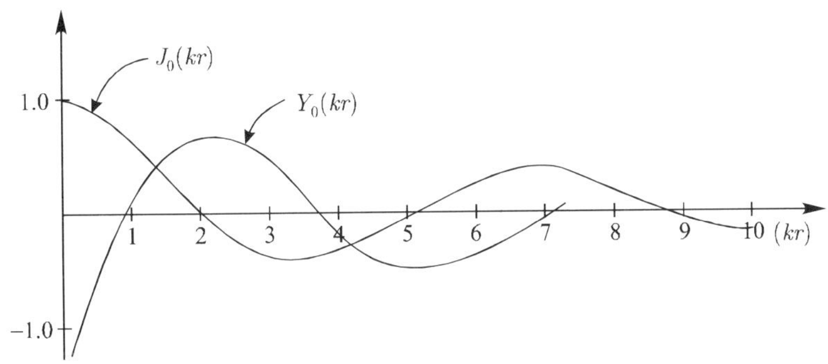

Figure 6.18: Bessel functions $J_0(kr)$ and $Y_0(kr)$

Applying the boundary condition (6.206) we have

$$C(a,t) = e^{-k^2Dt}[AJ_0(ka)] \tag{6.214}$$

For this condition to be satisfied for all t we require

$$J_0(ka) = 0 \tag{6.215}$$

From Figure 6.18, (6.215) is satisfied at the zero points of $J_0(ka)$, which are denoted by $k_n(n = 1, 2, 3, ...)$. The solution (6.214) can now be generalized to include all such values of k_n and the summation of such solutions gives the final solution as

$$C(r,t) = \sum_{n=1,2,...} A_n e^{-k_n^2 Dt} J_0(k_n r) \tag{6.216}$$

which is indeterminate to within the arbitrary constants A_n. These constants can be determined by invoking the initial condition, i.e.

$$C(r,0) = C_0 = \sum_{n=1,2,...} A_n J_0(k_n r) \tag{6.217}$$

We now need to expand the constant C_0 as a series of Bessel functions. This can be done in just the same way as a function can be expanded as a Fourier series. It can be shown that a function $F(\xi)$ in the interval $\xi = 0$ to $\xi = 1$ can be expressed in a series of the form

$$F(\xi) = \sum_{n=1,2..}^{\infty} E_n J_s(\alpha_n \xi) \tag{6.218}$$

where E_n are constants and α_n are the successive positive roots of

$$J_s(\alpha) = 0 \; ; \tag{6.219}$$

The coefficients E_n can be obtained by multiplying both sides of (6.218) by

$\xi J_s(\alpha_k\xi)d\xi$ and integrating between $\xi = 0$ and $\xi = 1$; i.e.

$$\int_0^1 \xi F(\xi) J_s(\alpha_k\xi)d\xi = E_k \int_0^1 \xi J_s^2(\alpha_k\xi)d\xi \tag{6.220}$$

which gives

$$E_k = \frac{2}{J_{s+1}^2(\alpha_k)} \int_0^1 \xi F(\xi) J_s(\alpha_k\xi)d\xi \tag{6.221}$$

Using this result we can obtain an integral expression for A_n in (6.216), i.e.

$$A_n = \frac{2}{a^2\{J_1(k_n a)\}^2} \int_0^a rF(r)J_0(k_n r)dr \tag{6.222}$$

for $n = 1, 2, 3, ...$ In the particular instance when $F(r) = C_0$, the integral in (6.222) gives

$$\int_0^a rJ_0(k_n r)dr = \frac{a}{k_n} J_1(k_n a) \tag{6.223}$$

The final result for the radially symmetric variation of chemical concentration within the cylindrical porous specimen is given by

$$C(r,t) = C_0 \sum_{n=1,2,} \frac{2e^{-k_n^2 Dt} J_o(k_n r)}{[k_n a J_1(k_n a)]} \tag{6.224}$$

and k_n are the successive roots of the equation (6.215).

Example 6.8

A circular wafer of a heat conducting material is placed between two perfectly insulating surfaces. The heat conduction pattern in the wafer is assumed to be radially symmetric. Initially the temperature varies linearly with zero temperature at the circular boundary and T_0 at the centre (Figure 6.19).

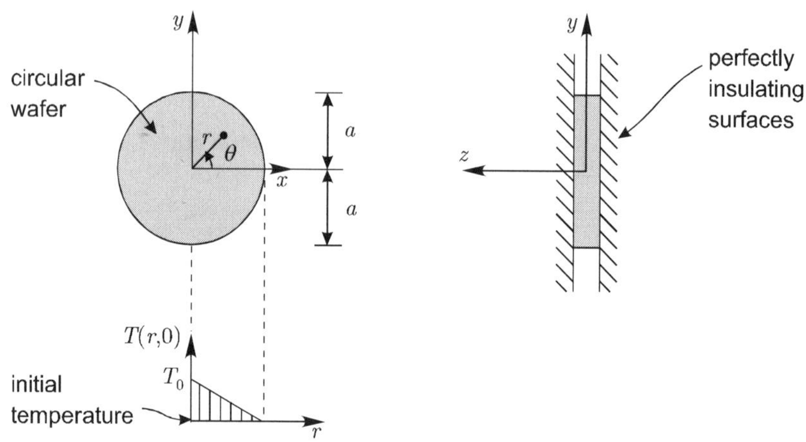

Figure 6.19: Heat conduction problem for a wafer

If the circular boundary of the wafer is perfectly insulated, derive an expression for the time-dependent radially symmetric distribution of temperature within the wafer.

Solution

The partial differential equation governing heat conduction is

$$\alpha\left(\frac{\partial^2 T}{\partial r^2}+\frac{1}{r}\frac{\partial T}{\partial r}\right)=\frac{\partial T}{\partial t} \tag{6.225}$$

The boundary conditions governing the problem is

$$\left(\frac{\partial T}{\partial r}\right) = 0 \quad \text{on} \quad r = a \quad ; \quad t > 0 \tag{6.226}$$

and the initial condition is

$$T(r, 0) = T_0 \left(1 - \frac{r}{a}\right) \tag{6.227}$$

Considering a variables separable solution of the form

$$T(r, t) = e^{-\lambda t} T(r) \tag{6.228}$$

it can be shown that the general solution which satisfies the partial differential equation (6.225) and gives finite temperatures within the wafer region takes the form

$$T(r, t) = Ae^{-k^2 \alpha t} J_0(kr) \tag{6.229}$$

and

$$\frac{\partial T}{\partial r} = -Ake^{-k^2 \alpha t} J_1(kr) \tag{6.230}$$

where J_1 is the Bessel function of the first kind of order unity. To satisfy the boundary condition (6.226) we require

$$J_1(ka) = 0 \tag{6.231}$$

The roots of (6.231) are denoted by $k_n (n = 1, 2, 3, ...)$. The solution (6.229) can now be generalized to include all values of k_n and the summation of all such solutions give

$$T(r,t) = \sum_{n=1,2,..} A_n e^{-k_n^2 \alpha t} J_0(k_n r) \tag{6.232}$$

In order to evaluate the constants A_n, we can use the initial condition

$$T(r,0) = T_0 \left(1 - \frac{r}{a}\right) = \sum_{n=1,2,...} A_n J_0(k_n r) \tag{6.233}$$

Using the results given in Example 6.6, equation (6.222), it can be shown that

$$A_n = \frac{2}{a^2 \{J_1(k_n a)\}^2} \int_0^a r T_0 \left(1 - \frac{r}{a}\right) J_0(k_n r) dr \tag{6.234}$$

The integrals occurring in (6.234) can be evaluated using *Tables of Definite Integrals* including the relationships of the form

$$\int x J_0(x) dx = x J_1(x) \tag{6.235}$$

$$\int x^2 J_0(x) dx = x^2 J_1(x) + x J_0(x) - \int J_0(x) dx \tag{6.236}$$

Avoiding details of the calculations it can be shown that the temperature distribution in the wafer takes the form

$$T(r,t) =$$

$$2T_0 \sum_{n=1,2,...} \frac{e^{-k_n^2 \alpha t} J_0(k_n r)}{\{k_n a J_1(k_n a)\}^2} \left[\frac{1}{a} \int_0^a J_0(k_n r) dr - J_0(k_n a)\right] \tag{6.237}$$

6.4 Some generalized results associated with the diffusion equation

In the previous section we have focused on applications of the diffusion equation to some specific problems which are spatially one-dimensional. A natural extension of the treatment would involve spatially two-dimensional initial boundary value problems associated with the diffusion equation. Before attempting to formulate such problems, it is instructive to consider some general results and theorems which will be of assistance in establishing the general character of the solutions of the diffusion equation.

6.4.1 Reduction to Helmholtz equation

In the diffusion equation, the time variable occurs in such a manner that it can be separated out giving an ordinary differential equation of the first-order. The remaining terms, which depend only on the spatial variables (x, y, z) always gives rise to a Helmholtz equation (named after Hermann von Helmholtz (1821-1894)).

Consider the diffusion equation

$$a^2\nabla^2\varphi = \frac{\partial\varphi}{\partial t} \tag{6.238}$$

Assume that φ admits a variables separable solution of the form

$$\varphi = F(x, y, z)T(t) \tag{6.239}$$

Substituting (6.239) in (6.238) we obtain

$$\frac{1}{F}\nabla^2 F = \frac{1}{a^2T}\frac{dT}{dt} = -\gamma^2 \tag{6.240}$$

Equation (6.240) is equivalent to

$$\frac{dT}{dt} + \gamma^2 a^2 T = 0 \tag{6.241}$$

$$\nabla^2 F + \gamma^2 F = 0 \tag{6.242}$$

The general solution of (6.241) is

$$T(t) = Ae^{-\gamma^2 a^2 t} \tag{6.243}$$

where A is an arbitrary constant. The solution of the *Helmholtz equation* (6.242) will depend on the form of Laplace's operator and can be handled via a variables separation technique. Specific solutions for the diffusion equation are obtained by combining the time-dependent solution which has an *exponential decay term* with solutions of the Helmholtz equation.

6.4.2 Product solutions for the diffusion equation

We develop here some general results which can assist the development of solutions to heat conduction problems involving spatially multi-dimensional regions. It can be shown that the solution to a multi-dimensional initial boundary value problem associated with the diffusion equation can be written as the product of the solutions for one dimensional initial boundary value problems. This is possible if the initial values in the multi-dimensional region can be expressed as a product of functions of the single space variables. It is convenient to illustrate the principle underlining the *product solution technique* by considering a specific diffusion problem referred to a rectangular region bounded by $0 \leq x \leq a$ and $0 \leq y \leq b$ (Figure 6.20).

The partial differential equation governing the diffusion problem is

$$\alpha \left(\frac{\partial^2 \varphi}{\partial x^2} + \frac{\partial^2 \varphi}{\partial y^2} \right) = \frac{\partial \varphi}{\partial t} \quad ; \quad 0 < x < a \quad ; \quad 0 < y < b \tag{6.244}$$

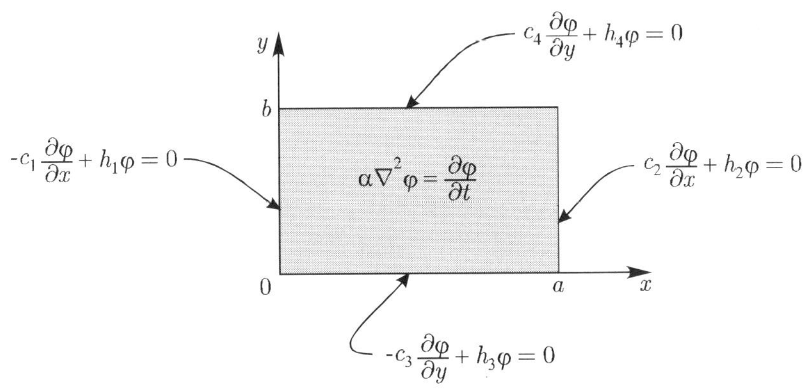

Figure 6.20: A diffusion problem referred to a rectangular region

We shall consider completely generalized boundary conditions of the type

$$c_i \frac{\partial \varphi}{\partial n} + h_i \varphi = 0 \tag{6.245}$$

where c_i and h_i are non-zero constants. The values of these constants can be suitably varied to recover boundary conditions either of the Dirichlet or Neumann-type or Robin-type if the transport processes are controlled (e.g. heat flow according to Newton's Law of cooling). We can state the boundary conditions (6.245) by considering the boundaries of the rectangular region shown in Figure 6.20. i.e.

$$-c_1 \frac{\partial \varphi}{\partial x} + h_1 \varphi = 0 \quad ; \quad x = 0 \quad ; \quad t > 0 \tag{6.246}$$

$$c_2 \frac{\partial \varphi}{\partial x} + h_2 \varphi = 0 \quad ; \quad x = a \quad ; \quad t > 0 \tag{6.247}$$

$$-c_3 \frac{\partial \varphi}{\partial y} + h_3 \varphi = 0 \quad ; \quad y = 0 \quad ; \quad t > 0 \tag{6.248}$$

$$c_4 \frac{\partial \varphi}{\partial y} + h_4 \varphi = 0 \quad ; \quad y = a \quad ; \quad t > 0 \tag{6.249}$$

The initial condition is assumed to be of the form

$$\varphi(x, y, 0) = \chi_1(x)\chi_2(y) \tag{6.250}$$

THEOREM 6.1

Consider $\varphi_1(x, t)$ a solution of the one-dimensional diffusion equation

$$\alpha \frac{\partial^2 \varphi_1}{\partial x^2} = \frac{\partial \varphi_1}{\partial t} \quad ; \quad 0 < x < a \quad ; \quad t > 0 \tag{6.251}$$

which satisfies the boundary conditions and initial conditions

$$-c_1 \frac{\partial \varphi_1}{\partial x} + h_1 \varphi_1 = 0 \quad ; \quad x = 0 \quad ; \quad t > 0 \tag{6.252}$$

$$c_2 \frac{\partial \varphi_1}{\partial x} + h_2 \varphi_1 = 0 \quad ; \quad x = a \quad ; \quad t > 0 \tag{6.253}$$

$$\varphi_1(x, 0) = \chi_1(x) \tag{6.254}$$

Consider $\varphi_2(y, t)$ a solution of the one-dimensional diffusion equation

$$\alpha \frac{\partial^2 \varphi_2}{\partial y^2} = \frac{\partial \varphi_2}{\partial t} \quad ; \quad 0 < y < b \quad ; \quad t > 0 \tag{6.255}$$

which satisfies the boundary conditions and initial conditions

$$-c_3 \frac{\partial \varphi_2}{\partial y} + h_3 \varphi_2 = 0 \quad ; \quad y = 0 \quad ; \quad t > 0 \tag{6.256}$$

$$c_4 \frac{\partial \varphi_2}{\partial y} + h_4 \varphi_2 = 0 \quad ; \quad y = b \quad ; \quad t > 0 \tag{6.257}$$

$$\varphi_2(y, 0) = \chi_2(y) \tag{6.258}$$

Then,

$$\varphi(x, y, t) = \varphi_1(x, t)\varphi_2(y, t) \tag{6.259}$$

is a solution of the initial boundary value problem defined by (6.244) to (6.250).

PROOF

Assuming that $\varphi(x, y, t)$ admits the representation (6.259); we substitute (6.259) into (6.244) and obtain

$$\varphi_2 \left(\frac{\partial^2 \varphi_1}{\partial x^2} - \frac{1}{\alpha} \frac{\partial \varphi_1}{\partial t} \right) + \varphi_1 \left(\frac{\partial^2 \varphi_2}{\partial y^2} - \frac{1}{\alpha} \frac{\partial \varphi_2}{\partial t} \right) = 0 \tag{6.260}$$

From (6.251) and (6.255) it is evident that (6.259) satisfies the governing differential equation (6.244). Similarly, by using (6.259) the boundary conditions (6.246) to (6.249) yield the following:

$$\varphi_2 \left\{ -c_1 \frac{\partial \varphi_1}{\partial x} + h_1 \varphi_1 \right\} = 0 \quad ; \quad x = 0 \quad ; \quad t > 0 \tag{6.261}$$

$$\varphi_2 \left\{ c_2 \frac{\partial \varphi_1}{\partial x} + h_2 \varphi_1 \right\} = 0 \quad ; \quad x = a \quad ; \quad t > 0 \tag{6.262}$$

$$\varphi_1 \left\{ -c_3 \frac{\partial \varphi_2}{\partial y} + h_3 \varphi_2 \right\} = 0 \quad ; \quad y = 0 \quad ; \quad t > 0 \tag{6.263}$$

$$\varphi_1 \left\{ c_4 \frac{\partial \varphi_2}{\partial y} + h_4 \varphi_2 \right\} = 0 \quad ; \quad y = b \quad ; \quad t > 0 \tag{6.264}$$

From (6.252), (6.253), (6.256) and (6.257) it is evident that the boundary conditions (6.246) to (6.249) will be identically satisfied. Finally, if we identify

$$\varphi_1(x,0) = \chi_1(x) \tag{6.265}$$

and

$$\varphi_2(y,0) = \chi_2(y) \tag{6.266}$$

then the initial condition (6.250) is identically satisfied. Thus, the validity of the product solution approach to the diffusion problem associated with spatially multi-dimensional regions is established.

6.4.3 Sturm-Liouville problems

At this stage, it is instructive to introduce aspects of a homogeneous one-dimensional boundary value problem which was originally investigated by J.C.F. Sturm (1803-1855) and J. Liouville (1809-1882) between 1836 - 1838. The general results associated with this problem have applications in the development of solutions to a wider class of partial differential equations involving *eigen-functions.*

Consider the ordinary differential equation having the form

$$\frac{d}{dx}\left[p(x)\frac{d\varphi}{dx}\right] + [q(x) + \lambda r(x)]\,\varphi = 0 \quad ; \quad a \le x \le b \tag{6.267}$$

subject to the boundary conditions

$$a_1\varphi(a) + a_2\left[\frac{d\varphi}{dx}\right]_{x=a} = 0 \tag{6.268}$$

$$b_1\varphi(b) + b_2 \left[\frac{d\varphi}{dx}\right]_{x=b} = 0 \tag{6.269}$$

where a_1, a_2, etc. are constants; $p(x), q(x)$ and $r(x)$ are given differentiable functions and λ is an unspecified parameter which is independent of x. The preceding constitutes a *Strum-Liouville boundary value problem.*

A non-trivial solution of this system exists only for a particular set of values of the parameter λ. These values are called the "*eigenvalues of the system*" and the corresponding solutions are called the "*eigenfunction solution*" of the system characterized by (6.267), (6.268) and (6.269). Some important attributes of a Sturm-Liouville problem can be summarized in following.

(i) In general, each eigenvalue is expected to provide one eigenfunction, although there can be exceptions to this rule.
(ii) When $p(x)$ and $q(x)$ are real, the eigenvalues are also real.
(iii) The eigenfunctions can form an orthonormal set with respect to the weighting function $r(x)$, which is assumed to be non-negative (i.e. $r(x) \geq 0$).
(iv) Through a normalization process the set of eigenfunctions can be made an orthonormal set with respect to the weighting function $r(x)$ in the range $a \leq x \leq b$.

Let λ_m and λ_n be two eigenvalues of (6.267) which provide eigenfunctions $\varphi(\lambda_m, x)$ and $\varphi(\lambda_n, x)$, which satisfy (6.267). For convenience let us denote the operator L by

$$L = \frac{d}{dx}\left[p(x)\frac{d}{dx}\right] \tag{6.270}$$

From (6.267), we have

$$L[\varphi_m(x)] + q(x)\varphi_m(x) + \lambda_m r(x)\varphi_m(x) = 0 \tag{6.271}$$

$$L[\varphi_n(x)] + q(x)\varphi_n(x) + \lambda_n r(x)\varphi_n(x) = 0 \tag{6.272}$$

where we have adopted the convention

$$\varphi_m(x) = \varphi(\lambda_m, x) \quad ; \quad \varphi_n(x) = \varphi(\lambda_n, x) \tag{6.273}$$

We multiply (6.271) by $\varphi_n(x)$ and (6.272) by $\varphi_m(x)$ and subtract one from the other to obtain the following:

$$\frac{d}{dx}\left[p(x)\left\{\varphi_n \frac{d\varphi_m}{dx} - \varphi_m \frac{d\varphi_n}{dx}\right\}\right] = (\lambda_n - \lambda_m) r(x) \varphi_m(x) \varphi_n(x) \tag{6.274}$$

When both sides of the equation (6.274) are integrated with respect to x from $x = a$ to $x = b$, the L.H.S. vanishes in view of the boundary conditions (6.268) and (6.269) and the remaining result is

$$(\lambda_n - \lambda_m) \int_a^b r(x) \varphi_m(x) \varphi_n(x) dx = 0 \tag{6.275}$$

If $\lambda_m \neq \lambda_n$, then the equation (6.275) is satisfied only if

$$\int_a^b r(x) \varphi_m(x) \varphi_n(x) dx = 0 \tag{6.276}$$

which proves that eigenfunctions of the Sturm-Liouville system are *orthogonal* with respect to the weighting function $r(x)$ in the interval (a, b).

Example 6.9

Examine the Sturm-Liouville problem resulting from the study of the one-dimensional problem in heat conduction

$$\alpha \frac{\partial^2 T}{\partial x^2} = \frac{\partial T}{\partial t} \tag{6.277}$$

which is subject to the Newton-cooling-type boundary conditions

$$-k_1 \frac{\partial T}{\partial x} + h_1 T = 0 \quad ; \quad x = 0 \quad ; t > 0 \tag{6.278}$$

$$k_2 \frac{\partial T}{\partial x} + h_2 T = 0 \quad ; \quad x = L \quad ; t > 0 \tag{6.279}$$

where k_1, k_2, etc. are arbitrary constants, and the initial condition

$$T(x, 0) = \Omega(x) \quad ; \quad 0 \leq x \leq L \tag{6.280}$$

Solution

A variables separable solution of the form

$$T(x, t) = T(t)X(x) \tag{6.281}$$

gives $T(t) = e^{-\alpha\beta^2 t}$ and $X(x)$ satisfies the reduced problem

$$\frac{d^2 X}{dx^2} + \beta^2 X = 0 \quad ; \quad 0 < x < L \tag{6.282}$$

and the corresponding boundary conditions applicable to (6.282) are

$$-k_1 \frac{dX}{dx} + h_1 X = 0 \quad ; \quad x = 0 \quad ; \quad t > 0 \tag{6.283}$$

$$k_2 \frac{dX}{dx} + h_2 X = 0 \quad ; \quad x = L \quad ; \quad t > 0 \tag{6.284}$$

The equation (6.282) together with (6.283) and (6.284) constitutes a special case of the generalized Sturm-Liouville problem discussed previously with

$$p(x) = 1 \quad ; \quad r(x) = 1 \quad ; \quad q(x) = 0 \quad ; \quad \lambda = \beta^2 \tag{6.285}$$

If $X(\beta_m, x) = X_m(x)$ and $X(\beta_n, x) = X_n(x)$ are orthogonal eigenfunctions then

$$\int_0^L X_m(x) X_n(x) dx = \begin{cases} \Delta(\beta_m) & \text{for} \quad m = n \\ 0 & \text{for} \quad m \neq n \end{cases} \tag{6.286}$$

where $\Delta(\beta_m)$ is referred to as the *norm*; i.e.

$$\Delta(\beta_m) = \int_0^L [X(\beta_m, x)]^2 dx \tag{6.287}$$

Assuming that the eigenfunctions $X(\beta_m, x)$ are known, we can construct the general solution of the initial boundary value problem posed by (6.277) to (6.280) as follows:

$$T(x,t) = \sum_{m=1,2,}^{\infty} A_m \;\; X(\beta_m, x) e^{-\alpha\beta_m t} \tag{6.288}$$

where A_m are arbitrary constants. In order to determine the arbitrary constants A_m, it is necessary to use the initial condition (6.280), which gives

$$\Omega(x) = \sum_{m=1,2}^{\infty} A_m \;\; X(\beta_m, x) \quad ; \quad 0 \leq x < L \tag{6.289}$$

This is effectively the representation of $\Omega(x)$ in terms of the eigenfunctions

$X(\beta_m, x)$ of the eigenvalue problem defined by (6.282), (6.283) and (6.284). If this representation is admissible, the coefficients A_m can be obtained by multiplying both sides of the equation (6.289) by $X(\beta_m, x)$ and integrating between the limits 0 to L. Performing this operation we obtain

$$A_m = \frac{\int_0^L X(\beta_m, x)\Omega(x)dx}{\int_0^L [X(\beta_m, x)]^2 dx} \tag{6.290}$$

and the denominator of (6.290) is $\Delta(\beta_m)$ defined by (6.287). The general solution of the initial boundary value problem can be represented in the form

$$T(x,t) = \sum_{m=1,2,} e^{-\alpha\beta_m^2 t} \frac{X(\beta_m, x)}{\Delta(\beta_m)} \int_0^L X(\beta_m, \xi)\Omega(\xi)d\xi \tag{6.291}$$

where the integration parameter is changed from x to ξ. As $t \to 0$,

$$\Omega(x) = \sum_{m=1,2} \frac{X(\beta_m, x)}{\Delta(\beta_m)} \int_0^L X(\beta_m, \xi)\Omega(\xi)d\xi \tag{6.292}$$

It can be shown that the series (6.292) converges at each point x in the interval $0 \leq x \leq L$ where $\Omega(x)$ is continuous, provided that $\Omega(x)$ and $d\Omega/dx$ are sectionally continuous on the interval $x = 0$ to $x = L$.

In order to complete the solution to the initial boundary value problem defined by the (6.277) to (6.280) and reduced to the Sturm-Liouville problem (6.282) to (6.284) we need to utilize explicit expressions for $X(\beta_m, x)$. Considering (6.282), the solution $X(\beta_m, x)$ is chosen as

$$X(\beta_m, x) = A_m \cos(\beta_m x) + B_m \sin(\beta_m x) \tag{6.293}$$

To satisfy the boundary condition (6.283) at $x = 0$ and $x = L$ we require

$$h_1 A_m - k_1 \beta_m B_m = 0 \tag{6.294}$$

and

$$\begin{aligned} &\{h_2 \cos(\beta_m L) - \beta_m k_2 \sin(\beta_m L)\} A_m \\ &+\{h_2 \sin(\beta_m L) + \beta_m k_2 \cos(\beta_m L)\} B_m = 0 \end{aligned} \tag{6.295}$$

For a non-trivial solution (i.e. non-zero A_m and B_m) we require

$$\begin{vmatrix} h_2 \cos(\beta_m L) - \beta_m k_2 \sin(\beta_m L) & h_2 \xi \sin(\beta_m L) + \beta_m k_2 \cos(\beta_m L) \\ h_1 & -k_1 \beta_m \end{vmatrix} = 0 \tag{6.296}$$

Alternatively we require

$$\tan(\beta_m L) = \frac{\beta_m (h_2 k_1 + h_1 k_2)}{(\beta_m^2 k_1 k_2 - h_1 h_2)} \tag{6.297}$$

The solution of the transcendental equation (6.297) can be used to determine the positive roots which correspond to β_m. Once the roots β_m are known, it is possible to evaluate other parameters associated with the solution.

6.5 Separation of variables method for spatially two-dimensional problems

The basic procedures involving the separation of variables technique can be readily adopted to examine spatially two-dimensional problems involving the diffusion equation. Spatially two-dimensional problems can be formulated both in relation to rectangular Cartesian coordinates and plane polar coordinates depending upon the geometry of the problem under consideration.

6.5.1 Spatially two-dimensional problems - Cartesian coordinates

We consider spatially two-dimensional problems associated with the diffusion equation which are formulated in relation to a system of rectangular Cartesian coordinates (x, y). The basic type of problem associated with the diffusion equation can involve heat conduction in plates, pressure transients in porous layers in which the thickness is considerably smaller than the lateral dimensions and chemical diffusion in similar regions. We shall apply the separation of variables technique and the product of solutions method to obtain solutions to certain elementary problems which are formulated in relation to *rectangular regions* of finite extent and *quadrant regions* of infinite extent.

Example 6.10

Consider the problem of heat conduction in a thin rectangular plate which is kept between perfectly insulating surfaces to prevent any heat losses normal to the plane of the plate (Figure 6.21).

The plate is maintained at an arbitrary initial temperature distribution $T^*(x, y, 0)$ through insulation provided along the edges $x = 0, a$ and $y = 0, b$. At time $t = 0$, the free edges are exposed and maintained at zero temperature. Formulate the initial boundary value problem and derive an expression for the time-dependent variation of temperature within the plate.

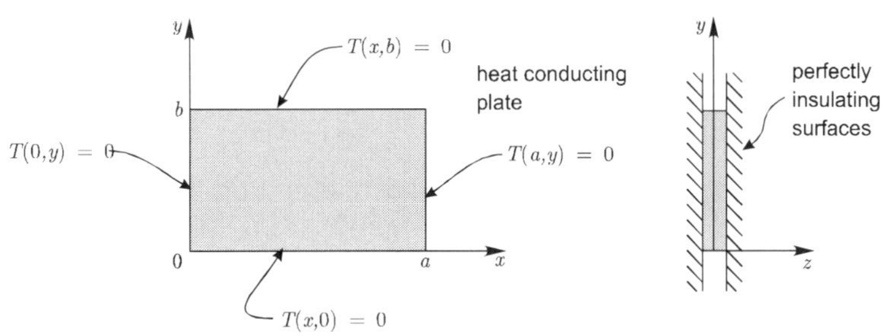

Figure 6.21: Heat conduction problem for an insulated plate

Solution

The partial differential equation governing the problem is

$$\alpha\left(\frac{\partial^2 T}{\partial x^2}+\frac{\partial^2 T}{\partial y^2}\right)=\frac{\partial T}{\partial t} \quad ; \qquad 0\le x\le a \quad ; \qquad 0\le y\le b \quad ; \qquad t>0 \tag{6.298}$$

The boundary conditions governing the problem are

$$T(0,y,t)=0 \quad ; \quad T(a,y,t)=0 \quad ; \quad t>0 \tag{6.299}$$

$$T(x,0,t)=0 \quad ; \quad T(x,b,t)=0 \quad ; \quad t>0 \tag{6.300}$$

The initial condition is

$$T(x,y,0)=T^*(x,y) \quad ; \quad 0\le x\le a \quad ; \quad 0\le y\le b \tag{6.301}$$

In addition, the temperature field $T(x,y,t)$ should reduce to zero as $t\to\infty$.

We seek a solution of the form

$$T(x,y,t)=e^{-\lambda t}F_1(x)F_2(y) \tag{6.302}$$

where λ is an undetermined positive parameter and $F_1(x)$ and $F_2(y)$ are arbitrary functions. Upon substituting (6.302) in (6.298) we obtain a representation in a variables separable form

$$\frac{d^2F_1/dx^2}{F_1}+\frac{\lambda}{\alpha}=-\frac{d^2F_2/dy^2}{F_2}=k^2 \tag{6.303}$$

This separates into two equations of the form

$$\frac{d^2F_1}{dx^2} + \xi^2 F_1 = 0 \tag{6.304}$$

$$\frac{d^2F_2}{dy^2} + k^2 F_2 = 0 \tag{6.305}$$

where

$$\xi^2 = \frac{\lambda}{\alpha} - k^2 \tag{6.306}$$

The equations (6.304) and (6.305) have solutions of the form

$$F_1(x) = A_1 \sin(\xi x) + B_1 \cos(\xi x) \tag{6.307}$$

$$F_2(y) = A_2 \sin(ky) + B_2 \cos(ky) \tag{6.308}$$

where A_1, A_2, etc., are arbitrary constants. Since the temperatures are zero on the boundaries of the plate, we require

$$B_1 = B_2 = 0 \tag{6.309}$$

and for a non-trivial solution, we require

$$\sin(\xi a) = 0 \tag{6.310}$$

and

$$\sin(kb) = 0 \tag{6.311}$$

These equations can be satisfied provided

$$\xi a = m\pi \quad \text{or} \quad \xi = \frac{m\pi}{a} \quad ; \quad m = 0, 1, 2, ... \tag{6.312}$$

$$kb = n\pi \quad \text{or} \quad k = \frac{n\pi}{b} \quad ; \quad n = 0, 1, 2, ... \tag{6.313}$$

and from (6.306) we have

$$\lambda_{mn} = \alpha \left[\left(\frac{m\pi}{a} \right)^2 + \left(\frac{n\pi}{b} \right)^2 \right] \tag{6.314}$$

For each value of λ_{mn} we have a solution which is now indeterminate to within an arbitrary constant C_{mn}; i.e.

$$T(x, y, t) = C_{mn} e^{-\lambda_{mn} t} \sin\left(\frac{m\pi x}{a}\right) \sin\left(\frac{n\pi y}{b}\right) \tag{6.315}$$

A summation of all such solutions over all possible values of m and n constitutes the complete solution, i.e.

$$T(x, y, t) = \sum_{m=1,}^{\infty} \sum_{n=1,}^{\infty} C_{mn} e^{-\lambda_{mn} t} \sin\left(\frac{m\pi x}{a}\right) \sin\left(\frac{n\pi y}{b}\right) \tag{6.316}$$

The arbitrary constants C_{mn} can now be determined by making use of the initial conditions (6.301); we have

$$T^*(x, y) = \sum_{m=1,}^{\infty} \sum_{n=1,}^{\infty} C_{mn} \sin\left(\frac{m\pi x}{a}\right) \sin\left(\frac{n\pi y}{b}\right) \tag{6.317}$$

The initial temperature can now be expanded in a double *sine* series. We multiply both sides (6.317) by $\sin(u\pi x/a)\sin(v\pi y/b)$ and integrate between the limits $x = 0, x = a$ and $y = 0, y = b$. Making use of the orthogonality properties of *sine* series, all the terms of the summation vanish except for the terms for which $m = u$ and $n = v$. We obtain

$$C_{uv} = \frac{4}{ab}\int_0^a \int_0^b T^*(x,y)\sin\left(\frac{u\pi x}{a}\right)\sin\left(\frac{v\pi y}{b}\right) dx dy \tag{6.318}$$

The solution of the heat conduction problem for the plate is now formally complete. The time-dependent distributions of the temperature within the plate can be determined from (6.316) and (6.318).

As an example consider the case when $T^*(x,y) \;=\; \text{const.} \;=\; T_0$. Using (6.318) we obtain

$$C_{mn} = 4T_0 \frac{[1-\cos(m\pi)]}{m\pi}\frac{[1-\cos(n\pi)]}{n\pi} \tag{6.319}$$

and the temperature distribution within the plate takes the form

$$T(x,y,t) = 4T_0 \sum_{m=1,}^{\infty}\sum_{n=1,}^{\infty}\left\{\frac{(1-\cos m\pi)(1-\cos n\pi)}{mn\pi^2}\right\}\sin\left(\frac{m\pi x}{a}\right)$$

$$.\sin\left(\frac{n\pi y}{b}\right)\exp\left\{-\alpha\left[\left(\frac{m\pi}{a}\right)^2+\left(\frac{n\pi}{b}\right)^2\right]t\right\} \tag{6.320}$$

Example 6.11

The plane faces of a particle board are provided with an impermeable membrane to prevent excessive drying due to changes in relative humidity. The edges of the particle board are also provided with impervious barriers and the initial distribution of volumetric moisture content within the particle board is $\Omega_0(x, y)$. The edges of the board are now removed such that the boundary conditions governing volumetric moisture content at the boundaries are as shown in Figure 6.22.

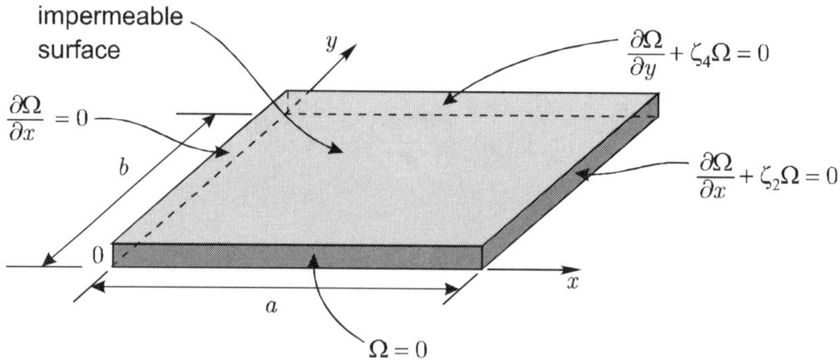

Figure 6.22: Moisture re-distribution problem for a particle board

If the partial differential equation governing the in-plane moisture transport in the particle board is given by

$$D\nabla^2\Omega = \frac{\partial\Omega}{\partial t} \tag{6.321}$$

where ∇^2 is Laplace's operator in two-dimensions and D is a moisture diffusivity parameter derive an expression for the time-dependent variation of $\Omega(x, y, t)$ within the board.

Solution

The initial boundary value problem governing the moisture transport problem can be formulated as follows: the partial differential equation governing the moisture transport problem is

$$D\left(\frac{\partial^2\Omega}{\partial x^2} + \frac{\partial^2\Omega}{\partial y^2}\right) = \frac{\partial\Omega}{\partial t} \quad ; \quad 0 \le x \le a \quad ; \quad 0 \le y \le b\,;\, t > 0 \tag{6.322}$$

The boundary conditions associated with the problem are

$$\frac{\partial \Omega}{\partial x} = 0, \quad \text{at} \quad x = 0 \quad ; \quad t > 0 \tag{6.323}$$

$$\frac{\partial \Omega}{\partial x} + \zeta_2 \Omega = 0, \quad \text{at} \quad x = a \quad ; \quad t > 0 \tag{6.324}$$

$$\Omega = 0, \quad \text{at} \quad y = 0 \quad ; \quad t > 0 \tag{6.325}$$

$$\frac{\partial \Omega}{\partial y} + \zeta_4 \Omega = 0, \quad \text{at} \quad y = b \quad ; \quad t > 0 \tag{6.326}$$

where ζ_2 and ζ_4 are constants and the initial condition is

$$\Omega(x, y, 0) \;=\; \Omega_0(x, y) \tag{6.327}$$

The boundary conditions (6.324) and (6.326) can be identified as surfaces which allow partial moisture efflux. This is analogous to the Newton-cooling-type boundary condition encountered in heat conduction problems. We can recover from these boundary conditions the limiting solutions to cases involving both impermeable boundaries ($\zeta_2 = \zeta_4 = 0$) or boundaries which maintain zero moisture conditions ($\zeta_2, \zeta_4 \to \infty$).

We assume a variables separable solution of the form

$$\Omega(x, y, t) = T(t)F(x)G(y) \tag{6.328}$$

Substituting (6.328) into (6.322) we obtain the separation equations for $F(x)$ and $G(y)$ as

$$\frac{d^2 F(x)}{dx^2} + \beta^2 F(x) = 0 \quad ; \quad 0 < x < a \tag{6.329}$$

with boundary conditions

$$\frac{dF}{dx} = 0 \quad \text{at} \quad x = 0 \quad ; \quad t > 0 \tag{6.330}$$

$$\frac{dF}{dx} + \zeta_2 F = 0 \quad \text{at} \quad x = a \quad ; \quad t > 0 \tag{6.331}$$

and

$$\frac{d^2G(y)}{dy^2} + \delta^2 G(y) = 0 \quad ; \quad 0 < y < b \tag{6.332}$$

with boundary conditions

$$G = 0 \quad \text{at} \quad y = 0 \quad ; \quad t > 0 \tag{6.333}$$

$$\frac{dG}{dy} + \zeta_4 G = 0 \quad \text{at} \quad y = b \quad ; \quad t > 0 \tag{6.334}$$

where β^2, δ^2 are positive parameters. The solution for $T(t)$ is given by

$$T(t) = e^{-D(\beta^2+\delta^2)t} \tag{6.335}$$

The general solution for $\Omega(x, y, t)$ can be constructed by summing up all the solutions associated with β and δ, i.e.

$$\Omega(x, y, t) = \sum_{m=1}^{\infty}\sum_{n=1}^{\infty} \tilde{A}_{mn} e^{-D(\beta_m^2+\delta_n^2)t} F(\beta_m, x) G(\delta_n, y) \tag{6.336}$$

Let us assume that we are able to determine the appropriate forms of the functions $F(\beta_m, x)$ and $G(\delta_n, y)$ by considering the solution of the reduced boundary value problems defined by (6.329) to (6.331) and (6.332) to (6.334) respectively. These boundary value problems are identical to the *Sturm-Liouville problems* discussed in Section 6.4.3.

If $F(\beta_m, x)$ and $G(\delta_n, y)$ are known, then the formal solution of the moisture diffusion problem is complete if we can determine the arbitrary constants $\tilde{A}_{mn}$. Invoking the initial condition (6.327) we have

$$\Omega(x, y, 0) = \Omega_0(x, y) = \sum_{m=1,}^{\infty} \sum_{n=1,}^{\infty} \tilde{A}_{mn} F(\beta_m, x) G(\delta_n, y) \tag{6.337}$$

The unknown coefficients $\tilde{A}_{mn}$ are determined by operating both sides of the equation (6.337) successively by

$$\int_0^a F(\beta_m, x) dx$$

and

$$\int_0^b G(\delta_n, y) dy$$

and by utilizing the orthogonality properties of eigen-functions. We thus obtain, after a change in the integration variables,

$$\tilde{A}_{mn} = \frac{1}{\Delta(\beta_m)\Delta(\delta_n)} \int_0^a \int_0^b F(\beta_m, \xi) G(\delta_n, \eta) \Omega_0(\xi, \eta) d\xi d\eta \tag{6.338}$$

where

$$\Delta(\beta_m) = \int_0^a [F(\beta_m, x)]^2 dx \tag{6.339}$$

$$\Delta(\delta_n) = \int_0^b [G(\delta_n, y)]^2 dy \tag{6.340}$$

The formal solution (6.337) now can be represented in the form

$$\Omega(x,y,t) = \sum_{m=1,}\sum_{n=1,} \frac{e^{-D(\beta_m^2+\delta_n^2)t} F(\beta_m,x)G(\delta_n,y)}{\Delta(\beta_m)\Delta(\delta_n)}.$$

$$\cdot \int_0^a \int_0^b F(\beta_m,\xi)G(\delta_n,\eta)\Omega_0(\xi,\eta)d\xi d\eta \tag{6.341}$$

It may be noted that the solution (6.341) is fairly general, in the sense that, the *eigen-functions are as yet unspecified* and can be chosen to reflect the boundary conditions applicable to the Sturm-Liouville-type problems defined by (6.329) to (6.331) and (6.332) to (6.334). Considering the ordinary differential equation (6.329), a solution of the form

$$F(\beta_m,x) = \cos(\beta_m x) \tag{6.342}$$

satisfies the differential equation and the boundary condition (6.330). Using the boundary condition (6.331) we obtain the transcendental equation

$$\beta_m \tan(\beta_m a) = \zeta_2 \tag{6.343}$$

from which the positive roots can be used to determine β_m.

Similarly, considering the ordinary differential equation (6.332), a solution of the form

$$G(\delta_n,y) = \sin(\delta_n y) \tag{6.344}$$

satisfies the differential equation and the boundary condition (6.333). Using the boundary condition (6.334) we obtain the transcendental equation

$$\delta_n \cot(\delta_n b) = -\zeta_4 \tag{6.345}$$

from which the positive roots can be used to determine δ_n. Also, we can evaluate the norms $\Delta(\beta_m)$ and $\Delta(\delta_n)$ by making use of (6.339), (6.342), (6.343) and (6.340), (6.344), (6.345) respectively. It can be shown that

$$\Delta(\beta_m) = \frac{1}{2}\left[\frac{a(\beta_m^2 + \zeta_2^2) + \zeta_2}{(\beta_m^2 + \zeta_2^2)}\right] \tag{6.346}$$

$$\Delta(\delta_n) = \frac{1}{2}\left[\frac{b(\delta_n^2 + \zeta_4^2) + \zeta_4}{(\delta_n^2 + \zeta_4^2)}\right] \tag{6.347}$$

It should be noted that the arbitrary functions for $F(\beta_m, x)$ and $G(\delta_n, y)$ defined by (6.342) and (6.344) will have arbitrary constants as multipliers. These can be suppressed by assuming that they are already incorporated in the definition of $\tilde{A}_{mn}$ given by (6.336). (Furthermore they are simply *eigenfunctions*)

Finally, the solution for the moisture distribution within the particle board region is given by

$$\begin{aligned}\Omega(x, y, t) = 4 \sum_{m=1,}^{\infty} \sum_{n=1,}^{\infty} e^{-D(\beta_m^2+\delta_n^2)t} \frac{(\beta_m^2 + \zeta_2^2)(\delta_n^2 + \zeta_4^2)}{[a(\beta_m^2 + \zeta_2^2) + \zeta_2][b(\delta_n^2 + \zeta_4^2) + \zeta_4]} \\ \cdot \cos(\beta_m x) \sin(\delta_n y) \int_0^a \int_0^b \cos(\beta_m \xi) \sin(\delta_n \eta) \Omega_0(\xi, \eta) d\xi d\eta\end{aligned} \tag{6.348}$$

6.5.2 Spatially two-dimensional problems - plane polar coordinates

The use of plane polar coordinates in the formulation of diffusion-type problems becomes more advantageous in situations where the natural boundaries of the domain of interest can be described in terms of the plane polar coordinate system. Such applications are largely restricted to circular, annular, sector regions and infinite regions bounded internally or externally by circular boundaries. In section 6.3.4 we discussed diffusion problems involving radial symmetry, which essentially involves a single spatial variable. The formulation and solution of problems which involve both coordinate variables r and θ, is in principle, quite straightforward and effectively utilizes the methodologies presented in Section 6.5.1 involving spatially two dimensional problems formulated in relation to the rectangular Cartesian coordinate system. The algebraic manipulation involving polar coordinates can, however, be a little more involved than it is with the Cartesian coordinate equivalent. The development of solutions can be greatly enhanced by the use of symbolic mathematical manipulations developed in mathematical software such as MAPLE®, MATHEMATICA® and others.

In the following, we shall develop certain solutions to spatially one-dimensional and spatially two-dimensional problems which are formulated in relation to the plane polar coordinate system (r, θ). As a preamble, however, we shall examine the structure of the variables separable solution in plane polar coordinates. We consider the diffusion equation

$$a^2\nabla^2\Psi = \frac{\partial \Psi}{\partial t} \tag{6.349}$$

where

$$\nabla^2 = \frac{\partial^2}{\partial r^2} + \frac{1}{r}\frac{\partial}{\partial r} + \frac{1}{r^2}\frac{\partial^2}{\partial \theta^2} \tag{6.350}$$

is Laplace's operator in plane polar coordinates. As indicated in section (6.4.1), the variables separable form

$$\Psi(r, \theta, t) = \chi(r, \theta)T(t) \tag{6.351}$$

leads to the Helmholtz equation for $\chi(r, \theta)$; i.e.,

$$\left(\frac{\partial^2\chi}{\partial r^2}+\frac{1}{r}\frac{\partial\chi}{\partial r}+\frac{1}{r^2}\frac{\partial^2\chi}{\partial\theta^2}\right)+\lambda^2\chi=0 \tag{6.352}$$

and

$$\frac{dT}{dt}+a^2\lambda^2T=0 \tag{6.353}$$

where λ^2 is a positive parameter. A repeated application of the variables separable form can be used to generate the solution of $\chi(r,\theta)$. We assume that

$$\chi(r,\theta)=R_\nu(r)\Theta(\theta) \tag{6.354}$$

Substituting (6.354) in (6.352) we obtain

$$\frac{1}{\Theta}\frac{d^2\Theta}{d\theta^2}=-\nu^2 \tag{6.355}$$

$$\frac{1}{R_\nu}\left[\frac{d^2R_\nu}{dr^2}+\frac{1}{r}\frac{dR_\nu}{dr}\right]-\frac{\nu^2}{r^2}=-\beta^2 \tag{6.356}$$

where ν^2 and β^2 are positive parameters. The elementary solutions associated with (6.353), (6.355) and (6.356) take the forms

$$R_\nu(\beta,r)\sim J_\nu(\beta r)\ \text{ and }\ Y_\nu(\beta r) \tag{6.357}$$

$$\Theta(\nu,\theta)\sim\sin(\nu\theta)\ \text{ and }\ \cos(\nu\theta) \tag{6.358}$$

$$T(t)\sim e^{-a^2\lambda^2t} \tag{6.359}$$

where

$$\lambda^2 = \beta^2 + \nu^2 \tag{6.360}$$

From a practical point of view, we cannot proceed to make use of these solutions in a general sense without examining specific initial boundary value problems. We can, however, investigate some general properties of a *Sturm-Liouville* problem resulting from the differential equation for $R_\nu(\beta, r)$, generated from the variables separation technique.

We consider the ordinary differential equation

$$\frac{d^2\varphi_\nu(r)}{dr^2} + \frac{1}{r}\frac{d\varphi_\nu(r)}{dr} + \left(\beta^2 - \frac{\nu^2}{r^2}\right)\varphi_\nu(r) = 0 \quad ; \quad a \leq r \leq b \tag{6.361}$$

which is subject to the generalized boundary conditions

$$\frac{-d\varphi_\nu}{dr} + \zeta_a\varphi_\nu = 0 \quad ; \quad r = a \tag{6.362}$$

$$\frac{d\varphi_\nu}{dr} + \zeta_b\varphi_\nu = 0 \quad ; \quad r = b \tag{6.363}$$

The generalized nature of the boundary conditions (6.362) and (6.363) allows consideration of both Dirichlet ($\zeta_i \to \infty \; ; \; i = a, b$) and Neumann ($\zeta_i \to 0 \; ; \; i = a, b$) boundary conditions as limiting cases.

The eigen-value problem defined by (6.361), (6.362) and (6.363) represents a *Sturm-Liouville* problem where the eigen-functions $R_\nu(\beta_m, r)$ are orthogonal in the interval $a \leq r \leq b$, with respect to the weighting function $w(r) = r$. Hence the eigen-functions have the following orthogonality properties:

$$\int_a^b r\varphi_\nu(\beta_m, r)\varphi_\nu(\beta_n, r)dr = \begin{cases} \Delta(\beta_m) \; ; & m = n \\ 0 \; ; & m \neq n \end{cases} \tag{6.364}$$

where the *norm*, $\Delta(\beta_m)$, is given by

$$\Delta(\beta_m) = \int_a^b r[\varphi_\nu(\beta_m, r)]^2 dr \tag{6.365}$$

As indicated previously the solution for $\varphi_\nu(\beta_m, r)$ invariably involves Bessel functions; therefore we can make use of the following relationship to evaluate the *norm*: i.e.

$$\int r B_k^2(\beta r) dr = \frac{r^2}{2}\left[\{B_k'(\beta r)\}^2 + \left(1 - \frac{k^2}{\beta^2 r^2}\right) B_k^2(\beta r)\right] \tag{6.366}$$

where $B_k(\beta r)$ is any Bessel function of the *first* or *second kind* of order k and $(\ )'$ indicates the derivative of the function with respect to its argument.

Now, in applying the initial condition to evaluate certain arbitrary constants that are encountered in the general solution it becomes necessary to express the initial condition in terms of the eigen-functions.

Let us examine the specific case when the initial condition has a form

$$\Psi(r, \theta, 0) = \varphi_0^*(r) \tag{6.367}$$

In this case we can assume that $\varphi^*(r)$ admits a representation

$$\varphi_0^*(r) = \sum_{m=1,2}^{\infty} A_m \varphi_\nu(\beta_m, r) \quad ; \quad a \le r \le b. \tag{6.368}$$

The coefficients A_m of the series expansion can be determined by operating both sides of the equation (6.368) by

$$\int_a^b r\varphi_\nu(\beta_n, r) dr$$

and invoking the orthogonality condition (6.364). We thus obtain

$$A_m = \frac{1}{\Delta(\beta_m)} \int_a^b r\varphi_\nu(\beta_m, r)\varphi_0^*(r)dr \tag{6.369}$$

Therefore, $\varphi_0^*(r)$ can be written as

$$\varphi_0^*(r) = \sum_{m=1,2,}^{\infty} \frac{1}{\Delta(\beta_m)} \varphi_\nu(\beta_m, r) \int_a^b \xi\varphi_\nu(\beta_m, \xi)\varphi_0^*(\xi)d\xi \quad ; \quad a \leq r \leq b \tag{6.370}$$

Let us now focus attention on the solution of $\Theta(\theta)$, (6.358) derived via the variables separation technique. The general form of the solution can be written as

$$\Theta(\nu, \theta) = B_\nu \sin(\nu\theta) + C_\nu \cos(\nu\theta) \tag{6.371}$$

where B_ν and C_ν are arbitrary constants. Again, the application of the initial condition could yield the result

$$\Psi(r, \theta, 0) = \Theta_0^*(\theta) \tag{6.372}$$

We can consider the representation of $\Theta_0^*(\theta)$ that is periodic in θ, with period 2π, as a summation of functions $\Theta(\nu, \theta)$ defined by (6.371):

$$\Theta_0^*(\theta) = \sum_{\nu=0,1,}^{\infty} \{B_\nu \sin(\nu\theta) + C_\nu \cos(\nu\theta)\} \tag{6.373}$$

If $\Theta_0^*(\theta)$ is to remain periodic, with period 2π, the constants ν associated with the separation scheme should take integer values; i.e., $\nu = 0, 1, 2, 3....$ To determine the arbitrary constants B_ν, we operate both sides of (6.373) by

$$\int_0^{2\pi} \sin(\omega\theta) d\theta$$

and make use of the orthogonality properties. We obtain

$$B_\nu = \frac{1}{\pi} \int_0^{2\pi} \Theta_0^*(\theta) \sin(\nu\theta) d\theta \quad ; \quad \nu = 0, 1, 2, \tag{6.374}$$

Similarly, to determine C_ν we operate both sides of equation (6.373) by

$$\int_0^{2\pi} \cos(\omega\theta) d\theta$$

and make use of the orthogonality conditions. We obtain

$$C_\nu = \frac{1}{2\pi} \int_0^{2\pi} \Theta_0^*(\theta) d\theta \quad ; \quad \nu = 0 \tag{6.375}$$

and

$$C_\nu = \frac{1}{\pi} \int_0^{2\pi} \Theta_0^*(\theta) \cos(\nu\theta) d\theta \quad ; \quad \nu = 1, 2, 3, \tag{6.376}$$

We can now express $\Theta_0^*(\theta)$ as

$$\Theta_0^*(\theta) = \frac{1}{2\pi} \int_0^{2\pi} \Theta_0^*(\xi) d\xi + \frac{1}{\pi} \sum_{\nu=1,} \int_0^{2\pi} \Theta_0^*(\xi) [\sin(\nu\theta) \sin(\nu\xi) + \cos(\nu\theta) \cos(\nu\xi)] d\xi \tag{6.377}$$

or

$$\Theta_0^*(\theta) = \frac{1}{\pi} \sum_{\nu=0,1,} \int_0^{2\pi} [\Theta_0^*(\xi) \cos \nu(\theta - \xi)] d\xi \ ; \quad 0 \le \theta \le 2\pi \tag{6.378}$$

The result (6.378) is generally valid for $\nu = 0, 1, 2, ...$ *provided* we replace multiplication factor $(1/\pi)$ by $(1/2\pi)$ when $\nu = 0$.

Example 6.12

Consider the radially symmetric problem of a long annular tube of porous material which is initially dosed with a chemical at a concentration C_0.

The tube is placed in an ambient moving fluid region which maintains a zero chemical concentration. The fabrication of the tube introduces a "skin effect" which maintains chemical diffusion at the boundaries $r = a$ and $r = b$ according to the relationships

$$-\frac{\partial C}{\partial r} + \zeta_a C = 0 \quad \text{and} \quad \frac{\partial C}{\partial r} + \zeta_b C = 0 \tag{6.379}$$

respectively. Formulate the initial boundary value problem and derive a relationship for the time-dependent radially symmetric variation of chemical concentration within the annular porous region.

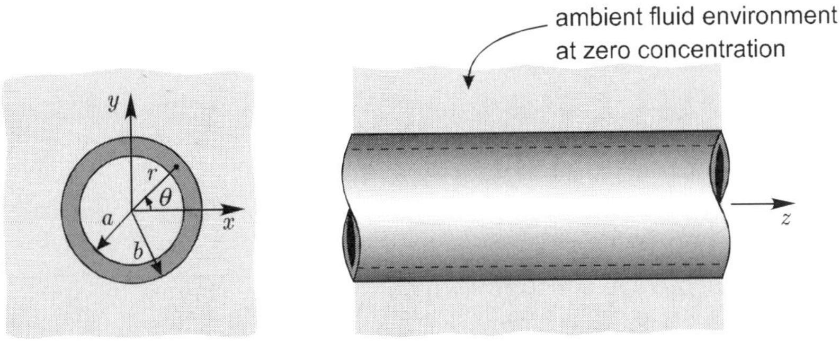

Figure 6.23: Diffusion of a chemical from an annular region

Solution

The partial differential equation governing the diffusion of the chemical is given by

$$D\left(\frac{\partial^2 C}{\partial r^2}+\frac{1}{r}\frac{\partial C}{\partial r}\right)=\frac{\partial C}{\partial t} \tag{6.380}$$

where D is a diffusion constant. The boundary conditions applicable to the inner and outer boundaries of the annular region are

$$-\frac{\partial C}{\partial r}+\zeta_a C=0 \quad ; \quad r=a \tag{6.381}$$

$$\frac{\partial C}{\partial r}+\zeta_b C=0 \quad ; \quad r=b \tag{6.382}$$

The initial condition is

$$C(r,0)=C_0 \tag{6.383}$$

Owing to the radial symmetry of the problem, the variables separable solution outlined in the previous section gives the following form for the general solution, i.e.

$$C(r,t) = \sum_{m=1,2}^{\infty} A_m e^{-D\beta_m^2 t} R_0(\beta_m, r) \tag{6.384}$$

where A_m are arbitrary constants which can be determined by applying the initial condition (6.383). We have

$$C(r,0) = C_0 = \sum_{m=1,2}^{\infty} A_m R_0(\beta_m, r) \quad ; \quad a \le r \le b \tag{6.385}$$

Using the result (6.369), the constants A_m are given by

$$A_m = \frac{C_0}{\Delta(\beta_m)} \int_a^b r R_0(\beta_m, r) dr \tag{6.386}$$

and the *norm*, $\Delta(\beta_m)$, is given by

$$\Delta(\beta_m) = \int_a^b r[R_0(\beta_m, r)]^2 dr \tag{6.387}$$

The general result for the time-dependent variation of chemical concentration within the annular region is given by

$$\frac{C(r,t)}{C_0} = \sum_{m=1,2}^{\infty} \frac{e^{-D\beta_m^2 t}}{\Delta(\beta_m)} R_0(\beta_m, r) \int_a^b \xi R_0(\beta_m, \xi) d\xi. \tag{6.388}$$

The explicit form of the eigen-function $R_0(\beta_m, r)$ now needs to be determined by selecting the functions in a form which satisfies the boundary conditions (6.381) and (6.382).

Considering the functions (6.357) we can show that

$$R_0(\beta_m, r) = B_0 J_0(\beta_m r) - C_0 Y_0(\beta_m r) \tag{6.389}$$

where

$$B_0 = \beta_m Y_0'(\beta_m a) - \zeta_a Y_0(\beta_m a) \tag{6.390}$$

$$C_0 = \beta_m J_0'(\beta_m a) - \zeta_a J_0(\beta_m a) \tag{6.391}$$

identically satisfies the differential equation governing the eigen-function, (6.361), and the boundary condition (6.381). If the solution is to satisfy the boundary condition (6.382), then the eigen-values β_m should be the positive roots of the transcendental equation

$$B_0 C_0^* - B_0^* C_0 = 0 \tag{6.392}$$

where

$$B_0^* = \beta_m Y_0'(\beta_m b) + \zeta_b Y_0(\beta_m b) \tag{6.393}$$

$$C_0^* = \beta_m J_0'(\beta_m b) + \zeta_b J_0(\beta_m b) \tag{6.394}$$

Also the *norm* can be evaluated as

$$\Delta(\beta_m) = \frac{2[(\zeta_b^2 + \beta_m^2) C_0^2 - (\zeta_a^2 + \beta_m^2) C_0^{*2}]}{\pi^2 \beta_m^2 C_0^2} \tag{6.395}$$

This, formally completes the solution of the problem. The solution of the transcendental equation (6.392) provides the values of β_m, which in turn provide explicit expressions for $R_0(\beta_m, r)$. These eigen-functions can be utilized in the series representation for $C(r,t)$ given by (6.388) to generate the final solution.

Example 6.13

The initial temperature in a flat heat conducting disc, which has a radius b (Figure 6.24), is given by $T(r,\theta,0) = T_0(r/a)\sin(\nu\theta)$. For time $t > 0$, the outer boundary of the disc is allowed to dissipate heat into the environment, which is maintained at zero temperature, through imperfect insulation.

If the temperature boundary condition at the boundary of the disc can be characterized by a Newton cooling-type condition, formulate the initial boundary value problem and derive an expression for the time-dependent variation of temperature within the disc.

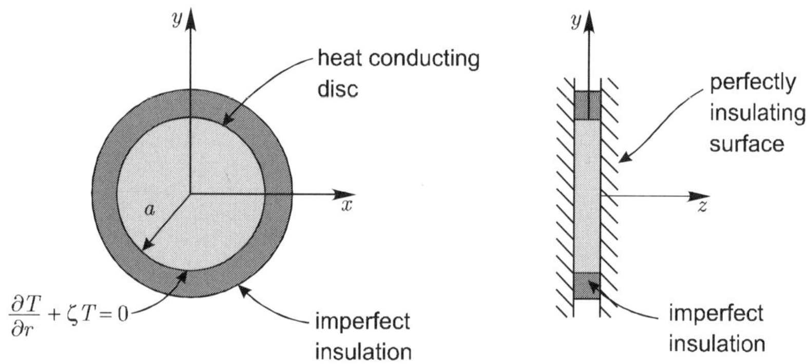

Figure 6.24: Heat conduction in a disc with an imperfect boundary insulation

Solution

The partial differential equation governing heat conduction in the disc is given by

$$\alpha\left(\frac{\partial^2 T}{\partial r^2} + \frac{1}{r}\frac{\partial T}{\partial r} + \frac{1}{r^2}\frac{\partial^2 T}{\partial \theta^2}\right) = \frac{\partial T}{\partial t} \tag{6.396}$$

The boundary condition governing the problem is

$$\frac{\partial T}{\partial r} + \zeta T = 0 \qquad \text{on} \qquad r = a \tag{6.397}$$

and the initial condition is

$$T(r,\theta,0) = T_0\left(\frac{r}{a}\right)\sin(\nu\theta) = T^*(r,\theta) \tag{6.398}$$

In addition to these conditions, the temperature field within the plate should be *finite* and *bounded*. Using the variables separation technique, the solution can be constructed by making use of the basic results (6.357), (6.358) and (6.359). We can write the solution in the most general form

$$T(r,\theta,t) = \sum_{m=1,}^{\infty}\sum_{\nu=0,}^{\infty} e^{-\alpha\beta_m^2 t}\left[B_{m\nu}\sin(\nu\theta) + C_{m\nu}\cos(\nu\theta)\right]R_\nu(\beta_m,r) \tag{6.399}$$

If we now apply the initial condition (6.398) we obtain

$$\begin{aligned} T(r,\theta,0) &= T_0\frac{r}{a}\sin(\nu\theta) \\ &= \sum_{m=1,}^{\infty}\sum_{\nu=0,}^{\infty}\left[B_{m\nu}\sin(\nu\theta) + C_{m\nu}\cos(\nu\theta)\right]R_\nu(\beta_m,r) \end{aligned} \tag{6.400}$$

If we operate both sides of (6.400) by

$$\int_0^a rR_\nu(\beta_n,r)dr$$

and make use of the orthogonality properties of the functions $R_\nu(\beta_m, r)$ we obtain

$$\int_0^a rR_\nu(\beta_m, r)T^*(r,\theta)dr = \Delta(\beta_m)\sum_{\nu=0}^{\infty}[B_{m\nu}\sin(\nu\theta) + C_{m\nu}\cos(\nu\theta)] \quad ; \quad 0 \le \theta \le 2\pi \tag{6.401}$$

and $\Delta(\beta_m)$ is the *norm.*

To determine the coefficients $B_{m\nu}$, we operate on both sides of the equation (6.401) by

$$\int_0^{2\pi} \sin(\nu^*\theta)d\theta$$

and make use of the orthogonality properties of function $\sin(\nu\theta)$. Hence we have

$$B_{m\nu} = \frac{1}{\pi\Delta(\beta_m)}\int_0^{2\pi}\left[\int_0^a rR_\nu(\beta_m, r)T_0\left(\frac{r}{a}\right)\sin(\nu\theta)dr\right]\sin(\nu\theta)d\theta \tag{6.402}$$

or

$$B_{m\nu} = \frac{T_0}{a\Delta(\beta_m)}\int_0^a \xi^2 R_\nu(\beta_m, \xi)d\xi \tag{6.403}$$

Also

$$\begin{aligned} C_{m\nu} &= \frac{1}{\pi\Delta(\beta_m)}\int_0^{2\pi}\left[\int_0^a rR_\nu(\beta_m, r)T_0\left(\frac{r}{a}\right)\sin(\nu\theta)dr\right]\cos(\nu\theta)d\theta \\ &\equiv 0 \end{aligned} \tag{6.404}$$

Hence the general solution (6.399) can now be written as

$$\frac{T(r,\theta,t)}{T_0} = \sum_{m=1,}^{\infty} \sum_{\nu=0,}^{\infty} \frac{e^{-\alpha\beta_m^2 t}}{a\Delta(\beta_m)} \left[\int_0^a \xi^2 R_\nu(\beta_m,\xi)d\xi\right] \sin(\nu\theta) R_\nu(\beta_m,r) \tag{6.405}$$

The eigen-functions $R_\nu(\beta_m, r)$ can be constructed by making use of the Bessel function solutions $J_\nu(\beta_m, r)$ and $Y_\nu(\beta_m r)$. If the temperature field is to be bounded and finite in the region $0 < r < a$, then $Y_\nu(\beta_m r)$ is an *inadmissible* solution. Hence, the appropriate eigen-function is

$$R_\nu(\beta_m, r) = J_\nu(\beta_m r) \tag{6.406}$$

In order to satisfy the boundary conditions (6.397) for all choices of $J_\nu(\beta_m r)$ we require

$$\beta_m J_\nu'(\beta_m a) + \zeta J_\nu(\beta_m a) = 0 \tag{6.407}$$

Equation (6.407) serves as the *transcendental equation* from which the positive roots are selected as values of β_m. From a knowledge of $R_\nu(\beta_m, r)$, the *norm* can be evaluated in an explicit form: i.e.

$$\Delta(\beta_m) = \int_0^a r[J_\nu(\beta_m r)]^2 dr = \frac{J_\nu^2(\beta_m a)[a^2(\zeta^2 + \beta_m^2) - \nu^2]}{2\beta_m^2} \tag{6.408}$$

6.5.3 Product solutions and solutions for infinite domains

The product solutions approach discussed in Section 6.4.2 can be utilized quite conveniently to develop solutions to diffusion problems associated with both infinite domains and finite domains. Although the application of the product solution technique discussed in the ensuing example is restricted to formulations involving rectangular Cartesian coordinates, the methodology can be easily applied to problems formulated in relation to other orthogonal curvilinear coordinate systems.

Example 6.14

A heat conducting plate has the shape of a *semi-infinite quarter plane* as shown in Figure 6.25. The plate is initially maintained at a constant temperature T^* and at time $t > 0$, the edges $x = 0$ and $y = 0$ of the plate are reduced to zero temperature.

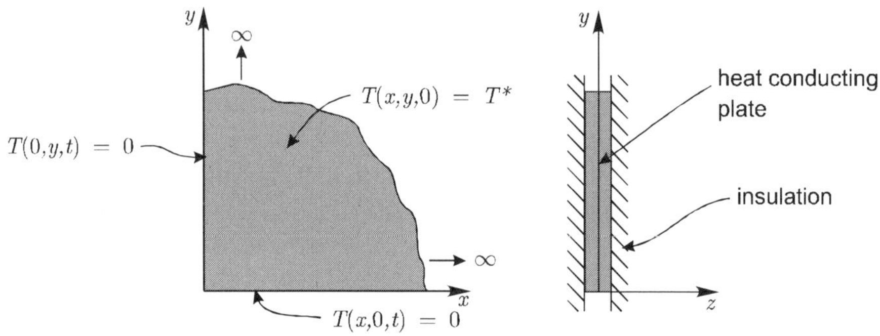

Figure 6.25: Heat conduction in a semi-infinite quarter plane

Formulate the initial boundary value problem and derive an expression for the time-dependent temperature distribution within the plate.

Solution

The partial differential equation governing heat conduction in the plate is

$$\alpha\left(\frac{\partial^2 T}{\partial x^2} + \frac{\partial^2 T}{\partial y^2}\right) = \frac{\partial T}{\partial t} \quad ; \quad 0 < x < \infty \quad ; \quad 0 < y < \infty \tag{6.409}$$

The temperature boundary conditions are

$$T(x,0,t) = 0 \quad ; \quad t > 0 \tag{6.410}$$

$$T(0,y,t) = 0 \quad ; \quad t > 0 \tag{6.411}$$

and the initial condition is

$$T(x,y,0) = T^* \tag{6.412}$$

The product solution has to be constructed by considering two, one-dimensional problems referred to semi-infinite plate regions which are governed by the following initial boundary value problems. These initial boundary value problems are formulated in relation to the semi-infinite heat conducting regions $0 \leq x < \infty$ (Figure 6.26a) and $0 \leq y < \infty$ (Figure 6.26b).

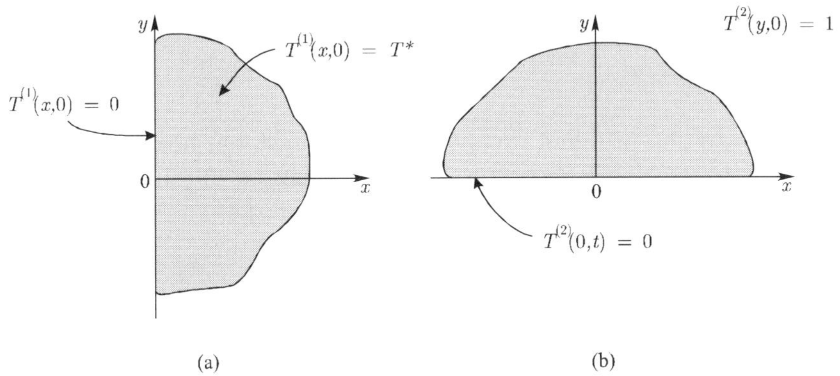

Figure 6.26: Sub-problems associated with the product solutions approach

Problem (1)

$$\alpha \frac{\partial^2 T^{(1)}}{\partial x^2} = \frac{\partial T^{(1)}}{\partial t} \quad ; \quad 0 \leq x < \infty \tag{6.413}$$

$$T^{(1)}(0,t) = 0 \quad ; \quad t > 0 \tag{6.414}$$

$$T^{(1)}(x,0) = T^{\,*} \quad ; \quad 0 \leq x < \infty \tag{6.415}$$

Problem (2)

$$\alpha \frac{\partial^2 T^{(2)}}{\partial y^2} = \frac{\partial T^{(2)}}{\partial t} \quad ; \quad 0 \leq y < \infty \tag{6.416}$$

$$T^{(2)}(0,t) = 0 \quad ; \quad t > 0 \tag{6.417}$$

$$T^{(2)}(y,0) = 1 \quad ; \quad 0 \leq y < \infty \tag{6.418}$$

The solution to the one-dimensional problem governed by (6.413) to (6.415) was presented in Example 6.1 involving heat conduction in a semi-infinite region which is maintained at a constant initial temperature (T_0) and subjected to a zero temperature boundary condition for $t > 0$. Using the solution (6.123) we obtain

$$T^{(1)}(x,t) = T^* \mathrm{erf}\left(\frac{x}{\sqrt{4\alpha t}}\right) \tag{6.419}$$

The solution to the problem involving one-dimensional heat conduction in the semi-infinite region $0 \leq y < \infty$, can similarly be obtained by a simple change the initial value and in the spatial variable; i.e.

$$T^{(2)}(y,t) = 1 \ \mathrm{erf}\left(\frac{y}{\sqrt{4\alpha t}}\right) \tag{6.420}$$

From the Theorem 6.1 involving product solutions, the time-dependent temperature distribution within the plate is given by

$$T(x,y,t) = T^{(1)}(x,t)\ T^{(2)}(y,t) \tag{6.421}$$

or

$$T(x, y, t) = T^* \text{erf}\left(\frac{x}{\sqrt{4\alpha t}}\right) \text{erf}\left(\frac{y}{\sqrt{4\alpha t}}\right) \tag{6.422}$$

which satisfies the boundary conditions on $x = 0$ and $y = 0$ and the initial condition (6.412).

• • •

We now consider the problem of diffusion in an infinite medium which is subjected to an initial condition which is radially symmetric. This a generalization of the problem considered in Example 6.3, to a radially symmetric one-dimensional domain.

Example 6.15

An infinite region of a saturated porous medium has a radially symmetric initial contaminant concentration $C(r, 0) = C^*(r)$ (Figure 6.27)

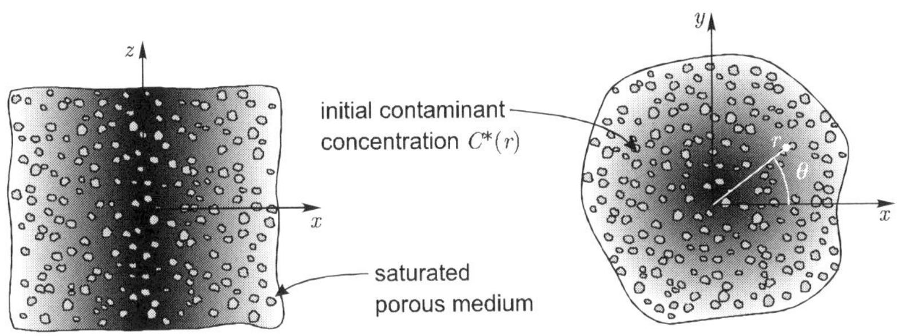

Figure 6.27: Diffusion of a contaminant in a porous medium

If the spatial movement of the contaminant is radially symmetric and governed by a diffusive-type process, develop an expression for the time-dependent variation of contaminant concentration as a function of the initial concentration $C^*(r)$.

Solution

The partial differential equation governing the radially symmetric diffusive transport of the contaminant is

$$D\left(\frac{\partial^2 C}{\partial r^2}+\frac{1}{r}\frac{\partial C}{\partial r}\right)=\frac{\partial C}{\partial r} \quad ; \quad 0\leq r<\infty \quad ; \quad t>0 \tag{6.423}$$

The initial condition for the problem is

$$C(r,0)=C^*(r) \tag{6.424}$$

Since the medium is of infinite extent, there are no "*boundary conditions*" associated with the problem. We can, however, impose some physically plausible constraints on the solution. The solution should be finite and bounded in the region $0 \leq r < \infty$, and should decay to zero as $r \to \infty$ and as $t \to \infty$. A separation of variables technique can be used to show that the time-dependency has a form similar to (6.359), i.e. $\exp(-D\beta^2 t)$ where β^2 is the separation constant. The spatial variable component of the solution $R_0(\beta, r)$ is governed by the differential equation

$$\frac{d^2 R_0}{dr^2}+\frac{1}{r}\frac{dR_0}{dr}+\beta^2 R_0=0 \tag{6.425}$$

If $R_0(\beta, r)$ is to remain finite as $r \to 0$, the only admissible solution is

$$R_0(\beta,r)=A(\beta)J_0(\beta r) \tag{6.426}$$

where $A(\beta)$ is an undetermined function. Since β^2 is an arbitrary parameter, we can construct a solution in an integral form as follows:

$$C(r,t)=\int_0^\infty A(\beta)e^{-D\beta^2 t}J_0(\beta r)d\beta \tag{6.427}$$

This general solution is complete to within an arbitrary function $A(\beta)$. We can determine this function by applying the known initial condition (6.424). i.e.

$$C(r,0) = C^*(r) = \int_0^\infty A(\beta) J_0(\beta r) d\beta \quad ; \quad 0 \le r < \infty \tag{6.428}$$

This is an integral representation of an arbitrary function $C^*(r)$, which is defined in the interval $0 \le r < \infty$, in terms of Bessel functions $J_0(\beta r)$. Using integral relationships involving Bessel functions, it can be shown that

$$C^*(r) = \int_0^\infty \beta J_0(\beta r) d\beta \int_0^\infty \xi J_0(\beta\xi) C^*(\xi) d\xi \tag{6.429}$$

From (6.428) and (6.429) it is evident that

$$A(\beta) = \beta \int_0^\infty \xi J_0(\beta\xi) C^*(\xi) d\xi \tag{6.430}$$

In obtaining this relationship it is implicitly assumed that the initial variation of the chemical concentration $C^*(r)$ is such that the integral in (6.430) is bounded. The general expression for the chemical concentration in the medium is given by

$$C(r,t) = \int_0^\infty e^{-D\beta^2 t} \beta J_0(\beta r) d\beta \int_0^\infty \xi C^*(\xi) J_0(\beta\xi) d\xi \tag{6.431}$$

A change in the order of integration gives

$$C(r,t) = \int_0^\infty \xi C^*(\xi) \left[\int_0^\infty \beta e^{-D\beta^2 t} J_0(\beta r) J_0(\beta\xi) d\beta \right] d\xi \tag{6.432}$$

We can obtain the inner integral of (6.432) in explicit form, i.e.

$$\int_0^\infty \beta e^{-D\beta^2 t} J_0(\beta r) J_0(\beta \xi) d\beta = \frac{1}{2Dt} \exp\left(-\frac{[r^2+\xi^2]}{4Dt}\right) I_0\left(\frac{\xi r}{2Dt}\right) \tag{6.433}$$

where I_0 is the *zeroth-order modified Bessel function of the first kind.* The result (6.432) can be written as

$$C(r,t) = \frac{1}{2Dt} \int_0^\infty \xi \exp\left(-\frac{[r^2+\xi^2]}{4Dt}\right) I_0\left(\frac{\xi r}{2Dt}\right) C^*(\xi) d\xi \tag{6.434}$$

In the particular instance when the contaminant is initially concentrated over a circular patch of radius a, i.e.

$$C(r,0) = C^*(r) = \begin{cases} C_0 & ; \quad 0 \le r \le a \\ 0 & ; \quad r > 0 \end{cases} \tag{6.435}$$

the equation (6.434) gives

$$\frac{C(r,t)}{C_0} = \frac{1}{2Dt} \exp\left(-\frac{r^2}{4Dt}\right) \int_0^b \xi \exp\left(-\frac{\xi^2}{4Dt}\right) I_0\left(\frac{\xi r}{2Dt}\right) d\xi \tag{6.436}$$

This result can be numerically evaluated to generate the contaminant concentration profiles in both space and time.

6.6 Uniqueness theorem

The proof of uniqueness of a solution is an important consideration in ensuring the well-posedness of the initial boundary value problem. The proof of the uniqueness of solution of an initial boundary value problem posed by a diffusion problem can be established by considering a relatively generalized initial value problem which can include Dirichlet, Neumann and Robin-type boundary conditions on separate regions of a closed boundary. From the point of view of the discussion of diffusion problems the time variable should also be considered in the domain of interest. The procedure adopted in proving the uniqueness theorem is again by a *reductio ad absurdum* approach (proof by contradiction).

THEOREM 6.2

Consider a region V with a closed boundary S made up of three parts S_D, S_N and S_M. In the region V, the function $\varphi(\mathbf{x}, t)$ satisfies the diffusion equation

$$a^2\nabla^2\varphi = \frac{\partial\varphi}{\partial t} \quad ; \quad \mathbf{x} \in V \quad ; \quad t > 0 \tag{6.437}$$

and the initial condition

$$\varphi(\mathbf{x}, 0) = f_0(\mathbf{x}) \quad ; \quad \mathbf{x} \in V \quad ; \quad t = 0 \tag{6.438}$$

The function $\varphi(\mathbf{x}, t)$ satisfies boundary conditions

$$\varphi = f_D(\mathbf{x}) \quad \text{on} \quad \mathbf{x} \in S_D \quad ; \quad t > 0 \tag{6.439}$$

$$\frac{\partial\varphi}{\partial n} = f_N(\mathbf{x}) \quad \text{on} \quad \mathbf{x} \in S_N \quad ; \quad t > 0 \tag{6.440}$$

$$\frac{\partial \varphi}{\partial n} + \zeta\varphi = f_M(\mathbf{x}) \qquad \text{on} \qquad \mathbf{x} \in S_M \quad ; \quad t > 0 \tag{6.441}$$

where $f_D(\mathbf{x}), f_N(\mathbf{x})$ and $f_M(\mathbf{x})$ are prescribed functions and ζ is a positive constant. The equations (6.439), (6.440) and (6.441) are, respectively, Dirichlet, Neumann and Robin or Newton-type boundary conditions, applied to the separate sub-regions of S (Figure 6.28) i.e.:

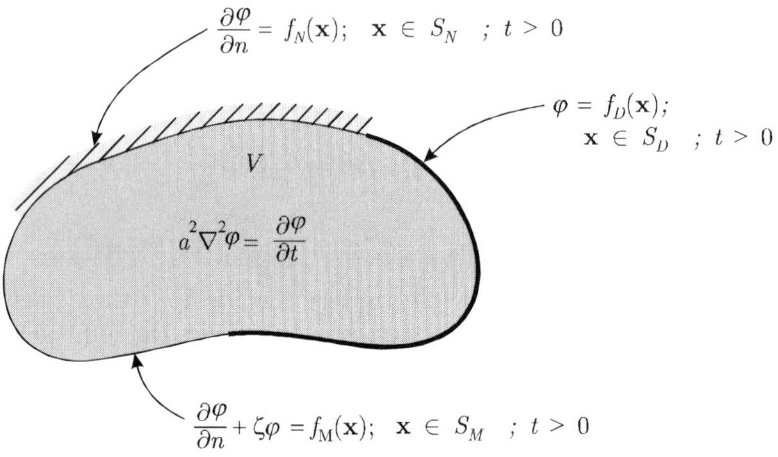

Figure 6.28: Spatial domain for the diffusion problem

$$S_D \cup S_N \cup S_M = S \qquad ; \qquad S_D \cap S_N \cap S_M = 0 \tag{6.442}$$

If a solution of $\varphi(\mathbf{x}, t)$ satisfying the diffusion equation (6.437), the initial condition (6.438) and the boundary conditions (6.439), (6.440) and (6.441) exists, then it is unique.

PROOF

Let us assume that there are two distinct solutions $\varphi_1(\mathbf{x}, t)$ and $\varphi_2(\mathbf{x}, t)$ satisfying the governing diffusion equation, the initial condition and boundary conditions i.e. if $\varphi_i(\mathbf{x}, t)$ $(i = 1, 2)$ are solutions, then

$$a^2\nabla^2\varphi_i = \frac{\partial \varphi_i}{\partial t} \quad ; \quad \mathbf{x} \in V \quad ; \quad t > 0 \tag{6.443}$$

$$\varphi_i = f_0(\mathbf{x}) \quad ; \quad \mathbf{x} \in V \quad ; \quad t = 0 \tag{6.444}$$

$$\varphi_i = f_D(\mathbf{x}) \quad ; \quad \mathbf{x} \in S_D \quad ; \quad t > 0 \tag{6.445}$$

$$\frac{\partial \varphi_i}{\partial n} = f_N(\mathbf{x}) \quad ; \quad \mathbf{x} \in S_N \quad ; \quad t > 0 \tag{6.446}$$

$$\frac{\partial \varphi_i}{\partial n} + \zeta\varphi_i = f_M(\mathbf{x}) \quad ; \quad \mathbf{x} \in S_M \quad ; \quad t > 0 \tag{6.447}$$

We now define a new solution

$$\varphi_3(\mathbf{x}, t) = \varphi_1(\mathbf{x}, t) - \varphi_2(\mathbf{x}, t) \tag{6.448}$$

This solution satisfies the following initial value problem

$$a^2\nabla^2\varphi_3 = \frac{\partial \varphi_3}{\partial t} \quad ; \quad \mathbf{x} \in V \quad ; \quad t > 0 \tag{6.449}$$

$$\varphi_3(\mathbf{x}, 0) = 0 \quad ; \quad \mathbf{x} \in V \quad ; \quad t = 0 \tag{6.450}$$

$$\varphi_3 = 0 \quad ; \quad \mathbf{x} \in S_D \quad ; \quad t > 0 \tag{6.451}$$

$$\frac{\partial \varphi_3}{\partial n} = 0 \quad ; \quad \mathbf{x} \in S_N \quad ; \quad t > 0 \tag{6.452}$$

$$\frac{\partial \varphi_3}{\partial n} + \zeta\varphi_3 = 0 \quad ; \quad \mathbf{x} \in S_M \quad ; \quad t > 0 \tag{6.453}$$

Now, from Green's first identity (1.45), we have (with $\varphi = \psi = \varphi_3$)

$$\int\int\int_V \varphi_3\nabla^2\varphi_3 \, dV + \int\int\int_V \mid \text{grad } \varphi_3 \mid^2 dV = \int\int_S \varphi_3 \frac{\partial \varphi_3}{\partial n} dS \tag{6.454}$$

Considering the right hand side of the equation (6.454) we have

$$\int\int_S \varphi_3 \frac{\partial \varphi_3}{\partial n} dS = \int\int_{S_D} \varphi_3 \frac{\partial \varphi_3}{\partial n} dS + \int\int_{S_N} \varphi_3 \frac{\partial \varphi_3}{\partial n} dS + \int\int_{S_M} \varphi_3 \frac{\partial \varphi_3}{\partial n} dS \tag{6.455}$$

Since either φ_3 and $\partial\varphi_3/\partial n$ are zero, respectively, on S_D and S_N, (6.453) and (6.455) give

$$\int\int_S \varphi_3 \frac{\partial \varphi_3}{\partial n} dS = -\zeta \int\int_{S_M} (\varphi_3)^2 dS \tag{6.456}$$

Also considering the partial differential equation (6.449) we can express this in a "*weak form*" as follows:

$$\int\int\int_V \varphi_3 \nabla^2 \varphi_3 \, dV = \frac{1}{a^2} \int\int\int_V \varphi_3 \frac{\partial \varphi_3}{\partial t} dV \tag{6.457}$$

Combining (6.454), (6.456) and (6.457) we obtain

$$\int\int\int_V \varphi_3 \frac{\partial \varphi_3}{\partial t} \, dV = -a^2 \left[\int\int\int_V \mid \text{grad } \varphi_3 \mid^2 dV + \zeta \int\int_{S_M} (\varphi_3)^2 dS \right] \tag{6.458}$$

If we denote by

$$F^*(t) = \frac{1}{2} \int\int\int_V (\varphi_3)^2 dV \tag{6.459}$$

then

$$\frac{dF^*}{dt} = \int\int\int_V \varphi_3 \frac{\partial \varphi_3}{\partial t} dV \tag{6.460}$$

We can write the result (6.458) in the form

$$\frac{dF^*}{dt} = -a^2 \left[\int\int\int_V \mid \text{grad } \varphi_3 \mid^2 dV + \zeta \int\int_{S_M} (\varphi_3)^2 dS \right] \tag{6.461}$$

An inspection of the above result indicates that F^* is always positive definite or zero (equation (6.459)) ; the time derivative of F^* is always negative or zero (equation (6.461) and at $t = 0, F^* = 0$ (equations (6.450) and (6.459)). The only way in which these constraints can be simultaneously satisfied is if $\varphi_3(\mathbf{x}, t) \equiv 0$.

Hence from (6.448)

$$\varphi_1(\mathbf{x}, t) = \varphi_2(\mathbf{x}, t) \tag{6.462}$$

and the solution to the initial boundary value problem or the Cauchy problem is unique.

6.7 A maximum principle

The application of the diffusion equation to a variety of problems encountered in heat conduction in solids, pressure transients in porous fluid saturated solids, drying of porous media and chemical diffusion in porous media was demonstrated by appeal to a number of examples. In developing these solutions, the assurances of a well-posed problem was the governing principle. Considerations of *boundedness* of the fields within spatial and temporal domains of interest was implicit in many of the solutions developed. These discussions were restricted to situations where singular fields were excluded and maximum values were usually prescribed at boundaries of the region. It is therefore natural to enquire whether a *maximum principle* can be developed to assure that the maximum value of the dependent variables, such as temperature, pressure, volumetric moisture content, chemical concentration etc., applicable to the region will always occur either within the region initially or on the boundary of the region. The notion that temperature, pressure, chemical concentration and other fields will "*diffuse*" within a medium is inherent in a maximum principle.

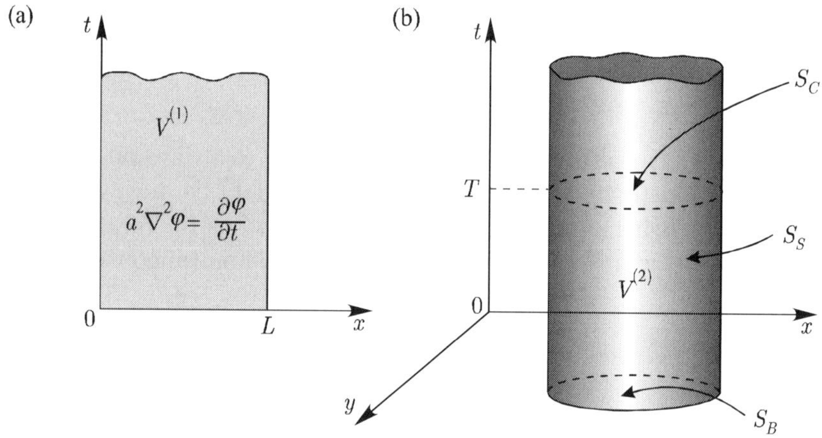

Figure 6.29: Domains for the applicability of the diffusion equation

We shall present the maximum principle along the general lines adopted for the development a maximum principle for Laplace's equation. We consider a bounded *open* region V in the $(\mathbf{x}, t)$ space. If we consider one spatial dimension, the region can correspond to the open-ended rectangular region $V^{(1)}$ shown in Figure 6.29(a). In two spatial dimensions, the region can correspond to the open ended cylindrical region $V^{(2)}$ shown in Figure 6.29(b). In three spatial dimensions, there is no simple visual representation of the domain V. We have to imagine a four dimensional space which consists of time and the three spatial dimensions.

For the purposes of illustration we shall consider Figure 6.29b and assume that $V^{(2)} \equiv V(\mathbf{x}, t)$. The boundary of the open region V, denoted by S_0, is obtained from the result

$$S_0 = S_B \cup S_S \tag{6.463}$$

We can consider a closed surface S by providing a closure S_C which is selected at $t = T$; i.e.

$$S = S_0 \cup S_C \tag{6.464}$$

The region contained with S is designated by $\overline{V}$.

THEOREM 6.3

The maximum principle states that if $\varphi(\mathbf{x}, t)$ is continuous in $\overline{V}$ and $(\partial\varphi/\partial t)$ and $\nabla^2\varphi$ which are continuous in V exist and satisfy

$$\frac{\partial \varphi}{\partial t} - a^2 \nabla^2 \varphi \leq 0 \tag{6.465}$$

then

$$\max_{(\mathbf{x},t)\in\overline{V}} [\varphi(\mathbf{x}, t)] = \max_{(\mathbf{x},t)\in S_0} [\varphi(\mathbf{x}, t)] \tag{6.466}$$

PROOF

We introduce an auxiliary function $\varphi^*(\mathbf{x}, t)$ defined by

$$\varphi^*(\mathbf{x}, t) = \varphi(\mathbf{x}, t) - \epsilon^* t \tag{6.467}$$

which is defined for $\epsilon^* > 0$, such that

$$\frac{\partial \varphi^*}{\partial t} - a^2 \nabla^2 \varphi^* = -\epsilon^* < 0 \tag{6.468}$$

Since $\varphi^*(\mathbf{x}, t)$ is continuous in $\overline{V}$, it attains its maximum value at some location $(\mathbf{x}_M, T_M)$. Let us assume that the point $(\mathbf{x}_M, T_M)$ is *not* located on S_0. Then, *either* $(\mathbf{x}_M, T_M)$ is in the interior of $\overline{V}$ so that it satisfies the conditions for local maxima; i.e.

$$\frac{\partial \varphi^*}{\partial t} = 0 \quad ; \quad \nabla^2 \varphi^* \leq 0 \quad \text{at} \quad (\mathbf{x}_M, T_M) \tag{6.469}$$

or, it is on S_C (see Figure 6.29b), so that it satisfies

$$\frac{\partial \varphi^*}{\partial t} \geq 0 \quad ; \quad \nabla^2 \varphi^* \leq 0 \quad \text{at} \quad (\mathbf{x}_M, T_M) \tag{6.470}$$

Both statements contradict (6.468), therefore we require

$$\max_{(\mathbf{x},t)\in\overline{V}} [\varphi^*(\mathbf{x}, t)] = \max_{(\mathbf{x},t)\in S_0} [\varphi^*(\mathbf{x}, t)] \tag{6.471}$$

for all $\epsilon^* > 0$.

From (6.467) we have

$$\max_{(\mathbf{x},t)\in\overline{V}} [\varphi(\mathbf{x}, t)] = \max_{(\mathbf{x},t)\in\overline{V}} [\varphi^*(\mathbf{x}, t)] + \epsilon^* t \tag{6.472}$$

From (6.471) and (6.472) we have

$$\max_{(\mathbf{x},t)\in\overline{V}} [\varphi(\mathbf{x}, t)] = \max_{(\mathbf{x},t)\in S_0} [\varphi^*(\mathbf{x}, t)] + \epsilon^* t \tag{6.473}$$

Considering the finite time T

$$\max_{(\mathbf{x},t)\in\overline{V}} [\varphi(\mathbf{x},t)] < \max_{(\mathbf{x},t)\in\overline{V}} [\varphi^*(\mathbf{x},t)] + \epsilon^* T \quad =$$

$$\max_{(\mathbf{x},t)\in S_0} [\varphi^*(\mathbf{x},t)] + \epsilon^* T \leq \max_{(\mathbf{x},t)\in S_0} [\varphi(\mathbf{x},t)] + \epsilon^* T \tag{6.474}$$

In the limit when $\epsilon^* \to 0$, we obtain the identity (6.466).

6.8 PROBLEM SET 6

6.1 The solutions to a one-dimensional diffusion problem is governed by the partial differential equation

$$\alpha \frac{\partial^2 U}{\partial x^2} = \frac{\partial U}{\partial t}.$$

The domain of applicability of the solution is $0 \leq x \leq L$ and $t > 0$. The solutions of the partial differential equation applicable to two problems are given by

(i) $U(x,t) = \dfrac{e^{-(x^2/4\alpha t)}}{\sqrt{t}} + e^{(x+\alpha t)}$

(ii) $U(x,t) = x^3 t + 3\alpha t^2 x + \dfrac{x^5}{20\alpha}$

Prove that these are solutions of the governing partial differential equation and establish the boundary value problem that is solved if either (a) Dirichlet or (b) Neumann boundary conditions are prescribed at the ends $x = 0$ and $x = L$.

6.2 A heat conducting plate of width $2a$ extends to infinity in one direction (Figure 6.30). This plate is incorporated with heating wire elements which induces an initial temperature distribution of the form

$$T(y,0) = T_0 \cos\left(\frac{\pi y}{2a}\right)$$

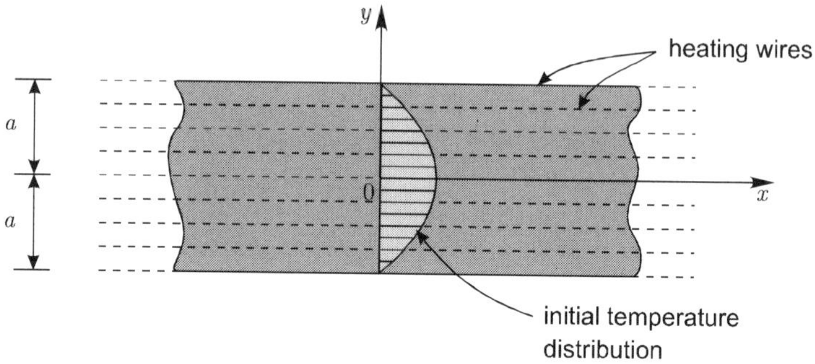

Figure 6.30: Internal heating of plate

(i) The power supply to the plate is terminated instantly and the boundaries $y = \pm a$, are maintained at the initial temperature. Determine an expression for the time-dependent variation of temperature within the plate.

(ii) The power supply to the plate is terminated instantly and the boundaries $y = \pm a$, are *insulated* to prevent any heat loss. Determine an expression for the time-dependent variation of temperature within the plate.

6.3 The Figure 6.31 shows a cross section through a composite heat conducting plate of large extent which contains heating wires and associated temperature controllers which induce a constant temperature T_0 over the region $0 < y < a$. The surface $y = 0$ is perfectly insulated and the surface $y = 2a$ is maintained at zero degrees to achieve a steady state temperature distribution over the thickness of the plate. The heating elements are switched off and the temperature at the surface $y = 2a$ is instantly raised to T_s. Formulate the initial boundary value problem and derive an expression for the time-dependent variation of temperature within the thickness of the plate.

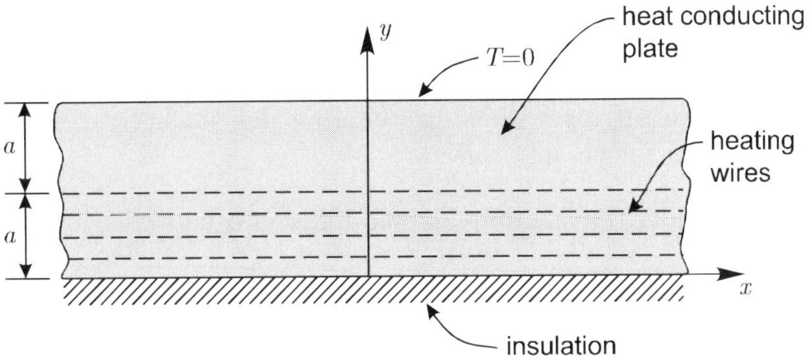

Figure 6.31: Heat conduction in a composite plate.

6.4 A heat conducting solid circular rod of length L and diameter D is insulated on the cylindrical surface and maintained at a uniform steady state temperature gradient with a constant temperature T_0 at the location $x = 0$ and at T_L at the location $x = L$ (Figure 6.32).

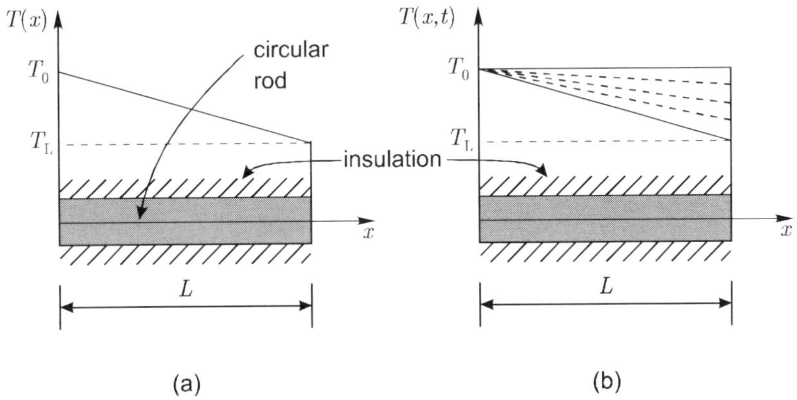

Figure 6.32: Heat conduction in an insulated circular rod

When the heat conduction processes have reached a steady state (Figure 6.32 a) the end $x = L$ is provided with insulation and the temperature at $x = 0$ is maintained at T_0. Formulate the initial boundary value problem giving the governing differential equation, boundary conditions and initial conditions.

Derive an expression for the time-dependent rise in temperature within the circular rod.

6.5 Consider the one-dimensional diffusion equation

$$\alpha \frac{\partial^2 \varphi}{\partial x^2} = \frac{\partial \varphi}{\partial t} \quad ; \quad 0 \leq x \leq L \quad ; \quad t > 0$$

subject to the boundary and initial conditions

$$\varphi(0, t) = 0 \quad ; \quad \left(\frac{\partial \varphi}{\partial x}\right)_{x=L} = 0 \quad ; \quad \varphi(x, 0) = \varphi_0 = \text{const.}$$

Formulate the Sturm-Liouville problem resulting from the development of a variables separable solution for $\varphi(x, t)$ of the form $\varphi(x, t) = X(x)T(t)$. Hence obtain the solution of $\varphi(x, t)$ in a series form.

6.6 Consider the generalized Sturm-Liouville problem defined by the differential equation

$$\frac{d}{dx}\left[p(x)\frac{d\varphi}{dx}\right] + [q(x) + \lambda r(x)]\,\varphi = 0 \quad ; \quad a \leq x \leq b$$

and subject to the boundary conditions

$$k_1 \varphi(a) + h_1 \left[\frac{d\varphi}{dx}\right]_{x=a} = 0$$

$$k_2 \varphi(b) + h_2 \left[\frac{d\varphi}{dx}\right]_{x=b} = 0$$

Show that the eigenvalues of the Sturm-Liouville system are real if $p(x)$, $q(x)$, $r(x)$, k_1, k_2, h_1 and h_2 are real.

6.7 Consider the eigenvalue problem associated with the Sturm-Liouville problem

$$\frac{d^2 X}{dx^2} + \beta^2 X = 0$$

subject to boundary conditions

$$k_1 X + h_1 \frac{dX}{dn} = 0 \qquad x = 0, L$$

Derive an expression for the *norm* which is defined by

$$\Delta(\beta_m) = \int_0^L [X(\beta_m, x)]^2 dx$$

6.8 A theory for the one-dimensional pore fluid pressure transient $p(x,t)$ in a fluid saturated porous geomaterial, first proposed by K. Terzaghi (1883-1963), is governed by a diffusion-type equation of the form

$$C_v \frac{\partial^2 p}{\partial y^2} = \frac{\partial p}{\partial t}$$

where C_v is a diffusion coefficient.

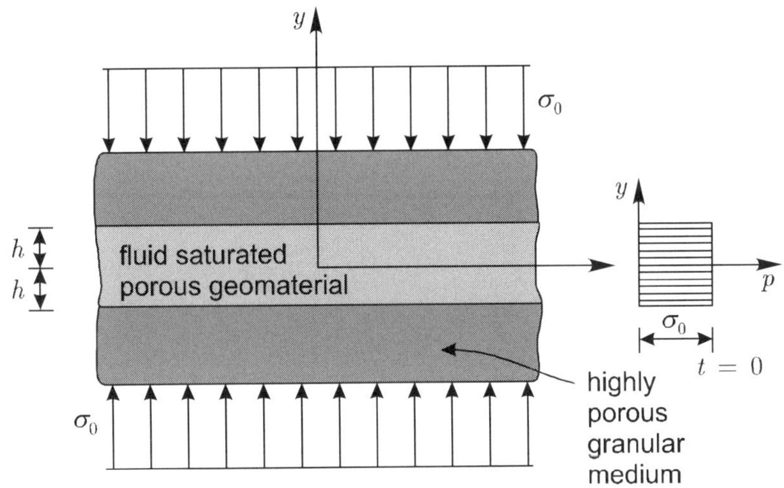

Figure 6.33: The loading of a fluid saturated porous geomaterial

In this one-dimensional diffusion model, the external stress σ_0 that is applied to the surface of the system (Figure 6.33) results in an immediate increase in fluid pressure in the fluid saturated medium. The excess fluid pressure will cause flow of water from the fluid saturated porous geomaterial to the highly porous granular medium. If the pressure at the boundaries of the fluid saturated porous geomaterial satisfy the boundary conditions

$$p(h,t) = 0 \quad ; \quad p(-h,t) = 0$$

derive an expression for time-dependent decay of the pressure.

6.9 The Figure 6.34 shows the surface of a fluid saturated porous medium of semi-infinite extent and occupying the region $0 \leq x \leq \infty$; $-\infty < y < \infty$; $-\infty < z < \infty$. The initial fluid pressure in the porous medium is given by

$$p(x,0) = \rho g(h+x)$$

The fluid layer of thickness h is suddenly removed. If the decay of pressure within the porous medium due to this perturbation is governed by a diffusion equation of the type

$$a^2 \frac{\partial^2 p}{\partial x^2} = \frac{\partial p}{\partial t}$$

where

$$a^2 = \frac{K}{\gamma(n^* C_f + C_s)}$$

K is the hydraulic conductivity of the porous medium; $\gamma(= \rho g)$ is the

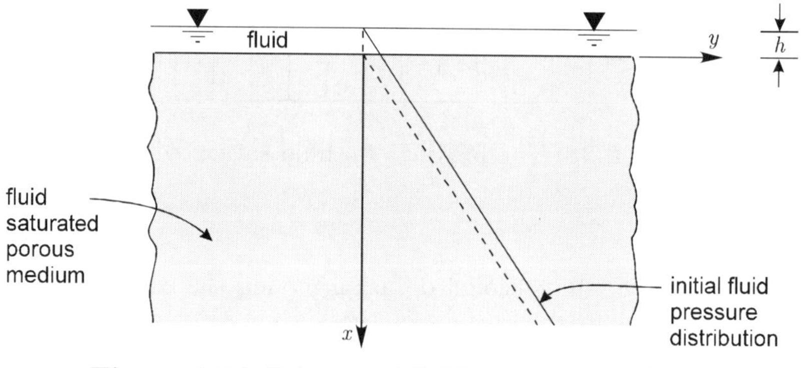

Figure 6.34: Submerged fluid saturated medium

unit weight of the fluid; n^* is the porosity of the porous medium and C_f and C_s are the compressibilities of the fluid and the solid fabric. Derive an expression for the time-dependent variation of pore fluid pressure in the porous medium.

6.10 The Figure 6.35 shows a submerged porous dyke of narrow width which is contained between basaltic impermeable rocks. Both the porous material (C_s) in the dyke and the pore fluid (C_f) are assumed to have finite compressibilities. The low permeability (K) of the porous material (porosity n^*) gives rise to time-dependent pore pressure diffusion which is governed by the partial differential equation

$$a^2 \nabla^2 p = \frac{\partial p}{\partial t};$$

where a^2 is a diffusivity coefficient for the pore pressure

$$a^2 = K/\gamma_w \left(n^* C_f + C_s\right)$$

Tidal fluctuations result in a variation of H in the form

$$H\left(t\right) = H_0 + H^* \sin\left(\omega t\right) \quad ; \quad H^* < H_0.$$

Derive an expression for the time-dependent variation of pore fluid pressure in the dyke.

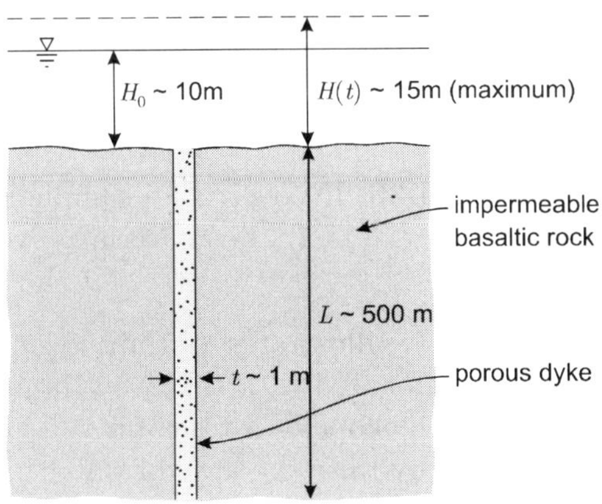

Figure 6.35: Porous dyke subjected to tidal actions.

6.11 The Figure 6.36 shows the cross section a saturated porous geomaterial from which a cylindrical sample is recovered by inserting a coring device which maintains perfect contact between the recovered core and the side wall of the sampling tube. The sample is sealed immediately and stored in a constant temperature chamber. Pore pressure redistribution takes place by pressure diffusion, primarily in the z-direction. The pore pressure diffusion is governed by the partial differential equation

$$a^2 \frac{\partial^2 p}{\partial z^2} = \frac{\partial p}{\partial t}$$

Determine the time-dependent redistribution of the fluid pressure within the sample if

(i) both ends of the sampling tube are perfectly sealed;

(ii) end A of the sampling tube is accidentally left open to the atmosphere.

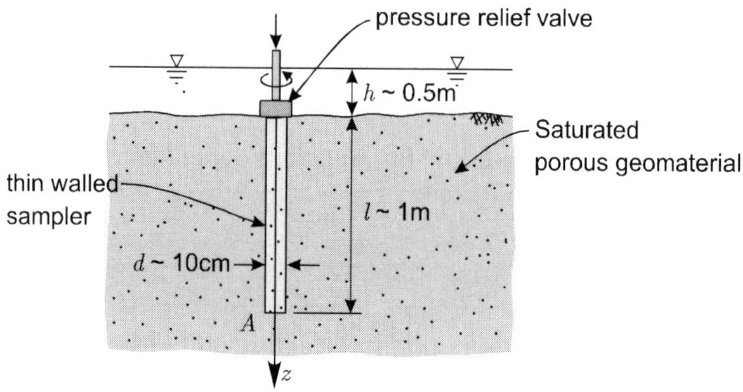

Figure 6.36: Recovery of a sample from a saturated geomaterial.

6.12 (i) The Figure 6.37 illustrates a thin layer of porous sandstone ($t << l$) which is located in a relatively impermeable parent rock, which forms the side slope of a reservoir. The geotechnical engineers intend to temporarily raise the level of water in the reservoir without affecting the stability of the weathered parent rock zone. To assess this, it is necessary to establish the rate of increase of pore fluid pressure in the porous sandstone region. The pore pressure diffusion in a porous region is governed by the general partial differential equation

$$a^2 \nabla^2 p = \frac{\partial p}{\partial t}$$

where

$$a^2 = \frac{K}{\gamma_w \left[n^* C_f + C_s \right]}$$

and K is the permeability of the porous sandstone; γ_w is the unit weight of water; n^* is the porosity; C_f is the compressibility of the pore water and C_s is the compressibility of the porous skeleton of the sandstone.

Derive an expression for the distribution of pore water pressure in the porous seam if the water level in the reservoir is suddenly raised to a height $D(t) = D_0 H(t)$ where $H(t)$ is the Heaviside step function.

(ii) Consider the idealized situation where the porous sandstone stratum is considered to be semi-infinite where $x \in (0, \infty)$. Derive an expression for the pore pressure distribution in the stratum if the water level $D(t)$ (measured above the initial water level h_0) is *gradually raised* according to the time-dependent variation

$$D(t) = h_0 \left[1 - e^{-\lambda t} \right]$$

where λ is a constant.

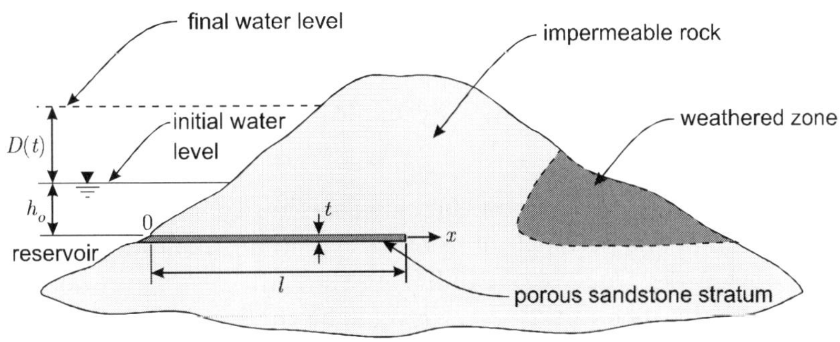

Figure 6.37: Porous sandstone seam embedded in impermeable rock.

6.13 The Figure 6.38 shows a cylindrical cavity which is drilled into a geological stratum. The stratum contains a saturated narrow porous seam which is contaminated with a chemical at a concentration C_0. The remediation of the porous seam involves the slow circulation of fresh water which will maintain the exposed cylindrical surface of the porous seam at zero concentration.

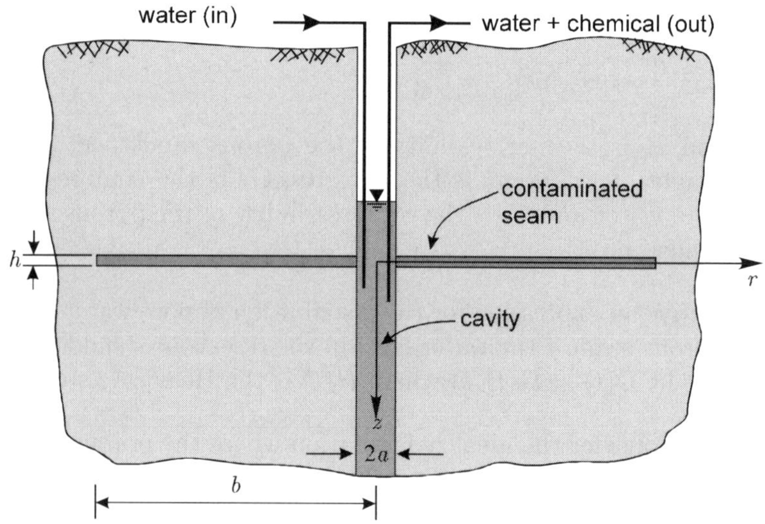

Figure 6.38: Radial diffusion in a contaminated porous seam

Assume that Fickian diffusive processes are present and that the problem is radially symmetric (i.e. the diffusion concentration is uniform over the thickness t of the contaminated seam.) Formulate the radially symmetric initial boundary value problem that should be solved to determine the radial distribution of chemical concentration $C(r,t)$. Derive an expression for the time-dependent radial distribution of the chemical along the porous seam. Hence obtain an expression for the chemical mass extracted with time. (The results can be presented in integral form).

6.14 The Figure 6.39 shows the cross section through a geological profile where a *low* permeability porous medium is contained within a *gravel* with a very *high* permeability. Both media rest on an *impervious* rock stratum. The initial water level is at the ground surface, $y = D$. The ground water level is instantly lowered to the level of the impervious rock stratum (Figure 6.39). The two-dimensional pore fluid pressure transient in the low permeability geological medium is governed by the partial differential equation

$$a^2\nabla^2 p = \frac{\partial p}{\partial t}$$

where a^2 is a pore fluid pressure diffusion coefficient which is positive and ∇^2 is Laplace's operator in rectangular Cartesian coordinates. Formulate the initial boundary value problem and derive an expression for the time-dependent variation of pore fluid pressure within the medium.

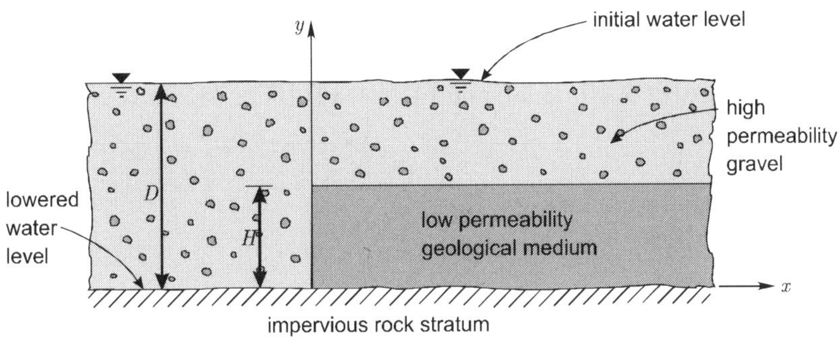

Figure 6.39: Pore fluid pressure transients in an embedded geological stratum

6.15 The Figure 6.40 shows the cross section a heat conducting solid region ($0 \leq x \leq L$; $0 \leq y \leq \infty$) where the boundary $y = 0$ is insulated and the temperatures on planes $x = 0$ and $x = L$ are maintained at *zero* and T_0, respectively, to achieve a steady state temperature distribution. At time $t = 0$ (Figure 6.40b) (i) the insulation at the boundary $y = 0$ is removed and this boundary is subjected to *zero* temperature, (ii) the boundary $x = 0$ is maintained at *zero* temperature, and (iii) the boundary $x = L$ is subjected to a *Robin*-type boundary condition

$$\frac{\partial T}{\partial n} + \zeta T = 0$$

where ζ is a positive constant.

Develop the solution for the time-dependent variation of temperature within the heat conducting solid.

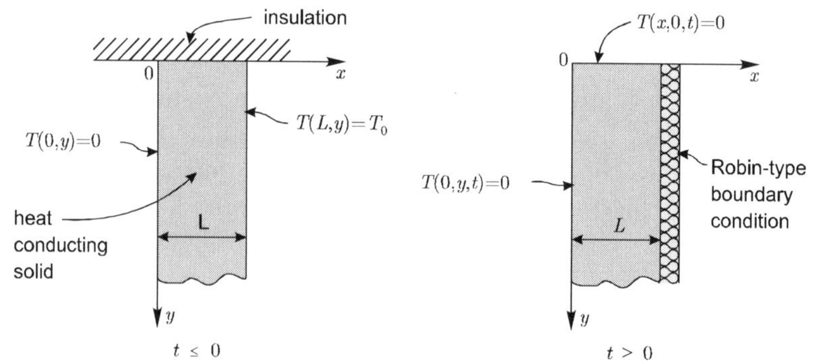

Figure 6.40: Heat conduction problem for a semi-infinite strip.

6.16 Use the *product solution* approach to derive an expression for the transient heat conduction in a plate with insulated plane faces which is at a uniform initial temperature T^* and at $t = 0$ subjected to the boundary conditions shown in Figure 6.41.

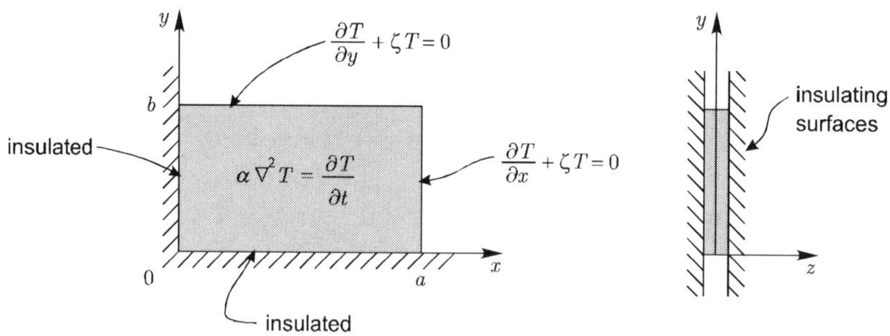

Figure 6.41: Heat conduction in a partly insulated plate

6.17 The plane surfaces of a square heat conducting plate and the edges are fully insulated (Figure 6.42a). The plate is maintained at zero initial temperature. At time $t = 0$ the surface BC is exposed and subjected to a temperature

$$T(x, a, t) = T_0 \cos\left(\frac{\pi x}{a}\right)$$

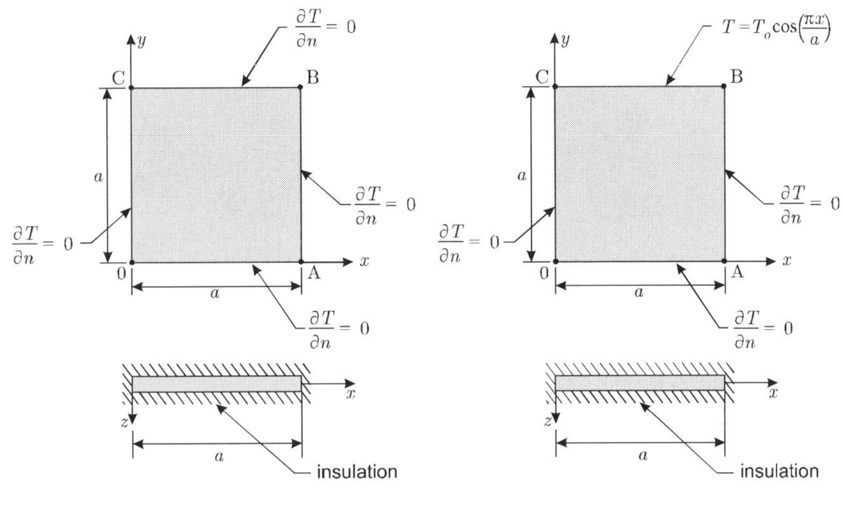

(a) Zero temperature initial state (b) Boundary conditions at $t = 0$

Figure 6.42: Heat conduction problem for a square plate.

where T_0 is a constant. Formulate the initial boundary value problem and derive an expression for the transient distribution of temperature in the plate if $T(x, y, t)$ satisfies

$$\alpha \nabla^2 T = \frac{\partial T}{\partial t}$$

6.18 A porous solid occupies the semi-infinite region $x \in (-\infty, \infty)$, $y \in (-\infty, \infty)$ and $z \in (-\infty, \infty)$. The entire surface of the semi-infinite region is subjected to a pollutant exposure with a time-dependent variation

$$C(x, y, 0, t) = C_0 e^{-\lambda t}$$

where λ is a constant (Figure 6.43). The migration of the pollutant in the porous medium is governed by the classical diffusion equation

$$D\nabla^2 C = \frac{\partial C}{\partial t}$$

where D is a diffusion coefficient. Use a Laplace transform technique to solve the initial boundary value problem governing $C(\mathbf{x}, t)$.

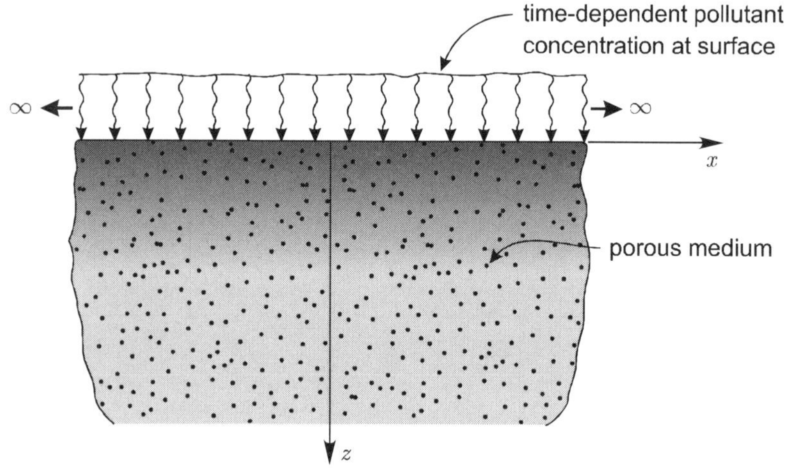

Figure 6.43: Pollutant migration in a semi-infinite porous medium.

6.19 The Figure 6.44 shows a semi-infinite porous region into which a chemical migrates. The initial concentration of the chemical in the region is *zero*, and a finite width of the surface $y = 0$ is subjected to a sudden increase in chemical concentration

$$C(x, 0, t) = C_0 H(t) \quad ; \quad -a < x < a$$

where C_0 is a constant and $H(t)$ is the Heaviside step function. The surface region $|x| > a$ is maintained at *zero* concentration. Use Laplace and Fourier transform techniques (with respect to t and x respectively) to develop a *formal integral expression* for the time-dependent variation of chemical concentration in the semi-infinite plane region due to surface chemical dosing.

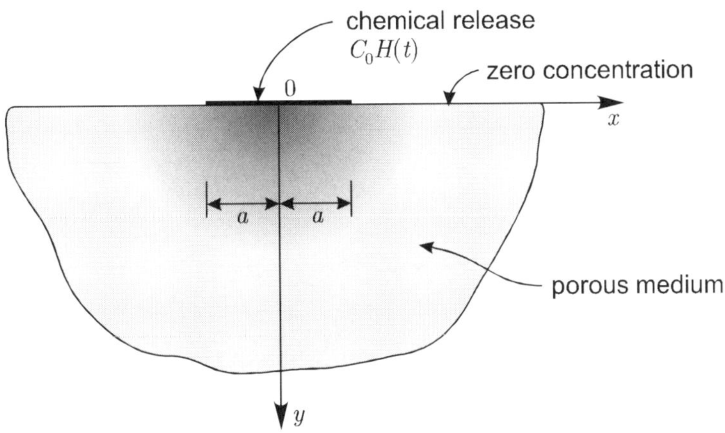

Figure 6.44: Chemical dosing of a porous half-plane by a surface release of chemical.

6.20 A magmatic vertical dyke has been intruded into a geological medium of semi-infinite extent (Figure 6.45). The dimensions of the dyke in the z-direction are such that the transient heat conduction process in the region is two-dimensional (i.e. $T = T(\mathbf{x}, t)$). The initial temperature distribution in the intruded dyke is constant; i.e.

$$T(x, y, 0) = T_0 \quad ; \quad |x| < b \quad ; \quad y \in (-\infty, \infty)$$

and, in the natural geological medium exterior to the dyke, the initial temperatures are *zero*; i.e.

$$T(x, y, 0) = 0 \quad ; \quad |x| > b \quad ; \quad y \in (-\infty, \infty)$$

The surface of the semi-infinite region $y = 0$ is maintained at *zero* temperature after the intrusion of the dyke.

(i) Formulate the initial boundary value problem assuming that the thermal diffusivity of the dyke material is the same as that for the surrounding geological medium

(ii) Use a double Fourier transform technique (with respect to the spatial variables x, y) to determine the transient temperature distribution in the semi-infinite medium.

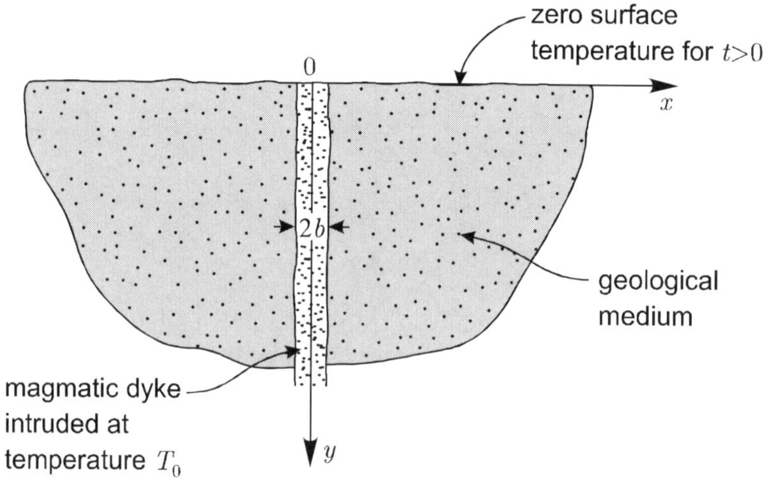

Figure 6.45: Heat conduction from an intruded magmatic dyke.

Chapter 7

The wave equation

The term 'wave' is used to signify an effect which propagates in a medium. The medium can be an object such as a string, cable, beam, plate, shell or a physical space consisting of fluids, solids, etc. The most common observation of a wave is that of the motion of the surface of still water when a pebble is thrown in it. The 'snap' at the end of a whip or a string is another form of a wave. The wind induced oscillations of power lines, telecommunication cables, suspension bridges and earthquake induced motion of structures represent other forms of wave motion. In contrast to these visible manifestations of waves, there are indirect consequences of waves which can include musical sounds generated by vocal cords, acoustic waves, heating effects of microwaves, waves that are generated by electrical oscillations of radio antennae, tidal movements due to gravitational effects, seismic waves in the earth's crust and tsunamis or tidal waves due to undersea earthquakes. There is a common theme associated with all of the physical phenomena associated with wave motion. It largely involves the notion of transmission of energy through a medium without any residual effect on the medium itself.

The study of waves in material bodies has been an important development in both engineering and mathematical physics. The origins of these developments can be traced to the Greek civilization (circa 500 BC) when the brotherhood of Pythagoras (c.580-500 BC) attempted to obtain the musical pitch of a vibrating string and its length. A mathematical treatment of waves in strings was presented in the 1747 classical work of Jean d'Alembert (1717-1783) dealing with the problem of the vibrating violin string. Much of d'Alembert's results were also given in an important paper "*On the vibration of strings*" by Leonhard Euler (1707-1783) and apparently after he had read d'Alembert's work on the subject. Although both d'Alembert and Euler agreed upon the structure of their equations, they disagreed on the nature of the functions that would qualify as solutions. A number of other distinguished mathematical physicists and engineers, including, Daniel Bernoulli (1700-1784), Joseph Lagrange (1736-1813), Lord Rayleigh (1842-1919), E.F.F. Chladni (1756-1829), Sophie Germain (1776-1831), C.L.M.H. Navier (1785-1836), Augustin Cauchy (1789-1857), Siméon-Denis Poisson (1781-1840), Heinrich Hertz (1857-1894), and Sir Horace Lamb (1849-1934) have investigated a variety of problems dealing with waves and vibrations in strings, beams and plates, wave propagation in the earth and the periodic motion of celestial bodies.

The study of waves and vibrations in media, be it a discrete structure or a continuum, have significant applications in a number of areas of importance to engineering, including waves in cable structures, vibrations of structures and structural elements, study of earthquake effects in buildings, materials evaluation and detection of defects in materials by ultrasonic pulse testing, wave phenomena in biomedical applications, including blood flow in deformable arteries, fragmentation of materials by acoustic waves (fragmentation of kidney stones and rocks), flow induced vibrations in pipelines, energy resource exploration through geophysical prospecting and in the study of nano-devices and components subjected to dynamic effects. These studies deal predominantly with wave phenomena in solids; however, historically important developments in the area of wave propagation in fluids and gases can also be included.

The study of wave phenomena in solid bodies has evolved through the examination of problems of specific interest, culminating in the development of general theories governing wave phenomena in solids of particular importance to scientific and engineering applications. In this chapter we will introduce the subject of wave motion by examining certain specific problems associated with spatially one-dimensional and two-dimensional problems associated with flexible strings and membranes. Other topics including longitudinal and torsional wave propagation in bars are also introduced briefly. Further aspects related to vibrations of plates and waves in elastic solids are discussed in the bibliography provided at the end of the text.

7.1 Wave motion in strings

The study of waves in a stretched string represents a physical problem that can be used to illustrate many important aspects of relevance to wave propagation. The stretched string problem also enables us to examine a solution to a classical problem in wave propagation developed by d'Alembert in 1747. Quite apart from the mathematical and historical interest, the stretched string problem has applications in the study of the vibration of vocal cords and stringed musical instruments such as the violin, harp, guitar and others. The mathematical modelling of the stretched string problem also has a variety of engineering applications including the study of dynamics of transmission lines and the dynamics of strings in the manufacture of fibres and textiles. In order to examine the problem of a stretched string, it is necessary to develop the equation governing its motion. This can be achieved by considering a differential element of a stretched string under a constant tension, T_0 (Figure 7.1).

We assume that during its motion, the stretched string displaces to a new position such that the segment of length dx acquires the displaced position as shown in Figure 7.1. The string is assumed to be stretched between its end points under a tension T_0 sufficiently large that gravity effects can be ignored. The stretched string is assumed to be elastic with no

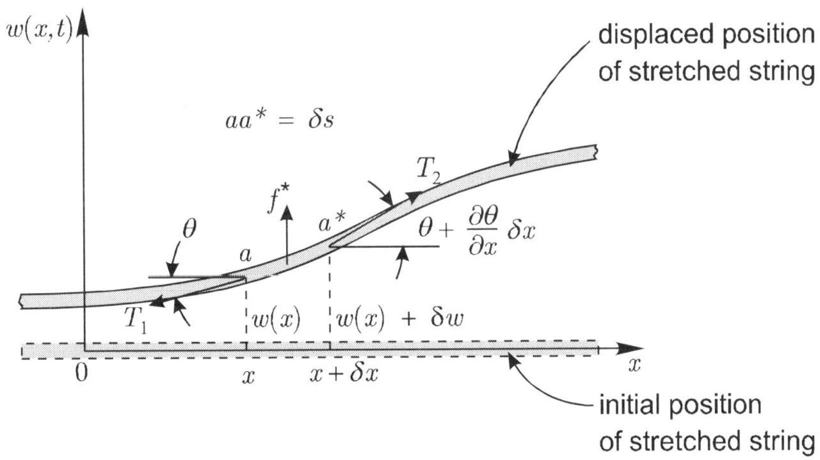

Figure 7.1: Deformed configuration of a stretched string.

resistance to bending. The tension in the string is assumed to remain constant throughout the wave motion. The *mass per unit length of the string* is denoted by ρ^* (i.e. $\rho^* = \rho\tilde{a}$ where ρ is the mass density and $\tilde{a}$ is the cross sectional area of the string). In its initial position the stretched string is located along the x-axis and during wave motion it displaces to the position indicated in Figure 7.1. We consider the equation of motion of the element aa^* shown in Figure 7.1. We have

$$-T_1 \sin\theta + T_2 \sin(\theta + \delta\theta) + f^*(\delta s) = (\rho^* \delta s)\frac{\partial^2 w}{\partial t^2} \tag{7.1}$$

$$-T_1 \cos\theta + T_2 \cos(\theta + \delta\theta) = 0 \tag{7.2}$$

From geometry we have

$$\delta s = \left[1 + \left(\frac{\partial w}{\partial x}\right)^2\right]^{1/2} \delta x \simeq \delta x \tag{7.3}$$

and

$$\sin\theta = \frac{\partial w}{\partial x}\left[1 + \left(\frac{\partial w}{\partial x}\right)^2\right]^{-1/2} \simeq \frac{\partial w}{\partial x} \tag{7.4}$$

$$\cos\theta = \left[1 + \left(\frac{\partial w}{\partial x}\right)^2\right]^{-1/2} \simeq 1 \tag{7.5}$$

$$\cos(\theta + \delta\theta) \simeq 1 \quad ; \quad \sin(\theta + \delta\theta) \simeq \frac{\partial w(x + \delta x, t)}{\partial x} \tag{7.6}$$

to within second degree terms in $(\partial w/\partial x)$.

Using the results (7.4) to (7.6) in (7.2), the equation gives

$$-T_1 + T_2 = 0 \tag{7.7}$$

Therefore we can set $T_1 = T_2 = T_0$. The first equation of motion can be written as

$$T_0 \left\{ \frac{\partial w(x+\delta x, t)}{\partial x} - \frac{\partial w(x,t)}{\partial x} \right\} + f^* \delta x = \rho^* \delta x \frac{\partial^2 w}{\partial t^2} \tag{7.8}$$

If we divide the equation (7.8) by δx and take the limit as $\delta x \to 0$ we obtain

$$T_0 \frac{\partial^2 w}{\partial x^2} + f^* = \rho^* \frac{\partial^2 w}{\partial t^2} \tag{7.9}$$

This represents a non-homogeneous second-order partial differential equation which describes the *forced vibration* of a stretched string. When the *forcing term* f^* vanishes, (7.9) reduces to

$$c_0^2 \frac{\partial^2 w}{\partial x^2} = \frac{\partial^2 w}{\partial t^2} \tag{7.10}$$

where

$$c_0 = \sqrt{\frac{T_0}{\rho^*}} \tag{7.11}$$

is the velocity of the small amplitude transverse waves in the stretched string (the letter 'c' after *celeritas* the Latin term for speed). The equation (7.10) is referred to as the *Wave Equation*, and describes the free vibrations of a stretched string. It is a linear partial differential equation of the second-order. By applying the criteria developed in Chapter 4 we can show that since c_0^2 is positive, the partial differential equation is *hyperbolic*.

In the ensuing we shall develop certain elementary solutions to the wave equation.

7.1.1 Harmonic waves

We first consider the problem of simple harmonic waves in the stretched string. Following the separation of variables technique, we assume a solution of the form

$$w(x,t) = W(x)\,T(t) \tag{7.12}$$

Substituting (7.12) into (7.10) we obtain

$$\frac{d^2W/dx^2}{W} = \frac{d^2T/dt^2}{c_0^2 T} = -\gamma^2 \tag{7.13}$$

where γ^2 is a positive constant. The resulting solution for $w(x,t)$ can be written as

$$w(x,t) = [A_1 \sin\gamma x + A_2 \cos(\gamma x)]\,[A_3 \sin(\omega t) + A_4 \cos(\omega t)] \tag{7.14}$$

where ω is the radial frequency given by

$$\omega = \gamma c_0 \tag{7.15}$$

We can rewrite the solution (7.14) as

$$\begin{aligned} w(x,t) = {} & A_1A_4 \sin(\gamma x)\cos(\omega t) + A_2A_3 \cos(\gamma x)\sin(\omega t) \\ & + A_2A_4 \cos(\gamma x)\cos(\omega t) + A_1A_3 \sin(\gamma x)\sin(\omega t) \end{aligned} \tag{7.16}$$

If we consider a typical term of this solution, which can be written in the form

$$w(x,t) = A^* \cos(\gamma x)\ \sin(\omega t) \tag{7.17}$$

the deflected shape of the harmonic motion of the stretched string can be as illustrated in Figure 7.2.

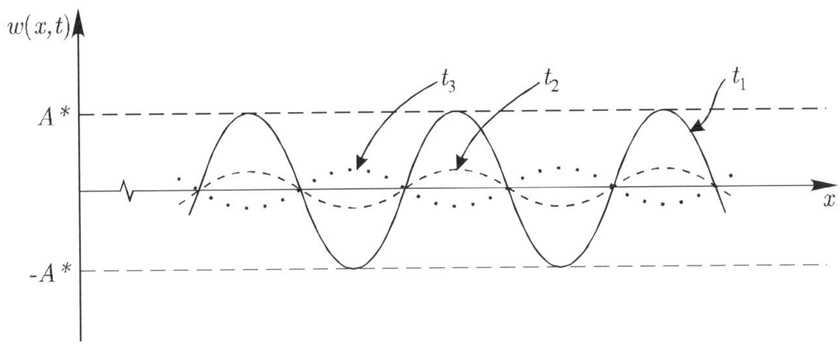

Figure 7.2: Deflected shapes of the stretched string.

The following definitions should be noted:

1. the points of *zero wave amplitude* are referred to as *nodes.*

2. the points of *maximum wave amplitude* are referred to as *antinodes.*

3. all free vibrations of this type and defined by result of the form (7.17) are called *stationary* or *standing waves.*

   The nodes and antinodes occur at regular intervals along the string and they are stationary with the passage of time. We can re-write (7.16) as

$$\begin{aligned} w(x,t) = & B_1 \sin(\gamma x + \omega t) + B_2 \sin(\gamma x - \omega t) \\ & + B_3 \cos(\gamma x + \omega t) + B_4 \cos(\gamma x - \omega t) \end{aligned} \tag{7.18}$$

Again if we consider a typical solution given by

$$w(x,t) = A^* \cos(\gamma x - \omega t) = A^* \cos \gamma (x - c_0 t) \tag{7.19}$$

This represents a wave propagating in the x-direction.

4. The argument of (7.19) is designated as the *phase* ϕ, where

$$\phi = \gamma (x - c_0 t) \tag{7.20}$$

In other words, for increasing time, increasing values of x are required to maintain the phase constant.

5. The deflection patterns at successive instances of time would be as shown in Figure 7.3. The propagation velocity of constant phase is c_0, which is defined as the *phase velocity.*

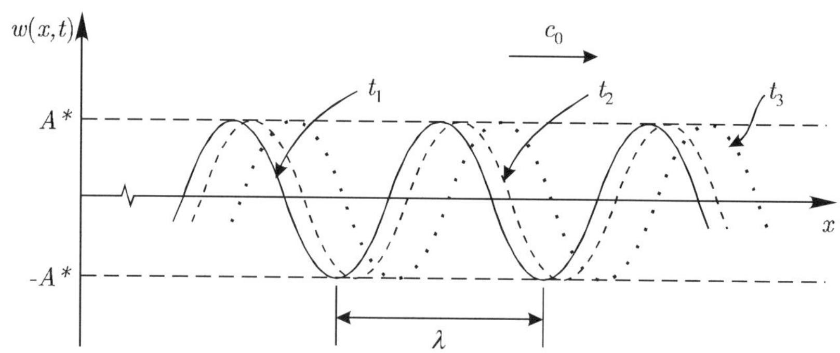

Figure 7.3: Transverse deflections at successive times resulting from propagating wave $w(x,t) = A^* \cos(\gamma x - \omega t)$.

The wave (7.19) represents a wave of infinite length; thus is has no *wave-front* or beginning. The consideration of the *phase* enables a propagation velocity to be identified with the motion.

6. We can define the distance between successive crests of constant phase as the *wave length* λ. Referring to Figure 7.3, this is shown as two successive minima; alternatively we could use successive maxima or intermediate points. The wavelength λ is related to γ by

$$\lambda = \frac{2\pi}{\gamma} \tag{7.21}$$

   and γ is designated the wave number.

7. Other quantities encountered in wave propagation problems include the following:

   A = amplitude (dimensions of length)
   ω = radial frequency of wave (dimensions of radians/time)
   f = cyclic frequency of wave (dimensions of cycles/time)
   T_P = period of wave (dimensions of time)

$$\omega = \gamma c_0 \quad ; \quad \omega = 2\pi f \quad ; \quad T_P = \frac{1}{f} \tag{7.22}$$

Having considered a typical term in (7.18) it is evident that the remaining terms are similar in character. The essential difference is that the terms having the argument $(\gamma x + \omega t)$ are propagating in the *negative x-direction.* Therefore, the general result (7.18) may be described as a *propagating wave solution* which has been derived from a *standing wave solution* (7.16). We can therefore interpret standing waves as the end result of the interference of *leftward* and *rightward* propagating waves e.g.

$$\begin{aligned} w(x,t) &= \frac{A^*}{2}\sin(\gamma x + \omega t) + \frac{A^*}{2}\sin(\gamma x - \omega t) \\ &= A^* \sin(\gamma x)\cos(\omega t) \end{aligned} \tag{7.23}$$

We can consider alternate forms for the mathematical representation of harmonic waves since

$$e^{i\omega t} = \cos(\omega t) + i \sin(\omega t) \quad ; \quad e^{-i\omega t} = \cos(\omega t) - i \sin(\omega t) \tag{7.24}$$

we can seek a solution of $w(x,t)$ of the form

$$w(x,t) = W(x) e^{i\omega t} \tag{7.25}$$

Substituting (7.25) in (7.10) and solving the resulting differential equation for $W(x)$ we can show that

$$w(x,t) = A_1 e^{i(\gamma x + \omega t)} + B_1 e^{-i(\gamma x - \omega t)} \tag{7.26}$$

These solutions represent harmonic waves propagating in the negative and positive x directions respectively. Similarly, considering an initial time behaviour of the form $e^{-i\omega t}$, we obtain the solutions

$$w(x,t) = A_2 e^{i(\gamma x - \omega t)} + B_2 e^{-i(\gamma x + \omega t)} \tag{7.27}$$

7.1.2 d'Alembert's solution

A classical analysis of the wave equation was given by d'Alembert in 1747. This classical nature of the solution stems not only from the simplicity of the analysis but also from its ability to provide insight into wave propagation phenomena. Considering the wave equation (7.10), we can utilize the results derived in Chapter 4, for purposes of reducing the partial differential equations to their canonical forms. Writing (7.10) as

$$c_0^2 \frac{\partial^2 w}{\partial x^2} - \frac{\partial^2 w}{\partial t^2} = 0 \tag{7.28}$$

and comparing (7.28) with (4.16) we have

$$A = c_0^2 \quad ; \quad 2B = 0 \quad ; \quad C = -1 \tag{7.29}$$

From (4.31) and (4.32) we have

$$\alpha = \frac{1}{A}\left[-B + \sqrt{B^2 - AC}\right] = \frac{1}{c_0} \tag{7.30}$$

$$\beta = \frac{1}{A}\left[-B - \sqrt{B^2 - AC}\right] = -\frac{1}{c_0} \tag{7.31}$$

and the differential equations corresponding to (4.36) and (4.37) give

$$\frac{dt}{dx} = \frac{1}{c_0} \quad ; \quad \frac{dt}{dx} = -\frac{1}{c_0} \tag{7.32}$$

Integrating these, we have

$$(x - c_0 t) = C_1 = \xi \quad ; \quad (x + c_0 t) = C_2 = \eta \tag{7.33}$$

Alternatively, we can rewrite the wave equation in the form

$$\left(\frac{\partial}{\partial x} + \frac{1}{c_0}\frac{\partial}{\partial t}\right)\left(\frac{\partial}{\partial x} - \frac{1}{c_0}\frac{\partial}{\partial t}\right) w\,(x,t) = 0 \tag{7.34}$$

which suggests that the combination of $(x - c_0 t)$ and $(x + c_0 t)$ could form the natural variables for the solution of the problem. We can write

$$\frac{\partial w}{\partial x} = \frac{\partial w}{\partial \xi}\frac{\partial \xi}{\partial x} + \frac{\partial w}{\partial \eta}\frac{\partial \eta}{\partial x} = \frac{\partial w}{\partial \xi} + \frac{\partial w}{\partial \eta} \tag{7.35}$$

$$\frac{\partial w}{\partial t} = \frac{\partial w}{\partial \xi}\frac{\partial \xi}{\partial t} + \frac{\partial w}{\partial \eta}\frac{\partial \eta}{\partial t} = -c_0\frac{\partial w}{\partial \xi} + c_0\frac{\partial w}{\partial \eta} \tag{7.36}$$

and the second derivatives give

$$\frac{\partial^2 w}{\partial x^2} = \frac{\partial^2 w}{\partial \xi^2} + 2\frac{\partial^2 w}{\partial \xi \partial \eta} + \frac{\partial^2 w}{\partial \eta^2} \tag{7.37}$$

$$\frac{\partial^2 w}{\partial t^2} = c_0^2\left(\frac{\partial^2 w}{\partial \xi^2} - 2\frac{\partial^2 w}{\partial \xi \partial \eta} + \frac{\partial^2 w}{\partial \eta^2}\right) \tag{7.38}$$

Substituting (7.37) and (7.38) in the wave equation we obtain its canonical form

$$\frac{\partial^2 w}{\partial \xi \partial \eta} = 0 \tag{7.39}$$

This can be integrated to give

$$w(\xi, \eta) = F(\xi) + G(\eta) \tag{7.40}$$

where $F(\xi)$ and $G(\eta)$ are arbitrary functions. In terms of the x, t variables we have

$$w(x, t) = F(x - c_0 t) + G(x + c_0 t) \tag{7.41}$$

D'Alembert's solution can be interpreted in the following way: if we consider the solution $w(x, t) = F(x - c_0 t)$, at time $t = 0$, $w(x) = F(x)$ and this is depicted in Figure 7.4(a). If we consider the solution at a subsequent time $t = t'$, the position of the wave front will appear as shown in Figure 7.4(b). This represents the wave form in Figure 7.4(a), $w = F(x)$ moved a distance

c_0t'. This implies that the solution $w(x,t) = F(x - c_0t)$ represents a wave profile $w(x) = F(x)$ which moves in the *positive* x-direction at a uniform speed c_0, and *without distortion.* Similarly $w(x,t) = G(x + c_0t)$ represents a wave which moves in the *negative* x-direction with speed c_0 and *without distortion.*

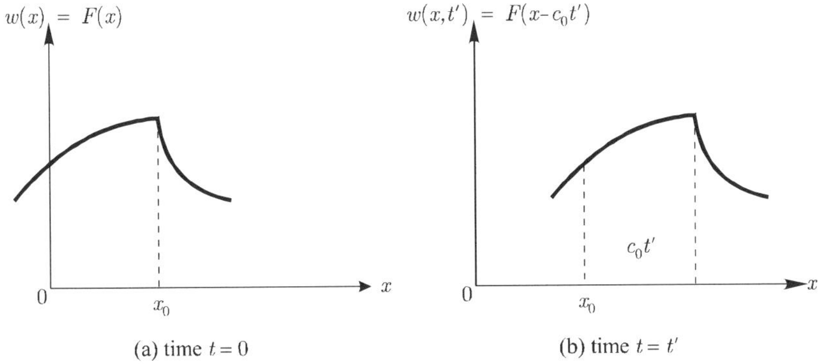

Figure 7.4: A moving wave.

Therefore d'Alembert's solution for the wave equation represents a combination of two arbitrary wave profiles $F(x)$ and $G(x)$ which move, respectively, in the *positive* and *negative* x-directions with speed c_0. *They can pass through each other without distortion or interference*. This solution is a convenient representation for progressive waves which travel large distances (Figure 7.5).

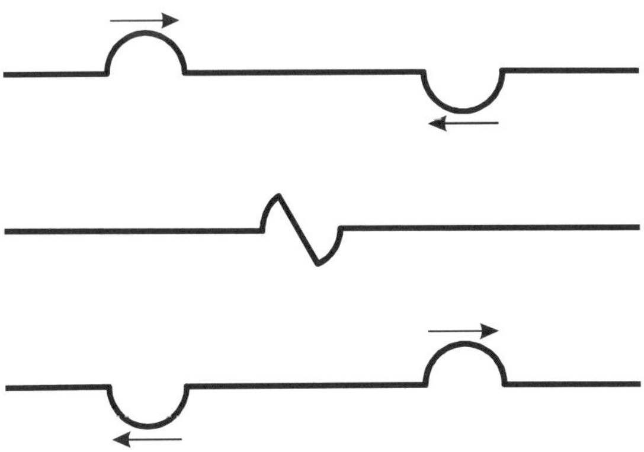

Figure 7.5: Interaction of travelling waves.

Example 7.1

Consider a stretched string of infinite length $-\infty < x < \infty$ which is prescribed with a displacement $w(x,0) = f(x)$ and a velocity $[\partial w/\partial t]_{t=0} = g(x)$. Pose the initial value problem and derive a general result for the deflected shape of the stretched string.

Solution

The partial differential equation governing the motion of the stretched infinite string problem is

$$c_0^2 \frac{\partial^2 w}{\partial x^2} = \frac{\partial^2 w}{\partial t^2} \quad ; \quad -\infty < x < \infty \quad ; \quad c_0^2 = \frac{T_0}{\rho^*} \tag{7.42}$$

with initial conditions

$$w(x,0) = f(x) \tag{7.43}$$

$$\left(\frac{\partial w}{\partial t}\right)_{t=0} = g(x) \tag{7.44}$$

D'Alembert's solution [(7.41)] is

$$w(x,t) = F(x - c_0 t) + G(x + c_0 t) \tag{7.45}$$

Using the initial conditions (7.43) and (7.44) we have

$$F(x) + G(x) = f(x) \quad ; \quad -\infty < x < \infty \tag{7.46}$$

$$-c_0 F'(x) + c_0 G'(x) = g(x) \quad ; \quad -\infty < x < \infty \tag{7.47}$$

Integrating (7.47) we have

$$-F(x)+G(x)=\frac{1}{c_0}\int_b^x g(\tau)\,d\tau \tag{7.48}$$

where the arbitrary lower limit $\tau = b$ simply represents the constant of integration. Using (7.46) and (7.48) we have

$$F(x)=\frac{1}{2}\left[f(x)-\frac{1}{c_0}\int_b^x g(\tau)\,d\tau\right] \tag{7.49}$$

$$G(x)=\frac{1}{2}\left[f(x)+\frac{1}{c_0}\int_b^x g(\tau)\,d\tau\right] \tag{7.50}$$

Using (7.49) and (7.50) we can represent the general solution (7.45) as

$$w(x,t)=\frac{1}{2}\left[f(x-c_0t)+f(x+c_0t)\right]+\frac{1}{2c_0}\int_{x-c_0t}^{x+c_0t} g(\tau)\,d\tau \tag{7.51}$$

Equation (7.51) indicates that the initial conditions are incorporated in the solution through $f(x)$ and $g(x)$. At the location $x = x^*$ at time $t = t^*$, the displacement $w(x^*, t^*)$ depends on

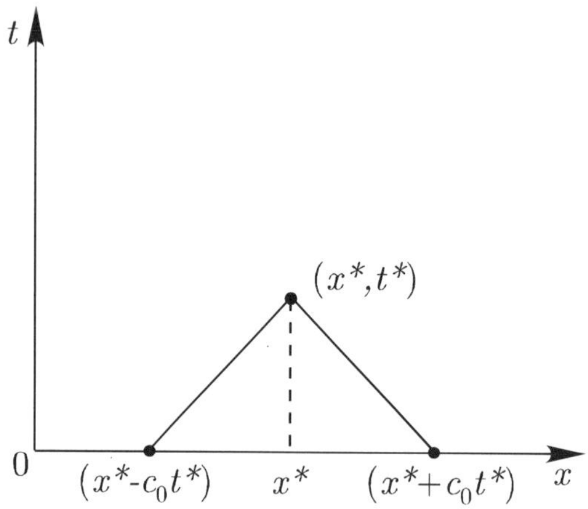

Figure 7.6: The influence of the initial conditions governing the solution of the wave equation (7.51).

$f\left(x^{*}-c_{0}t^{*}\right)$, $f\left(x^{*}+c_{0}t^{*}\right)$ which are the initial displacements at the points $x=\left(x^{*}-c_{0}t^{*}\right)$ and $x=x^{*}+c_{0}t^{*}$ and on

$$\int_{(x^{*}-c_{0}t^{*})}^{(x^{*}+c_{0}t^{*})} g\left(\tau\right) d\tau$$

This integral represents the velocity over the range $\left(x^{*}-c_{0}t^{*}\right)\leq x\leq\left(x^{*}+c_{0}t^{*}\right)$. Thus the initial conditions prescribed by (7.43) and (7.44) propagate with speed c_0 and at time $t=t^{*}$, the location $x=x^{*}$ has received the effects of the wave applicable only to the range

$$\left(x^{*}-c_{0}t^{*}\right)\leq x\leq\left(x^{*}+c_{0}t^{*}\right) \tag{7.52}$$

The influence of the initial condition is indicated in Figure 7.6.

7.1.3 Fourier analysis of the stretched string

In the section 7.1.1 we proved that *harmonic waves* satisfy the wave equation. We can therefore consider constructing a general solution consisting of superposition of such *harmonic waves.* For convenience, we prescribe initial conditions which will limit the harmonic solutions. The initial conditions are

$$w\left(x,0\right)=U\left(x\right) \tag{7.53}$$

$$\left[\frac{\partial w}{\partial t}\right]_{t=0}=0 \tag{7.54}$$

From the theory of Fourier integrals, it can be shown that, if $\varOmega\left(x\right)$ is defined in the interval $-\infty<x<\infty$, then

$$\varOmega\left(x\right)=\frac{1}{\pi}\int_0^\infty d\alpha\int_{-\infty}^\infty \varOmega\left(\kappa\right)\cos\left\{\left(x-\kappa\right)\alpha\right\}d\kappa \tag{7.55}$$

Therefore we can represent $U\left(x\right)$ in (7.53) in the form

$$U\left(x\right)=\frac{1}{\pi}\int_0^\infty d\gamma\int_{-\infty}^\infty U\left(\zeta\right)\cos\left\{\left(x-\zeta\right)\gamma\right\}d\zeta \tag{7.56}$$

We can interpret the result (7.56) with the statement that, at $t=0$, $U\left(x\right)$ has been represented by superimposing waves of different wave lengths and hence different wave number γ. The various harmonic components also differ in the phase angle given by $\zeta\gamma$.

For $t>0$, the various Fourier components begin to move to the right (positive x) and left (negative x) (see e.g. (7.51) with $g(\tau)=0$) i.e.

$$\cos\gamma\left(x-\zeta\right)\rightarrow\frac{1}{2}\left[\cos\gamma\left(x-c_0t-\zeta\right)+\cos\gamma\left(x+c_0t-\zeta\right)\right] \tag{7.57}$$

The resulting wave system is

$$w(x,t) = \frac{1}{\pi} \int_0^{\infty} d\gamma \int_{-\infty}^{\infty} U(\zeta) \left[\cos\gamma (x - c_0 t - \zeta) \right. \\ \left. + \cos\gamma (x + c_0 t - \zeta) \right] d\zeta \tag{7.58}$$

If we replace $(x - c_0 t)$ by ξ and $(x + c_0 t)$ by η in (7.58) and compare the result with the Fourier integral expression (7.56) we can write

$$w(x,t) = \frac{1}{2} \left[U(\xi) + U(\eta) \right] \tag{7.59}$$

Reverting back to the x, t variables, we have

$$w(x,t) = \frac{1}{2} \left[U(x - c_0 t) + U(x + c_0 t) \right] \tag{7.60}$$

We can apply the Fourier transform technique to recover the same result.

7.1.4 Reflection and transmission at boundaries

The general procedures described in connection with the study of waves in infinite strings are useful for the purposes of illustrating phenomena associated with stationary or standing waves. The domain of the stretched string occupied the region $-\infty < x < \infty$ and the interaction of waves with boundaries was not relevant. We shall now consider the processes that occur when such waves encounter boundaries. The basic problem involves a stretched string of mass per unit length ρ^* which is attached to a boundary and kept in tension by a tensile force T_0. We will consider waves incident on the boundary of the string. In order to examine this problem we need to specify boundary conditions which describe the nature of the fixity of the stretched string at the rigid boundary. This boundary condition can vary depending upon the restraint, ranging from zero displacement to an elastically restrained support.

Example 7.2

Consider the problem of a semi-infinite stretched string which is attached to the rigid boundary. The wave propagation in the string is governed by the partial differential equation

$$c_0^2 \frac{\partial^2 w}{\partial x^2} = \frac{\partial^2 w}{\partial t^2} \quad ; \quad 0 < x < \infty \tag{7.61}$$

with initial conditions

$$w\,(x, 0) = f\,(x) \tag{7.62}$$

$$\left[\frac{\partial w}{\partial t}\right]_{t=0} = g\,(x) \tag{7.63}$$

and the necessary boundary condition is

$$w\,(0, t) = 0 \tag{7.64}$$

Solution

Before developing a solution to this problem it is important to note the following. If we consider the solutions (7.49) and (7.50), these were developed in connection with the *infinite string problem* where $-\infty < x < \infty$. If we choose the same form of the solution, we need to recognize the restrictions placed by the domain $0 < x < \infty$ associated with the semi-infinite stretched string (Figure 7.7). The arbitrary functions $F(x - c_0 t)$ and $G(x + c_0 t)$ must be determined for positive arguments since $x > 0$ and $t > 0$. With the function $G(x + c_0 t)$ the argument is always positive for $x > 0$ and $t > 0$. On the other hand, with $F(x - c_0 t)$ we have to distinguish the combination of values of x and t which will render the *argument of* F either *positive* or *negative*.

The initial conditions (7.62) and (7.63) are identically satisfied if we select

$$\left.\begin{matrix} G(x) \\ F(x) \end{matrix}\right\} = \frac{1}{2} f(x) \pm \frac{1}{2c_0} \int_0^x g(\tau) d\tau \quad ; x > 0 \tag{7.65}$$

If $x > c_0 t$, the wave is unaware of the boundary at $x = 0$ and the solution is given by

$$w(x,t) = \frac{1}{2} \left[f(x - c_0 t) + f(x + c_0 t) \right] + \frac{1}{2c_0} \int_{x - c_0 t}^{x + c_0 t} g(\tau) d\tau \quad ; \quad x > c_0 t \tag{7.66}$$

which is d'Alembert's classical solution. This will not be valid if $x < c_0 t$. Now since for $x < c_0 t$ (7.65) is valid we obtain

$$G(x + c_0 t) = \frac{1}{2} f(x + c_0 t) + \frac{1}{2c_0} \int_0^{x + c_0 t} g(\tau) d\tau \tag{7.67}$$

The expression for F for negative arguments can be obtained by using the boundary condition (7.64). Hence from general result (7.41) we have

$$\begin{aligned} w(0,t) &= \left[F(x - c_0 t) + G(x + c_0 t) \right]_{x=0} \\ &= F(-c_0 t) + G(c_0 t) = 0 \quad ; \quad t > 0 \end{aligned} \tag{7.68}$$

Therefore F for negative arguments is $-G$ for the corresponding positive arguments, i.e.

$$F(\Omega) = -G(-\Omega) \quad ; \quad \Omega < 0 \tag{7.69}$$

Hence the solution for $(x < c_0 t)$ is

$$w(x,t) = F(x - c_0 t) + G(x + c_0 t) = -G(c_0 t - x) + G(x + c_0 t) \quad (7.70)$$

or

$$w(x,t) = \frac{1}{2}\left[f(x + c_0 t) - f(c_0 t - x)\right] + \frac{1}{2c_0}\int_{c_0 t - x}^{x + c_0 t} g(\tau)\, d\tau \quad (7.71)$$

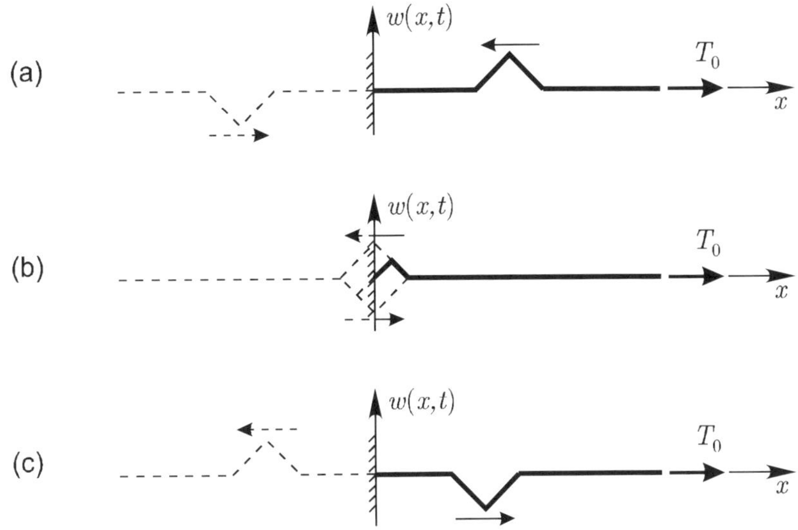

Figure 7.7: Reflection of a travelling wave at a rigid boundary.

We can interpret this solution in the following way. Consider a pulse $f(x + c_0 t)$ propagating to the 'left' and incident on the fixed boundary to which the stretched string is attached. We can "*remove*" the boundary at $x = 0$ and extend the string to $-\infty$. We can construct an image pulse to $f(x + c_0 t)$ which is symmetrically located, is opposite in sense to $f(x + c_0 t)$ and propagating to the right as shown in Figure 7.7. As these standing waves approach the origin, they will interact as shown in Figure 7.7(c). As they pass each other, the displacements cancel each other giving $w(0,t) = 0$. Hence the fixed end boundary condition for a semi-infinite stretched string is always satisfied by the image pulse system associated with the infinite string. With progress of time, the original pulse enters the region $x < 0$ and the "image" wave enters the region $x > 0$. The reflected wave is an original pulse with the sign reversed.

• • •

A similar but related problem occurs at the junction between two connected strings which are subjected to the same tension but have different mass densities per unit length.

Example 7.3

Consider a stretched string consisting of two materials each with mass per unit length ρ_A^* and ρ_B^* occupying the regions $-\infty < x \leq 0$ and $0 \leq x < \infty$ respectively (Figure 7.8). Examine the process of reflection and transmission of an incident wave.

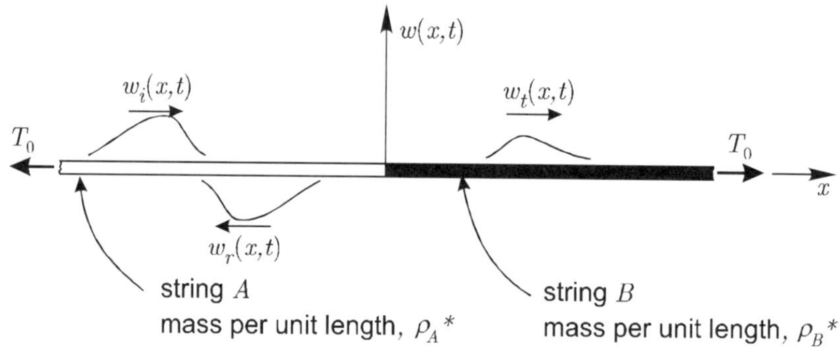

Figure 7.8: Incident, reflected and transmitted waves at a material discontinuity.

Solution

To examine this problem, we need to identify the conditions that are applicable to the interface or junction between the two semi-infinite strings. Since the wave equation is of second-order there should be at least two conditions that are applicable to the *interface*, which is in effect a boundary for *either* string.

(i) The first condition relates to the *continuity of transverse displacement* or alternatively the transverse velocity during the motion of the string.
(ii) The second condition relates to the *continuity of the vertical component of the tension* in the string at the interface. If this force was not continuous at the junction, an infinitesimal mass there would be subjected to a finite force resulting in an infinite acceleration.

If we denote the displacements corresponding to the string located to the left of the junction ($-\infty < x \leq 0$) by $w_A(x,t)$ and the displacements corresponding to right of the junction ($0 \leq x < \infty$) by $w_B(x,t)$, then these two conditions imply that

$$\left[\frac{\partial w_A}{\partial t}\right]_{x=0} = \left[\frac{\partial w_B}{\partial t}\right]_{x=0} \tag{7.72}$$

and

$$\left[-T_0\frac{\partial w_A}{\partial x}\right]_{x=0} = \left[-T_0\frac{\partial w_B}{\partial x}\right]_{x=0} \tag{7.73}$$

Let us now consider a wave travelling in string A (Figure 7.8) towards the interface between the two strings. For example, considering (7.27) we can take the *incident wave* as the *real part* of

$$w_I(x,t) = A_I e^{i(\gamma_A x - \omega t)} \quad ; \quad -\infty < x < 0 \tag{7.74}$$

which is specified by the amplitude A_I and $\gamma_A = \omega/c_A$; $c_A = (T_0/\rho^*_A)^{\frac{1}{2}}$ is the wave speed in the section A, and ρ^*_A is the mass per unit length in section A.

The *wave transmitted* through the interface is assumed to be the real part of

$$w_T(x,t) = \tilde{A}_T e^{i(\gamma_B x - \omega t)} \qquad ; \qquad 0 < x < \infty \tag{7.75}$$

where $\tilde{A}_T$ is a complex amplitude which is to be determined and a wave speed (c_B) and wave number (γ_B) which are different from the incident wave. (Both waves must necessarily have the same *frequency* unless there are dissipative phenomena at the junction).

It becomes evident that the wave forms (7.74) and (7.75) cannot satisfy the interface conditions (7.72) and (7.73). The conditions (7.72) and (7.73), respectively, would require

$$A_I = \tilde{A}_T \qquad ; \qquad \frac{A_I}{c_A} = \frac{\tilde{A}_T}{c_B} \tag{7.76}$$

This suggests the existence of a third wave which is a *reflected wave*; we assume that the reflected wave is the real part of

$$w_R(x,t) = \tilde{A}_R e^{i(\gamma_A x + \omega t)} \qquad ; \qquad -\infty < x < 0 \tag{7.77}$$

where $\tilde{A}_R$ is an undetermined complex amplitude and the wave number is the relevant value for the region A. The conditions at the junction between the string can now be expressed in terms of $w_I(x,t), w_T(x,t)$ and $w_R(x,t)$ as follows:

$$\frac{\partial w_I}{\partial t} + \frac{\partial w_R}{\partial t} = \frac{\partial w_T}{\partial t} \qquad ; \qquad \text{at} \quad x = 0 \tag{7.78}$$

and, since the tension T_0 is continuous

$$\frac{\partial w_I}{\partial x} + \frac{\partial w_R}{\partial x} = \frac{\partial w_T}{\partial x} \qquad ; \qquad \text{at} \quad x = 0 \tag{7.79}$$

Substituting (7.74), (7.75) and (7.77) in the equations (7.78) and (7.79) we obtain

$$A_I + \tilde{A}_R = \tilde{A}_T \quad ; \quad \frac{A_I}{c_A} - \frac{\tilde{A}_R}{c_A} = \frac{\tilde{A}_T}{c_B} \tag{7.80}$$

These equations can be solved to evaluate the amplitudes of the transmitted and reflected components of the incident waves we have

$$\tilde{A}_T = \frac{2c_B}{(c_A + c_B)} A_I \quad ; \quad \tilde{A}_R = \frac{(c_B - c_A)}{(c_A + c_B)} A_I \tag{7.81}$$

7.1.5 Energy in a string

The process of propagation of waves invariably involves the propagation of energy. In the case of a stretched string, energy can be transmitted as *kinetic energy* and *potential energy.* The transmission of energy by waves is an important aspect of the theory of wave motion. In its stationary position, the stretched string has *zero potential energy.* When a wave traverses, it acquires both potential energy and kinetic energy. In the formulation of the stretched string problem described in section 7.1, there are no processes which account for the dissipation of energy. Therefore, the principle of conservation of mechanical energy requires that the *rate at which energy enters a section* of the string through its ends or boundaries must balance *the rate of energy increase* in the section. This aspect of the energy associated with wave motion has important consequences in the theory of wave propagation and vibrations, including the consideration of the uniqueness of the solution to problems in wave propagation.

Considering the differential element of the stretched string shown in Figure 7.1, the kinetic energy increment is

$$dK = \frac{\rho^*}{2} \left(\frac{\partial w}{\partial t} \right)^2 \delta s \tag{7.82}$$

We can compute the total kinetic energy over a section located between $x = x_1$ and $x = x_2$ from the result

$$K = \frac{1}{2} \int_{x_1}^{x_2} \rho^* \left(\frac{\partial w}{\partial t} \right)^2 \left[1 + \left(\frac{\partial w}{\partial x} \right)^2 \right]^{\frac{1}{2}} dx \tag{7.83}$$

Neglecting higher-order quantities we obtain

$$K = \frac{1}{2} \int_{x_1}^{x_2} \rho^* \left(\frac{\partial w}{\partial t} \right)^2 dx \tag{7.84}$$

Alternatively, the *specific kinetic energy* or *kinetic energy density* is given by

$$k(x,t) = \frac{dK}{dx} = \frac{1}{2} \rho^* \left(\frac{\partial w}{\partial t} \right)^2 \tag{7.85}$$

In order to compute the *potential energy* V, we need to establish how much a finite region of the taut string is further stretched when it experiences small transverse deflections during the propagation of a wave. Firstly, the stretched string must possess *potential energy* due to the *work done* in achieving a deflected shape corresponding to the wave form. Since this *potential energy* must exist as *elastic energy* of the deflected shape, a more comprehensive calculation of the potential energy must involve the *elastic stiffness* of the string and the tension. The complexity associated with the incorporation of "*elasticity*" of the string is avoided by invoking the assumption that the *tension* T_0 *is independent of the deflected shape of the string.* In effect, in these elementary formulations of wave motion in strings, we are neglecting the axial stiffness of the string which does not permit the transmission of any longitudinal waves along the stretched string. Therefore, *regardless of the configuration of the wave, the tension* T_0 *is assumed to remain constant.*

To evaluate the potential energy of the deflected form of the string we consider the element aa^* shown in Figure 7.1. The change in length of the element by virtue of its deformed configuration is given by

$$\delta\ell = \int_{x_1}^{x_2} ds \; - \int_{x_1}^{x_2} dx \tag{7.86}$$

Using (7.3), we can rewrite (7.86) as

$$\delta\ell = \int_{x_1}^{x_2} \left[1 + \left(\frac{\partial w}{\partial x}\right)^2\right]^{\frac{1}{2}} dx \; - (x_2 - x_1) \tag{7.87}$$

Expanding the integrand of (7.87) in powers of $(\partial w/\partial x)^2$ and neglecting terms higher than the second, we have

$$\delta\ell = \frac{1}{2}\int_{x_1}^{x_2} \left(\frac{\partial w}{\partial x}\right)^2 dx \tag{7.88}$$

The work done by stretching the string by an amount $\Delta\ell$ against a constant tension T_0 is $(\Delta\ell)T_0$. Since the string is assumed to possess no elasticity, the work of the axial forces is stored as potential energy of the string in its displaced configuration, i.e.

$$V = \int_{x_1}^{x_2} \left\{\frac{1}{2}T_0 \left(\frac{\partial w}{\partial x}\right)^2\right\} dx \tag{7.89}$$

Hence the specific potential energy or potential energy density is given by

$$v(x,t) = \frac{1}{2}T_0 \left(\frac{\partial w}{\partial x}\right)^2 \tag{7.90}$$

Example 7.4

Consider the problem of an arbitrary travelling wave

$$w(x,t) = f(x - c_0 t) \tag{7.91}$$

moving in the positive x-direction. Derive an expression for the kinetic energy density of the wave and the power that is transmitted at any location x.

Solution

The specific kinetic energy is given by

$$k(x,t) = \frac{1}{2}\,\rho^* \left(\frac{\partial w}{\partial t}\right)^2 = \frac{1}{2}\,\rho^* c_0^2 \{f'(x - c_0 t)\}^2 \tag{7.92}$$

where f' is the ordinary derivative of the arbitrary function. Using the relationship $c_0^2 = T_0/\rho^*$, we can write (7.92) as

$$k(x,t) = \frac{1}{2}\,T_0 \{f'\}^2 \tag{7.93}$$

Similarly, the specific potential energy is given by

$$v(x,t) = \frac{1}{2}\,T_0 \left(\frac{\partial w}{\partial x}\right)^2 = \frac{1}{2}\,T_0 \{f'\}^2 \tag{7.94}$$

Since the waves move with velocity c_0, the total energy density (e^*) is given by

$$e^*(x,t) = k(x,t) + v(x,t) = T_0 \{f'\}^2 \tag{7.95}$$

The power carried by the wave at any location is

$$p^*(x,t) = c_0\; e^*(x,t) = T_0\; c_0 \{f'\}^2 \tag{7.96}$$

If we consider the power flow or the rate of transfer of energy past a given point we have

$$p^*(x_0, t) = -T_0 \left[\frac{\partial w}{\partial x} \frac{\partial w}{\partial t} \right]_{x=x_0} = T_0 \; c_0 \{f'\}^2_{x=x_0} \tag{7.97}$$

Hence the power is *entering* the section of the string $x \geq x_0$. Similarly, if we consider a wave travelling to the left (i.e. negative x-direction) with the wave form

$$w(x, t) = g(x + c_0 t) \tag{7.98}$$

we can show that

$$p^*(x_0, t) = -T_0 \; c_0 \{g'\}^2_{x=x_0} \tag{7.99}$$

Hence, power is leaving the section of the stretched string $x \geq x_0$.

7.1.6 Forced motion of a semi-infinite string

We now consider the problem of a stretched semi-infinite string which is subjected to *forced motion*. We note that the equilibrium of the stretched semi-infinite string is preserved by the provision of a rigid boundary to which the string is attached. With forced motions, the wave equation in its generalized form can be written as (see e.g. equation (7.9))

$$\frac{\partial^2 w}{\partial x^2} - \frac{1}{c_0^2} \frac{\partial^2 w}{\partial t^2} = q^*(x, t) \tag{7.100}$$

where

$$q^*(x,t) = -\frac{f^*(x,t)}{T_0} \tag{7.101}$$

As an example of forced motion, let us consider the motion of a semi-infinite stretched string occupying the region $0 \leq x < \infty$ (Figure 7.9). The string is *unloaded* in the region $0 \leq x < \infty$, but the rigid boundary is subjected to the harmonic displacement

$$w(0,t) = w_0 e^{i\omega t} \tag{7.102}$$

where w_o is a constant. Note that, although this represents a forced motion of the string, $q^*(x,t) = 0$.

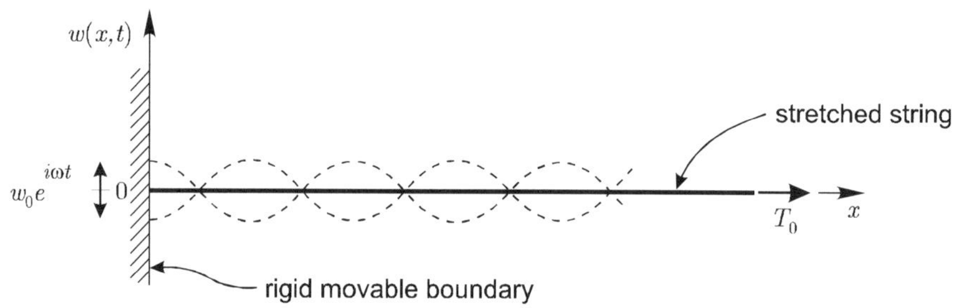

Figure 7.9: Forced motion of a semi-infinite string

A method of obtaining a solution to this problem is to assume a solution of the form

$$w(x,t) = W(x)e^{i\omega t} \tag{7.103}$$

Substituting (7.103) in (7.100), setting $q^*(x,t) = 0$, we obtain a second-order ordinary differential equation for $W(x)$. The solution of this equation can be expressed as

$$W(x) = A_1 e^{i\gamma x} + B_1 e^{-i\gamma x} \tag{7.104}$$

where $\gamma = \omega/c_0$ and A_1 and B_1 are arbitrary constants. Using (7.104) in (7.103) we have

$$w(x,t) = A_1 e^{i(\gamma x + \omega t)} + B_1 e^{-i(\gamma x - \omega t)} \tag{7.105}$$

(see e.g. equation (7.26)). The constants A_1 and B_1 can be determined by making use of the boundary conditions associated with the problem. One boundary condition is that specified by the forced motion, (7.102). Using this condition we obtain

$$A_1 + B_1 = w_0 \tag{7.106}$$

The resulting expression for the deflection of the stretched semi-infinite string can be written as

$$w(x,t) = A_1 e^{i(\gamma x + \omega t)} + (w_0 - A_1) e^{-i(\gamma x - \omega t)} \tag{7.107}$$

Since there are no further conditions prescribed at $x = 0$, this still leaves one constant undetermined. There is, however, a condition that is indirectly prescribed by virtue of the fact that the *stretched string is semi-infinite*. We note that the solution (7.107) contains waves which propagate in *both* directions. Since the domain of the semi-infinite string is $0 \leq x < \infty$, unless there is an energy source radiating waves towards the origin, there is no physical basis for the leftward propagating component of the solution (7.105). Hence we can invoke the *radiation condition* and obtain

$$A_1 \equiv 0 \tag{7.108}$$

and the solution for the forced motion of the string takes the form

$$w(x,t) = w_0 e^{-i(\gamma x - \omega t)} \tag{7.109}$$

The *radiation condition* is, in effect, a regularity condition which allows the elimination of certain solutions of the wave equation.

The solution which is selected to satisfy the radiation condition will depend on the nature of the initial solution. If the time-dependence of the initial solution has a form

$$w(x,t) = W(x)e^{-i\omega t} \tag{7.110}$$

then the corresponding radiation boundary condition will yield $B_1 = 0$, instead of $A_1 = 0$.

In the case when the end of the semi-infinite stretched string is subjected to a general transient displacement

$$w(0,t) = \Omega(t) \tag{7.111}$$

we can use a Fourier transform approach to develop a solution to the problem. We can apply the exponential Fourier transform on the *time variable* defined by

$$\mathcal{F}\{w(x,t);\omega\} = \overline{w}\{x;\omega\} = \frac{1}{\sqrt{2\pi}} \int_{-\infty}^{\infty} e^{i\omega t} w(x,t)dt \tag{7.112}$$

Using (1.173) we have

$$\mathcal{F}\left\{\frac{\partial^2 w}{\partial t^2};\omega\right\} = -\omega^2\, \overline{w}(x;\omega) \tag{7.113}$$

The wave equation now gives

$$\frac{d^2\overline{w}}{dx^2}\{x;\omega\} + \frac{\omega^2}{c_0^2}\,\overline{w}\{x;\omega\} = 0 \tag{7.114}$$

The solution of (7.114) gives

$$\overline{w}\{x;\omega\} = A^* e^{i\omega x/c_0} + B^* e^{-i\omega x/c_0} \tag{7.115}$$

where A^* and B^* are arbitrary constants. The exponential Fourier transform of the boundary condition (7.111) can be obtained from the general result

$$\overline{\Omega}(\omega) = \frac{1}{\sqrt{2\pi}} \int_{-\infty}^{\infty} e^{i\omega t}\,\Omega(t)dt \tag{7.116}$$

Using the boundary condition (7.111) at $x = 0$ we obtain, from (7.115) and (7.116),

$$A^* + B^* = \overline{\Omega}(\omega) \tag{7.117}$$

The second boundary condition is obtained by satisfying the *radiation condition* which ensures that all the waves resulting from the forced motion at $x = 0$ are *outgoing* in relation to $x \to \infty$. In order to apply the *radiation related* boundary condition it is necessary to invert the Fourier transform solution (7.115) to obtain an explicit expression for $w(x,t)$. Using (7.115), (7.117) and the Fourier inversion theorem (1.187) we obtain an explicit expression for $w(x,t)$:

$$\begin{aligned} w(x,t) = \frac{1}{\sqrt{2\pi}} \int_{-\infty}^{\infty} \Big[\left\{ \overline{\Omega}(\omega) - B^* \right\} e^{i\omega(x - c_0 t)/c_0} \\ + B^* e^{-i\omega(x + c_0 t)/c_0} \Big] d\omega \end{aligned} \tag{7.118}$$

To eliminate any incoming waves, the term containing $\exp\{ i\omega(x + c_0 t)/c_0 \}$ must be set to zero. The resulting solution is obtained by setting $B^* = 0$. We have

$$w(x,t) = \frac{1}{\sqrt{2\pi}} \int_{-\infty}^{\infty} \overline{\Omega}(\omega) e^{i\omega(x - c_0 t)/c_0} \, d\omega \tag{7.119}$$

The result (7.119) can be used to evaluate the time-dependent motion of the semi-infinite stretched string which is subjected to arbitrary time-dependent motions of the rigid movable boundary (Figure 7.9).

7.1.7 Forced motion of an infinite string

The examples involving the motion of a semi-infinite stretched string treated previously were restricted primarily to the situation where only the rigid movable boundary was subjected to a transverse displacement constraint and the initial displacement of the remainder of the stretched string was left unspecified. Furthermore, the string itself was not laterally loaded, thereby ensuring that $f^*(x,t)$ in the partial differential equation (7.9) was zero. We shall now consider the problem of a stretched string of *infinite length* which is subjected to an initial load which has both a *temporal distribution* and a *spatial distribution* along the stretched string. Although a variety of such loading configurations can be considered, we shall restrict attention to the specific case involving the loading distribution which is applied in a concentrated fashion. Prior to examining the localized loading problem it is instructive to cast the governing partial differential equation (7.9) in a non-dimensional form. We introduce the *non-dimensional* independent variables X and T such that

$$x = \ell X \quad ; \quad t = \tau T \tag{7.120}$$

where ℓ has units of length and τ has units of time. Using these substitutions and denoting

$$w(x,t) = \ell \, G(X,T) \tag{7.121}$$

the governing partial differential equation can be written as

$$\frac{\partial^2 G}{\partial X^2} - \left(\frac{\ell^2}{\tau^2}\right)\left(\frac{\rho^*}{T_0}\right)\frac{\partial^2 G}{\partial T^2} = -\frac{f^*\ell}{T_0} \tag{7.122}$$

Note that ρ^* has dimensions M/L; T_0 has dimensions ML/T^2; as such $(\ell^2\rho^*/\tau^2 T_0)$ is non-dimensional. Also we can fix ℓ and select τ such that

$$\frac{\ell^2}{\tau^2} = \frac{T_0}{\rho^*} = c_0^2 \quad \text{or} \quad \tau = \frac{\ell}{c_0} \tag{7.123}$$

where c_0 is the wave speed with dimensions, L/T. We can also replace f^*, the forcing function which has dimensions M/T^2, by

$$f^* = -f_0 F(X,T) \tag{7.124}$$

where f_0 is a constant with dimensions M/T^2 and $F(X,T)$ is a non-dimensional function. Using (7.123) and (7.124), the partial differential equation (7.122) can be written as

$$\frac{\partial^2 G}{\partial X^2} - \frac{\partial^2 G}{\partial T^2} = \left(\frac{f_0\ell}{T_0}\right) F(X,T) \tag{7.125}$$

without loss of generality we can specify

$$f_0 = \frac{T_0}{\ell} \tag{7.126}$$

which reduces the governing partial differential equation (7.9) to its non-dimensional form

$$\frac{\partial^2 G}{\partial X^2} - \frac{\partial^2 G}{\partial T^2} = F(X,T) \tag{7.127}$$

In this reduction, we have assumed that there exists a length parameter ℓ (or a time parameter τ) which is associated with the problem. Since the stretched string is of infinite extent, there is no naturally occurring length parameter ℓ in the problem (except perhaps the cross sectional dimensions of the string). Furthermore, the dimensionless groupings which occur in the reduction of the governing partial differential equation to a non-dimensional form, do not , naturally lend themselves for separating out either a length parameter (ℓ) or a time parameter (τ). All we can assume is that, in the application of the solution for $G(X,T)$ to a practical problem, a length parameter will naturally exist in a direct and identifiable form.

Let us now focus attention on the localized loading problem where $F(X,T)$ has the form

$$F(X,T) = \delta(X - X^*)\ \delta(T - T^*) \tag{7.128}$$

where $\delta(\ )$ represents the Dirac delta function. The force $F(X,T)$ is directed opposite to $G(X,T)$ and applied at time $T = T^*$ at the location $X = X^*$ (Figure 7.10). The solution for the motion of the stretched string induced by the application of the impulse function (7.128) is referred to as the *Green's function for the stretched string.*

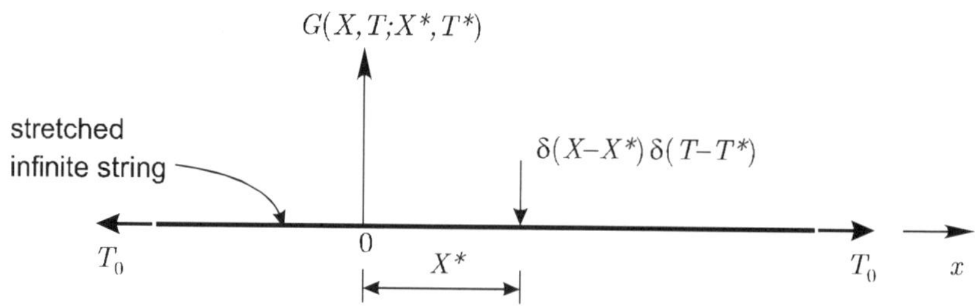

Figure 7.10: Localized loading of a stretched string

The solution for $G(X,T;X^*,T^*)$ should satisfy the governing partial differential equation

$$\frac{\partial^2 G}{\partial X^2} - \frac{\partial^2 G}{\partial T^2} = \delta(X - X^*)\,\delta(T - T^*) \tag{7.129}$$

and the *radiation conditions* applicable to *outgoing waves.* Since the stretched string is of infinite extent, there are no *boundary conditions* applicable to the problem.

The Green's function $G(X, T; X^*, T^*)$ for the problem should satisfy two initial conditions applicable to the problem; these are

$$[w(x, t; x^*, t^*)]_{t=0} = \ell\,[G(X, T; X^*, T^*)]_{T=0} = 0 \tag{7.130}$$

$$\left[\frac{\partial w}{\partial t}(x, t; x^*, t^*)\right]_{t=0} = c_0 \left[\frac{\partial G}{\partial T}(X, T; X^*, T^*)\right]_{T=0} = 0 \tag{7.131}$$

To solve the fundamental problem defined by (7.129), and subject to the initial conditions (7.130) and (7.131), we adopt a double integral transform approach; i.e. a *Fourier transform* with respect to the *non-dimensional spatial variable* X and a *Laplace transform* with respect to the *non-dimensional time variable* T.

We note that, from the definition of the exponential Fourier transform (see e.g. (1.168)), we have

$$\mathcal{F}\{\delta(X - X^*); \xi\} = \frac{e^{iX^*\xi}}{\sqrt{2\pi}} \tag{7.132}$$

and, from (1.109), the Laplace transform of $\delta(T - T^*)$ is given by

$$\mathcal{L}\{\delta(T - T^*)\} = e^{-T^* s} \tag{7.133}$$

First, we take the Fourier transform on the spatial variable (denoted by $(\tilde{\ })$)

which gives

$$-\xi^2\widetilde{G}(\xi,T;X^*,T^*) - \frac{d^2\widetilde{G}}{dT^2}(\xi,T;X^*,T^*) = \frac{e^{i\xi X^*}}{\sqrt{2\pi}}\delta(T-T^*) \qquad (7.134)$$

Taking the Laplace transform of (7.134) (denoted by ($\hat{\ }$)) and making use of the initial conditions (7.130) and (7.131) we obtain

$$-\xi^2\widehat{\widetilde{G}}(\xi,s;X^*,T^*) - s^2\widehat{\widetilde{G}}(\xi,s;X^*,T^*) = \frac{e^{i\xi X^*}e^{-T^*s}}{\sqrt{2\pi}} \qquad (7.135)$$

The transformed solution for $G(X,T;X^*,T^*)$ is given by

$$\widehat{\widetilde{G}}(\xi,s;X^*,T^*) = -\frac{e^{i\xi X^*}e^{-T^*s}}{\sqrt{2\pi}(s^2+\xi^2)} \qquad (7.136)$$

We can successfully invert the transforms to generate the *Green's function*. From *Tables of Laplace Transforms* we find that

$$\mathcal{L}^{-1}\left\{\frac{1}{s^2+\xi^2}\right\} = \frac{\sin(\xi T)}{\xi} \qquad (7.137)$$

Also, from Theorem II in Chapter 1 (equation (1.155)), we have

$$\mathcal{L}^{-1}\left\{e^{-T^*s}\left[\frac{1}{s^2+\xi^2}\right]\right\} = \frac{H(T-T^*)\ \sin(\xi[T-T^*])}{\xi} \qquad (7.138)$$

where $H(\ )$ is the Heaviside step function. Using the result (7.138), we can invert the Laplace transform of (7.136) to obtain

$$\widetilde{G}(\xi,T;X^*,T^*) = \frac{-1}{\sqrt{2\pi}}\left\{\frac{e^{i\xi X^*}}{\xi}\sin(\xi[T-T^*])\ H(T-T^*)\right\} \qquad (7.139)$$

Taking the inverse Fourier transform of (7.139) we have

$$G(X,T;X^*,T^*) = \frac{-H(T-T^*)}{2\pi}\int_{-\infty}^{\infty} \frac{e^{-i\xi(X-X^*)}}{\xi}\sin(\xi[T-T^*])d\xi \tag{7.140}$$

Introducing the substitutions

$$T' = (T-T^*) \quad ; \quad X' = (X-X^*) \tag{7.141}$$

we can write

$$G(X,T;X^*,T^*) = \frac{-H(T')}{2}\left\{\frac{1}{\sqrt{2\pi}}\int_{-\infty}^{\infty} \sqrt{\frac{2}{\pi}}\,\frac{\sin(\xi T')}{\xi}e^{-i\xi X'}d\xi\right\} \tag{7.142}$$

From *Tables of Fourier Transforms* we have

$$\frac{1}{\sqrt{2\pi}}\int_{\infty}^{\infty}\sqrt{\frac{2}{\pi}}\frac{e^{-i\xi X'}}{\xi}\,\sin(\xi T')d\xi = \begin{cases} 1 \quad ; \quad |\, X' \,| < T' \\ 0 \quad ; \quad |\, X' \,| > T' \end{cases} \tag{7.143}$$

We can now write (7.142) as

$$G(X,T;X^*,T^*) = \frac{-H(T')}{2}\left[H(X'+T') - H(X'-T')\right] \tag{7.144}$$

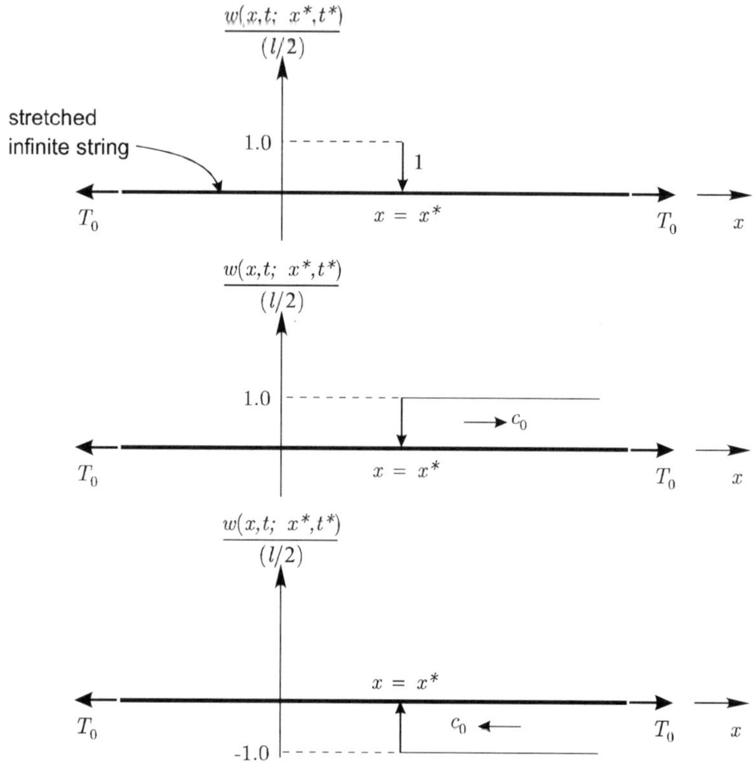

Figure 7.11: Heaviside function representation of the Green's function for a stretched infinite string.

We can express (7.144) in terms of the variables x, t, x^* and t^*as follows:

$$w(x,t;x^*,t^*) = -\frac{\ell}{2}H\left\{\frac{c_0}{\ell}(t-t^*)\right\}\left[H\left\{\frac{1}{\ell}[(x-x^*)+c_0(t-t^*)]\right\}\right.$$
$$\left. -H\left\{\frac{1}{\ell}[(x-x^*)-c_0(t-t^*)]\right\}\right] \tag{7.145}$$

The *impulse* or *Green's function* can be visualized as a combination of *two moving Heaviside step functions in space and time* which are initially applied at the location $x = x^*$ of the stretched infinite string at time $t = t^*$. The step function wave form $H\{[(x-x^*)+c_0(t-t^*)]/\ell\}$ is a wave which moves to the *left* (i.e. in the *negative* x-direction) and the step function wave form $H\{[(x - x^*) - c_0(t - t^*)]/\ell\}$ is a wave which moves to the *right* (i.e. in the *positive* x-direction), both wave forms starting precisely at $t = t^*$. (Figure 7.11) With progress of time, the deflected shape of the stretched string is obtained from the superposition of the two wave forms (Figure 7.12).

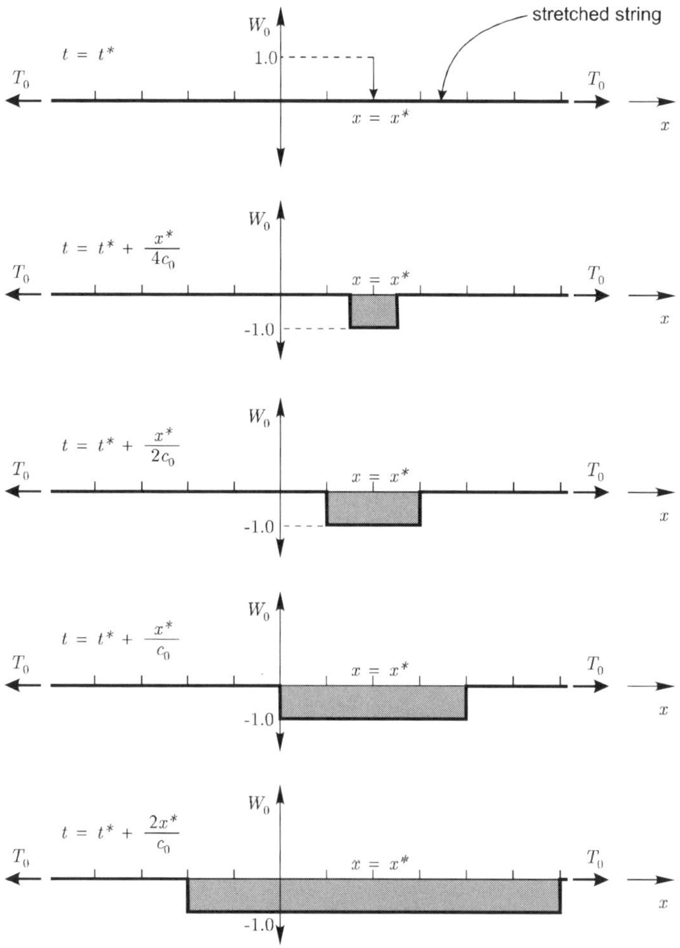

Figure 7.12: Wave profiles in the stretched string due to the application of the localized impulse loading.$[W_0 = w(x, t; x^*, t^*)/(\ell/2)]$

One of the distinct advantages of developing a *Green's function* for a particular partial differential equation is that the resulting solution can be effectively utilized to generate solutions to other types of distributions of *forcing functions* with spatial and temporal distributions. The *linearity* of the governing partial differential equation places the Green's function technique as an extremely effective method of obtaining solutions via a *superposition technique.*

Example 7.5

A stretched string of infinite extent is subjected to a time harmonic force such that

$$F(X,T) = \delta(X - X^*)e^{-i\Omega T} \tag{7.146}$$

where $X = X^*$ represents the point of application of the harmonic load (Figure 7.13). Examine the form of the deflection pattern of the stretched infinite string associated with this motion.

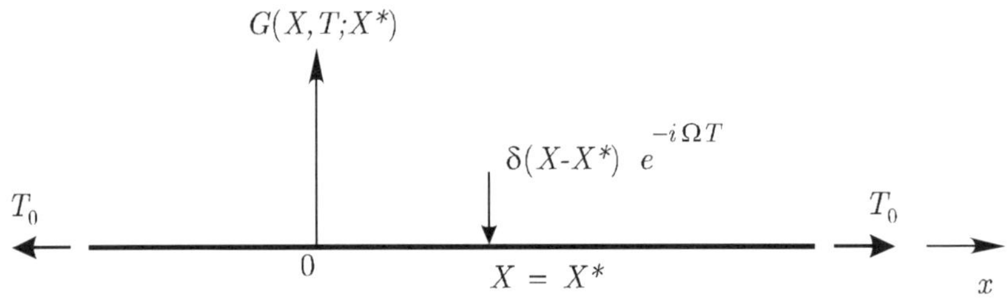

Figure 7.13: Time harmonic impulsive force acting on a stretched string.

Solution

The partial differential equation governing the motion of the string can be written in the non-dimensional form

$$\frac{\partial^2 G}{\partial X^2} - \frac{\partial^2 G}{\partial T^2} = \delta(X - X^*)\ e^{-i\Omega T} \tag{7.147}$$

where Ω is a non-dimensional form of the radial frequency of the forcing function with dimensions or radians per unit time. This pre-supposes that there is a normalizing parameter τ with dimensions of time.

Since no initial conditions are specified on the displacement and its derivatives, we shall assume that there are *no transient effects* associated with the time-harmonic forcing function and that the solution of (7.147) can be represented in the form

$$G(X,T;X^*) = \mathfrak{g}(X;X^*)e^{-i\Omega T} \tag{7.148}$$

substituting (7.148) in (7.147) we obtain

$$\frac{d^2\mathfrak{g}}{dX^2}(X;X^*) + \Omega^2\mathfrak{g}(X;X^*) = \delta(X - X^*) \tag{7.149}$$

If we take the Fourier transform of $\mathfrak{g}(X,X^*)$ with respect to the variable X we obtain

$$\widetilde{\mathfrak{g}}(\xi;X^*) = -\frac{e^{i\xi X^*}}{\sqrt{2\pi}(\xi^2 - \Omega^2)} \tag{7.150}$$

Inversion of (7.150) gives

$$\mathfrak{g}(X;X^*) = -\frac{1}{2\pi}\int_{-\infty}^{\infty}\frac{e^{-i\xi(X-X^*)}}{(\xi^2 - \Omega^2)}d\xi \tag{7.151}$$

Without loss of generality we can set $X^* = 0$, which implies that the forcing function is localized at the origin. The deflected form of the infinite string is given by

$$G(X,T) = \left[-\frac{1}{2\pi}\int_{-\infty}^{\infty}\frac{e^{-i\xi X}d\xi}{(\xi^2 - \Omega^2)}\right]e^{-i\Omega T} \tag{7.152}$$

The problem is now reduced to the evaluation of the integral in (7.152). If we assume that ξ is complex, then we have, from the theory of residues,

$$\int_C \frac{e^{-i\xi X} d\xi}{(\xi^2 - \Omega^2)} = 2\pi i \sum \text{Res.} \tag{7.153}$$

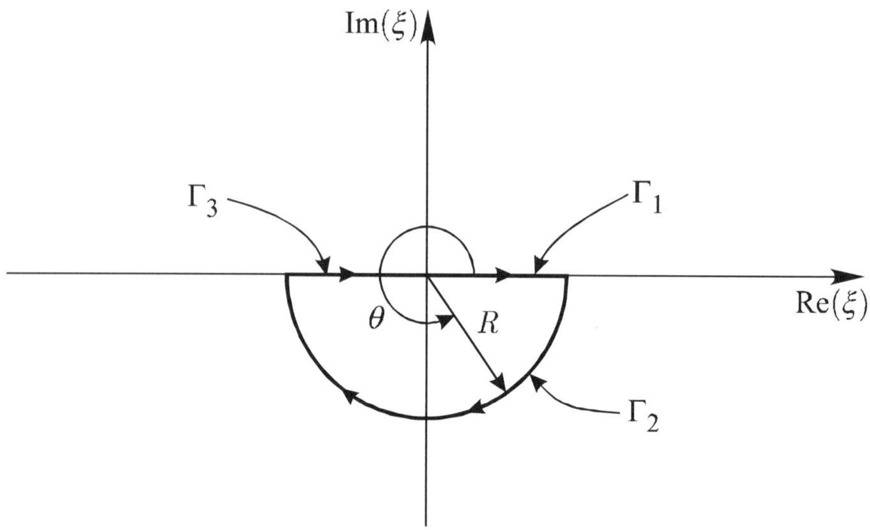

Figure 7.14: Contour integration path for evaluating the integral in (7.153).

where C is any closed contour in the complex plane. The contour C can be chosen as indicated in Figure 7.14. (Admittedly, this is not the only possible contour; an alternative form of the contour can be chosen as the circle located on the upper half of the complex plane.)

We note that in this contour integration, the poles at $\xi = \pm\Omega$ that are located on the real axis of ξ and have been ignored. We shall first deal with the contour integration which excludes the poles; i.e.

$$\int_C [\ ]d\xi = \int_{\Gamma_1} [\ ]d\xi + \int_{\Gamma_2} [\ ]d\xi + \int_{\Gamma_3} [\ ]d\xi \tag{7.154}$$

If we introduce the substitutions

$$\xi = R\,e^{i\theta} \quad \text{or} \quad d\xi = iR\,e^{i\theta}d\theta \tag{7.155}$$

the integral in (7.153) can be written as

$$I = \lim_{R\to\infty} \int_{\pi}^{2\pi} \frac{iR\,\mathrm{e}^{i\theta}}{[R^2 e^{2i\theta} - \Omega^2]} \left\{ e^{RX\sin\theta} \;\; e^{-iRX\cos\theta} \right\} d\theta \tag{7.156}$$

The dominant term in (7.156) is $\exp(RX\sin\theta)$. Along Γ_2 we have $\sin\theta < 0$;hence if $X > 0$ the exponential term decays rapidly for large values of R and as $R \to \infty, I \to 0$.

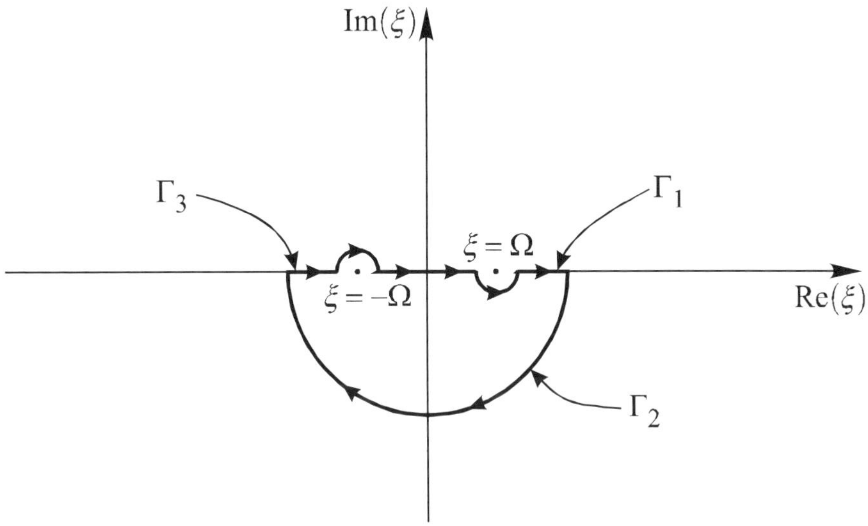

Figure 7.15: Contour integrations and indentations around the poles.

Let us now examine the poles that are located at $\xi = \pm\Omega$. The *poles* are singularities on the path of integration Γ_1 and Γ_3 (Figure 7.14) which should either be avoided or included by *indenting* the integration contour appropriately along a small semi-circular path around the poles (Figure 7.15).

These indentations will either *include* or *exclude* a singularity from the path of integration. The residues at the poles are given by

$$[\text{Res}]_{\xi=-\Omega} = -\frac{e^{i\Omega X}}{2\Omega} \tag{7.157}$$

$$[\text{Res}]_{\xi=\Omega} = \frac{e^{-i\Omega X}}{2\Omega} \tag{7.158}$$

The choice of which *residue* should either be *included* or *excluded* from the contour integration will depend on the physics of the problem and on the range of X. We recall that the wave propagation effects in the stretched infinite string are time-dependent with a time variation of $e^{-i\Omega T}$. When this time-dependency is combined with either (7.157) or (7.158), the resulting expression for $G(X,T)$ can give either *leftward* (negative X-direction) or *rightward* (positive X-direction) propagating waves. Therefore, the configuration of the contour integration path with the "*indentations*" should reflect the specific form of the "*radiation conditions*" applicable to the outgoing waves. If we select $X > 0$ and require outgoing waves in this region, we require that the waveform should be of the type

$$G(X,T) \sim e^{i(X-T)} \tag{7.159}$$

This will be obtained if we include the residue at $\xi = -\Omega$:i.e.

$$G(X,T) \sim \left[-\frac{e^{i\Omega X}}{2\Omega}\right]\left[e^{-i\Omega T}\right] \sim -\frac{e^{i\Omega(X-T)}}{2\Omega} \tag{7.160}$$

The residue at $\xi = \Omega$, on the other hand, will give

$$G(X,T) \sim \left[\frac{e^{-i\Omega X}}{2\Omega}\right]\left[e^{-i\Omega T}\right] \sim \frac{e^{-i\Omega(X+T)}}{2\Omega} \tag{7.161}$$

which represent *incoming waves* which will not satisfy the *radiation condi-*

tions. Therefore, the final form of the contour integration should be configured to include the pole at $\xi = -\Omega$ and to exclude the pole at $\xi = \Omega$ (Figure 7.14). Using these results, the final form of (7.152) can be written as

$$G(X,T) = \frac{1}{2\Omega}\left[\sin\Omega(X-T) - i\cos\Omega(X-T)\right] \tag{7.162}$$

7.2 Wave motion in stretched finite strings

The discussion which focused on the wave motion in an infinite string is, of course, a highly idealized situation which is useful from the point of view of explaining basic concepts applicable to wave motions in general. In most practical applications of the theory of wave motion, attention is restricted to stretched strings of *finite length*. The vibrations of stretched strings of musical instruments, vocal cords, power transmission cables, guy wires for antennae structures and spatially one-dimensional vibrations in highly flexible stretched surfaces can be examined by considering the basic form of the wave equation discussed in Section 7.1. The phrase '*waves in stretched strings*' is usually reserved to describe the movement of travelling waves in stretched strings of both infinite and finite length. The phrase, *vibrations in finite strings*, usually refers to the development of wave forms which usually occupy the entire interval of the stretched finite string. In the analysis of wave motion and vibrations in finite stretched strings, the governing wave equation has to satisfy both the initial conditions and boundary conditions applicable to the extremities of the stretched string.

7.2.1 Waves in a stretched finite string

We now consider the problem of a stretched string of finite length ℓ which is subjected to an initial disturbance, which can be an imposed displacement. We can visualize this string to be a part of an infinite string as shown in Figure 7.16. When the string is released from this initially displaced configuration, waves will propagate in *either* direction. When these separate waves encounter a boundary, a reflection will occur back into the region of the string. The *method of images* and the processes of reflection and transmission at the boundaries (see section 7.1.4) can be used to trace the generation of waves in the section of the string of length ℓ. Of course, it is necessary to construct an *infinite set of image strings* in order to characterize the time-dependent response of the stretched finite string for all time.

We can formulate the initial boundary value problem as follows: the governing partial differential equation is

$$\frac{\partial^2 w}{\partial x^2} - \frac{1}{c_0^2}\frac{\partial^2 w}{\partial t^2} = 0 \tag{7.163}$$

the initial conditions are

$$w(x,0) = W_0(x) \tag{7.164}$$

$$\left[\frac{\partial w}{\partial t}\right]_{t=0} = 0 \tag{7.165}$$

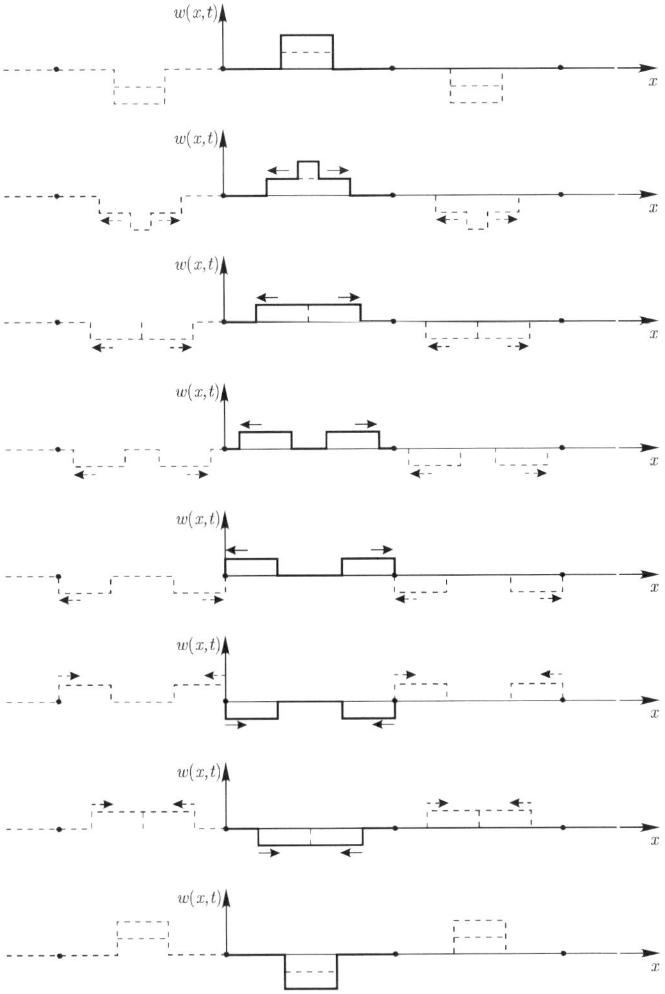

Figure 7.16: Travelling waves in a finite string.

and the boundary conditions are

$$w(0,t) = 0 \quad ; \quad w(\ell,t) = 0 \tag{7.166}$$

We can define $\overline{W}_0(x)$ such that

$$\overline{W}_0(x) = \begin{cases} W_0(x) & ; \quad 0 < x < \ell \\ \\ -W_0(-x) & ; \quad -\ell < x < 0 \end{cases} \tag{7.167}$$

and

$$\overline{W}_0(x) = \overline{W}_0(x + 2n\ell) \quad ; \quad n = 0, 1, 2... \tag{7.168}$$

From d'Alembert's solution, when the initial conditions (7.164) and (7.165) are prescribed, the general form of the solution of the wave equation is

$$w(x,t) = \frac{1}{2}[\ \overline{W_0}(x - c_0 t) + \overline{W_0}(x + c_0 t)\ ] \tag{7.169}$$

The primary aspect of the d'Alembert-type approach to the study of waves in stretched finite strings is that it provides a useful illustration of the basic periodic motion in such a system. The image approach to the study of waves in finite strings is effective when simple boundary conditions are involved. Other techniques prove to be more effective when examining problems related to vibrations of finite stretched strings.

7.2.2 Vibrations of a stretched finite string: trial function approach

In the previous chapters dealing with Laplace's equation and the diffusion equation, we were able to utilize the trial function approach to develop solutions to certain elementary problems. It is therefore natural to enquire whether such a procedure can be used to examine the spatially one-dimensional wave equation. Much of the success of the trial function-type approaches will of course depend on how successfully the trial function can be identified *a priori*.

Example 7.6

Consider the problem of a stretched string of finite length ℓ which is held fixed at the ends (Figure 7.17) and subjected to an initial displacement

$$w(x,0) = w_0 \sin\left(\frac{\pi x}{\ell}\right) \tag{7.170}$$

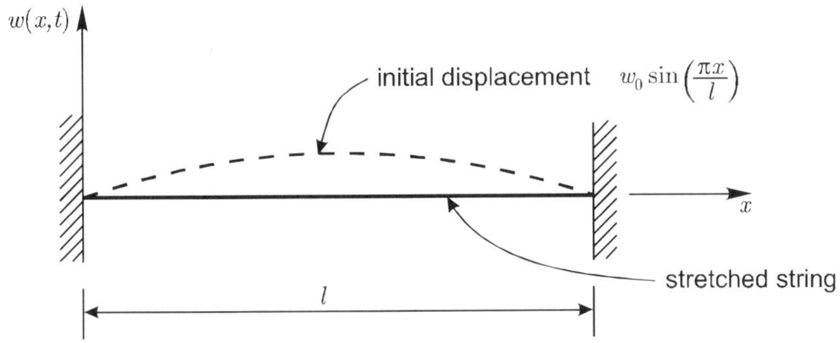

Figure 7.17: Initial displacement of stretched finite string.

The string is released from this position with *zero initial velocity*. Determine the resulting time-dependent motion of the stretched string.

Solution

The partial differential equation governing the problem is

$$\frac{\partial^2 w}{\partial x^2} - \frac{1}{c_0^2}\frac{\partial^2 w}{\partial t^2} = 0 \tag{7.171}$$

which is subject to boundary conditions

$$w(0,t) = 0 \quad ; \quad w(\ell,t) = 0 \tag{7.172}$$

The initial conditions applicable to the problem are

$$w(x,0) = w_0 \sin\left(\frac{\pi x}{\ell}\right) \tag{7.173}$$

$$\left[\frac{\partial w}{\partial t}\right]_{t=0} = 0 \tag{7.174}$$

If we compare the partial differential equation (7.171) with the general form of the second-order partial differential equation given in Chapter 4 (equation (4.2)) and identify y with t we have

$$A = 1 \quad ; \quad B = 0 \quad ; \quad C = -\frac{1}{c_0^2} \tag{7.175}$$

Hence

$$B^2 - AC = \frac{1}{c_0^2} > 0 \tag{7.176}$$

and the governing partial differential equation is *hyperbolic*. The characteristic equations are (equations (4.36) and (4.37)) are given by

$$\frac{dt}{dx} + \alpha(x,y) = 0 \tag{7.177}$$

$$\frac{dt}{dx} + \beta(x,y) = 0 \tag{7.178}$$

where

$$\alpha = \frac{1}{A}\left[-B + \sqrt{B^2 - AC}\right] = \frac{1}{c_0} \tag{7.179}$$

$$\beta = \frac{1}{A}\left[-B - \sqrt{B^2 - AC}\right] = -\frac{1}{c_0} \tag{7.180}$$

Substituting (7.179) and (7.180) respectively in (7.177) and (7.178) and performing the integrations, we have

$$t = \frac{x}{c_0} + C_1^* \quad ; \quad t = -\frac{x}{c_0} + C_2^* \tag{7.181}$$

The characteristic coordinates are

$$C_1 = (x - c_0 t) = \eta \ ; \tag{7.182}$$

$$C_2 = (x + c_0 t) = \xi \tag{7.183}$$

and the canonical form of the partial differential equation is

$$\frac{\partial^2 w}{\partial \xi \partial \eta} = 0 \tag{7.184}$$

Integrating (7.184) we have

$$w(\xi, \eta) = \varphi(\xi) + \psi(\eta) \tag{7.185}$$

where $\varphi(\xi)$ and $\psi(\eta)$ are arbitrary functions.

Alternatively, we can write

$$w(x,t) = \varphi(x + c_0 t) + \psi(x - c_0 t) \tag{7.186}$$

We can now choose a "*trial function*" of the form

$$w(x,t) = A\left[\sin\frac{\pi}{\ell}(x + c_0 t) + \sin\frac{\pi}{\ell}(x - c_0 t)\right] \tag{7.187}$$

where A is an arbitrary constant. If we expand the sine functions in (7.187) we obtain

$$w(x,t) = 2A\left[\sin\left(\frac{\pi x}{\ell}\right)\cos\left(\frac{c_0 \pi t}{\ell}\right)\right] \tag{7.188}$$

It is clear that (7.188) can satisfy the initial condition (7.173) (with $A = w_0/2$) and the second initial condition is identically satisfied. Hence, the final solution for the time-dependent motion of the stretched string is given by

$$w(x,t) = w_0 \sin\left(\frac{\pi x}{\ell}\right)\cos\left(\frac{c_0 \pi t}{\ell}\right) \tag{7.189}$$

This is a particularly simple example for which the canonical form can be effectively utilized to narrow down the type of solution and the initial conditions are such that the solutions can be obtained through a reasonable search for a trial function. It is somewhat more difficult to use a trial function approach if, for example, the initial conditions were posed as a general function $f(x)$ rather than a sine function. Other techniques prove to be more effective when examining generalized problems related to vibration of stretched strings.

7.2.3 Vibrations of a stretched finite string - variables separable solution

We now focus attention on the application of the method of separation of variables technique for the solution of the wave equation. With stretched strings of finite length, these solutions will yield the so called *standing wave* solutions. We seek variables separable solutions of (7.171), of the form

$$w(x,t) = X(x)T(t) \tag{7.190}$$

substituting (7.190) in (7.171) we obtain

$$\frac{d^2X/dx^2}{X} = \frac{1}{c_0^2}\frac{d^2T/dt^2}{T} = \text{const} \ = \lambda^2 \tag{7.191}$$

The constant λ^2 can either be *positive*, *negative* or *zero*. If the constant λ^2 is *positive*, the equation (7.191) yields the following:

$$\frac{d^2X}{dx^2} - \lambda^2 X = 0 \tag{7.192}$$

and

$$\frac{d^2T}{dt^2} - \lambda^2 c_0^2 T = 0 \tag{7.193}$$

These equations have exponential solutions of the form

$$X(x) = A^* e^{\lambda x} + B^* e^{-\lambda x} \tag{7.194}$$

$$T(t) = C^* e^{\lambda c_0 t} + D^* e^{-\lambda c_0 t} \tag{7.195}$$

where A^*, B^*, C^* and D^* are arbitrary constants. Although the resulting solution for $w(x,t)$ given by

$$w(x,t) = \left\{A^* e^{\lambda x} + B^* e^{-\lambda x}\right\}\left\{C^* e^{\lambda c_0 t} + D^* e^{-\lambda c_0 t}\right\} \tag{7.196}$$

is mathematically acceptable, it has limitations in that $w(x,t)$ has *exponential variations* in both *space* and *time*.

If the constant λ^2 is negative, the equation (7.191) gives

$$\frac{d^2X}{dx^2} + \lambda^2 X = 0 \tag{7.197}$$

and

$$\frac{d^2T}{dt^2} + \lambda^2 c_0^2 T = 0 \tag{7.198}$$

These have solutions

$$X(x) = A\sin(\lambda x) + B\cos(\lambda x) \tag{7.199}$$

$$T(t) = C\sin(c_0\lambda t) + D\cos(c_0\lambda t) \tag{7.200}$$

where A, B, C and D are arbitrary constants and the resulting expression of $w(x,t)$ is periodic in *both* x and t. The constant λ in (7.199) and (7.200) can be assigned a positive value without altering the generality of the solution.

When the constant λ^2 is zero, the integration of the differential equations (7.192) and (7.193) gives

$$X(x) = A'x + B' \tag{7.201}$$

$$T(t) = C't + D' \tag{7.202}$$

where A', B', C' and D' are arbitrary constants. To satisfy the boundary conditions applicable to a stretched string of length ℓ with fixed end (see e.g. Figure 7.17) we require

$$X(0) = 0 \quad ; \quad X(\ell) = 0 \tag{7.203}$$

(i) For the solution with $\lambda^2 > 0$ (i.e. equation (7.194)) we have

$$A^* + B^* = 0 \quad ; \quad A^* e^{\lambda \ell} + B^* e^{-\lambda \ell} = 0 \tag{7.204}$$

For a non-trivial solution we require $e^{\lambda \ell} - e^{-\lambda \ell} = 0$; this is, however, always non-zero for$\lambda \neq 0$. Hence $X(x)$ defined by (7.194) is identically zero for the case of the stretched string with fixed ends. (Also, note that the end fixity is necessary to maintain the tension T_0).

(ii) For the solution with $\lambda^2 = 0$ (i.e. equation (7.201)) we have

$$A' = B' \equiv 0 \tag{7.205}$$

Hence, the only applicable solution for $w(x,t)$ takes the form

$$w(x,t) = [A \sin(\lambda x) + B \cos(\lambda x)] \, [C \sin(c_0 \lambda t) + D \cos(c_0 \lambda t)] \tag{7.206}$$

To satisfy the boundary condition at $x = 0$ for all t we require

$$B = 0 \tag{7.207}$$

Similarly, to satisfy the boundary condition at $x = \ell$ for all t we require

$$A \sin(\lambda \ell) = 0 \tag{7.208}$$

The result $A = 0$ will render the third solution, trivial. Hence a *non-trivial* solution is obtained if we set

$$\sin(\lambda\ell) = 0 \tag{7.209}$$

The positive solutions of (7.209) are

$$\lambda_n \ell = n\pi \quad ; \quad n = 0, 1, 2, ... \tag{7.210}$$

where n is an integer. As observed in previous chapters, the result (7.209) is the *eigen-value equation* for the problem and the values of λ_n generated through the result (7.210) are referred to as the *eigen-values*, i.e.

$$\lambda_n = \frac{n\pi}{\ell} \quad ; \quad n = 0, 1, 2, ... \tag{7.211}$$

The solution of the wave equation which satisfies the boundary conditions at the ends of the stretched string and applicable to the eigen-value λ_n (defined by (7.211)) can be written as

$$w_n(x,t) = \sin\left(\frac{n\pi x}{\ell}\right)\left[C_n \sin\left(nc_0\frac{\pi t}{\ell}\right) + D_n \cos\left(nc_0\frac{\pi t}{\ell}\right)\right] \tag{7.212}$$

where C_n and D_n are arbitrary constants, implying that these arbitrary constants can take different values as n varies.

The following comments should be noted.

(i) The displaced form of the stretched string defined by (7.212) is referred to as the nth *eigenfunction* or the nth *normal mode* of vibration.

(ii) The normal modes corresponding to *negative integers* are identical to those which correspond to the *positive integers*.

(iii) The period of the nth *normal mode* is $2\ell/nc_0$; which translates to $nc_0/2\ell$ cycles per second.

(iv) The frequency can also be expressed as

$$f = \frac{n}{2\ell}\left(\frac{T_0}{\rho^*}\right)^{\frac{1}{2}} \tag{7.213}$$

The frequency can be increased by either reducing ℓ or increasing the tension T_0.

(v) For a given ℓ, T_0 and ρ^*, the first normal mode $n = 1$ vibrates with the *lowest frequency* and

$$f_0 = \sqrt{\frac{T_0}{4\ell^2\rho^*}} = \frac{c_0}{2\ell} \tag{7.214}$$

is called the *fundamental frequency.*

(vi) If the stretched string can be made to vibrate in a *higher normal mode*, the frequency is increased by an *integer multiple* (i.e. development of *harmonics*).

The deflected configuration of the stretched string corresponding to a given normal mode at a specific time $t = t^*$ can be obtained from (7.212). The deflected shapes associated with the first three normal modes and the associated frequencies are given in Figure 7.18.

If we consider the initial conditions that were prescribed for Example 7.6 (i.e. equations (7.173) and (7.174)) it is evident that the only eigenfunction required to satisfy the initial condition (7.173) is $w_1(x,t)$ which gives

$$w_1(x,0) = \sin\left(\frac{\pi x}{\ell}\right)[D_1] = w_0 \sin\left(\frac{\pi x}{\ell}\right) \tag{7.215}$$

To satisfy the second initial condition defined by (7.174) we require

$$C_1 = 0 \tag{7.216}$$

Again, the motion of the stretched string resulting from the prescribed initial displacement (7.173) is given by the first mode solution

$$w_1(x,t) = w_0 \sin\left(\frac{\pi x}{\ell}\right) \cos\left(c_0 \frac{\pi t}{\ell}\right) \tag{7.217}$$

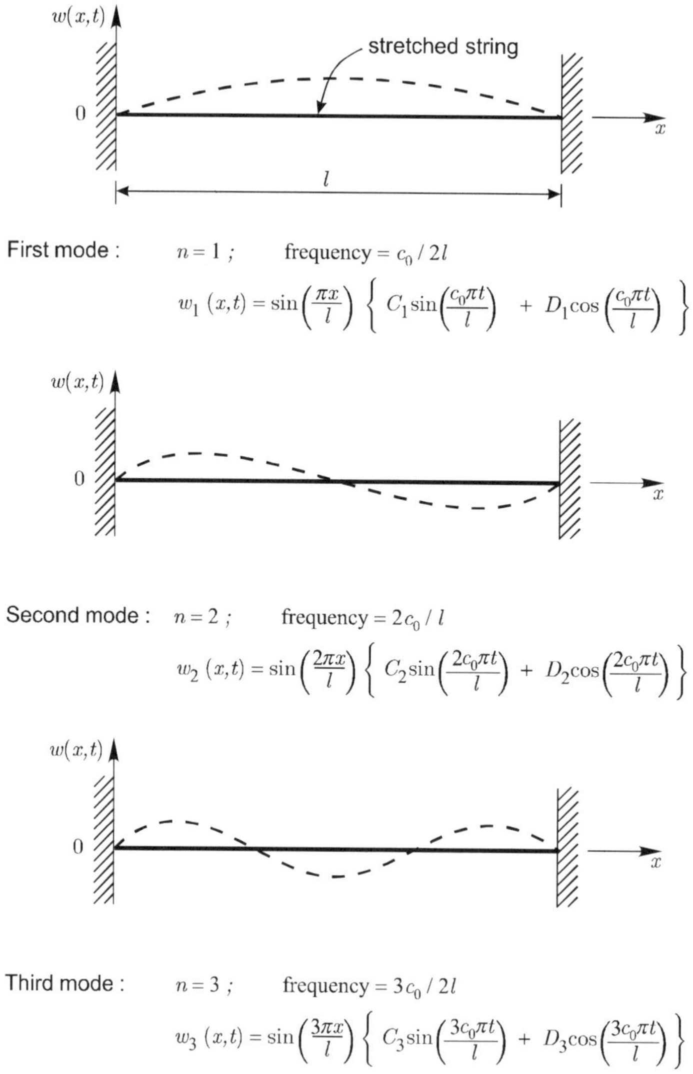

Figure 7.18: Normal modes of a vibrating stretched string.

In general, the *initial conditions* specified for a particular vibration problem can be more complicated than the idealized sinusoidal form discussed previously. In such cases, all normal modes need to be combined to satisfy the initial conditions. The infinite sum of all normal modes generates the deflected shape

$$w(x,t) = \sum_{n=1}^{\infty} \left\{ C_n \sin\left(c_0 \frac{n\pi t}{\ell} \right) + D_n \cos\left(c_0 \frac{n\pi t}{\ell} \right) \right\} \sin\left(\frac{n\pi x}{\ell} \right) \tag{7.218}$$

The constants C_n and D_n can be evaluated by using Fourier series techniques.

Example 7.7

A stretched string of finite length is held fixed at $x = 0$ and $x = \ell$ and subjected to an initial displacement

$$w(x,0) = w_0 \sin\left(\frac{\pi x}{\ell} \right) \tag{7.219}$$

and an initial velocity

$$\left[\frac{\partial w}{\partial t} \right]_{t=0} = v_0 \sin\left(\frac{\pi x}{\ell} \right) \tag{7.220}$$

as indicated in Figure 7.19. Formulate the initial boundary value problem and derive an expression for the time-dependent displacement of the stretched string.

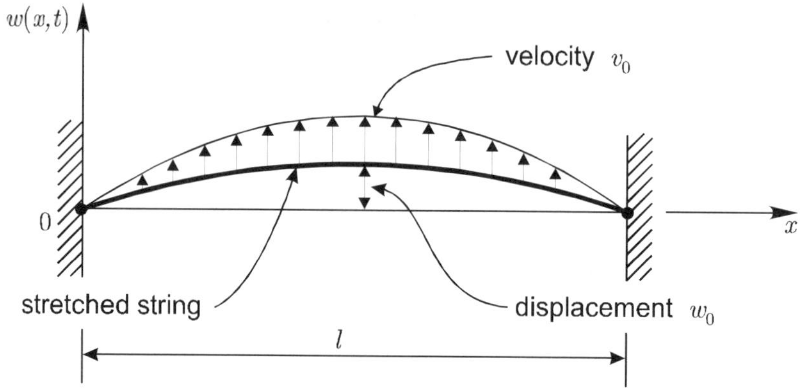

Figure 7.19: Stretched string subjected to initial displacement and initial velocity.

Solution

Since the ends of the stretched string are held fixed, and the initial displacement takes the form (7.219), the single eigenfunction solution required is given by

$$w(x,t) = \sin\left(\frac{\pi x}{\ell}\right)\left[C_1 \sin\left(c_0 \frac{\pi t}{\ell}\right) + D_1 \cos\left(c_0 \frac{\pi t}{\ell}\right)\right] \tag{7.221}$$

Using the initial condition (7.219) we have

$$D_1 = w_0 \tag{7.222}$$

Using the initial condition (7.220) we have

$$C_1 = \frac{v_0 \ell}{\pi c_0} \tag{7.223}$$

The time-dependent displacement of the stretched string is given by

$$w(x,t) = w_0 \sin\left(\frac{\pi x}{\ell}\right)\left[\frac{1}{\pi}\left(\frac{v_0 \ell}{c_0 w_0}\right)\sin\left(c_0\frac{\pi t}{\ell}\right) + \cos\left(c_0\frac{\pi t}{\ell}\right)\right] \quad (7.224)$$

7.2.4 Vibrations of a stretched string: variable boundary conditions

In previous sections dealing with wave propagation and vibrations in a stretched finite string, the boundary conditions applicable to the problems examined involved cases where the ends of the string were held stationary. This assumption is satisfied by the majority of problems involving the applications of the theory to practical problems. On occasions, the boundary constraints are different from the homogeneous constraints associated with stationary ends. In this section we shall examine, very briefly, certain cases where the homogeneous boundary conditions are relaxed.

(a) *A free end boundary*

In the context of a stretched finite string problem, a free end boundary condition involves a situation where the spatial gradient of the displacement is specified physically, this can involve unrestricted vertical movement of the end of the stretched string but preserving a *zero slope* at the boundary. Both ends of the stretched string cannot be prescribed with the "*free end*" boundary condition since this would lead to time-dependent displacements which are *indeterminate*. Also, despite the designation as a "*free end*", the string is capable of maintaining the tension constant during the free vertical movement of the boundary. We could visualize this boundary condition as a mechanical device such as a "*slider in a track*" (Figure 7.20) which can maintain the tension T_0, allows free movement and gives the boundary condition

$$\left[\frac{\partial w}{\partial x}\right]_{x=0} = 0 \quad (7.225)$$

It must be recognized that this boundary condition represents a highly idealized situation where the stretched string, which has no resistance to bending, is expected to maintain a zero slope at the boundary. Nonetheless, this boundary condition is utilized for the analysis of "stretched string" problems where the extensional stiffness is significantly higher than the flexural stiffness.

Example 7.8

A flexible stretched string is constrained to move with zero slope at the end $x = 0$ and held fixed against any movement at the end $x = \ell$ (Figure 7.20). Derive an expression for the time-dependent motion of the string if it is subjected to an initial displacement

$$w(x,0) = w_0 \cos\left(\frac{\pi x}{2\ell}\right) \tag{7.226}$$

and released from this position with zero velocity.

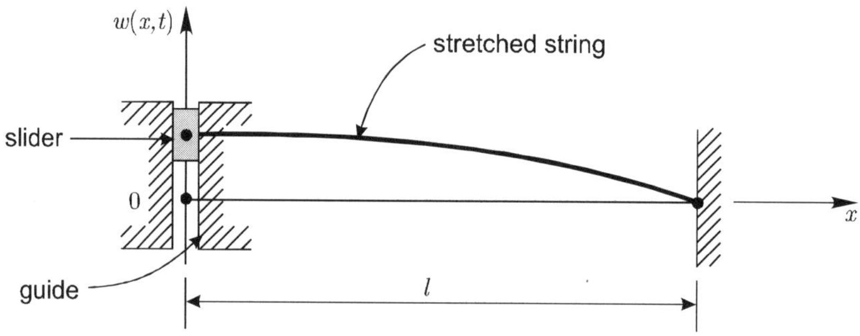

Figure 7.20: Free end boundary condition for a stretched string

Solution

Considering the separation of variables solution (7.206) for the governing partial differential equation

$$\frac{\partial^2 w}{\partial x^2} - \frac{1}{c_0^2}\frac{\partial^2 w}{\partial t^2} = 0 \tag{7.227}$$

we can write a generalized eigenfunction solution as

$$w(x,t) = \sum_{n=1,2,} \{A_n \sin(\lambda_n x) + B_n \cos(\lambda_n x)\}\{C_n \sin(c_0 \lambda_n t) + D_n \cos(c_0 \lambda_n t)\} \tag{7.228}$$

which should satisfy the boundary conditions

$$\left[\frac{\partial w}{\partial x}\right]_{x=0} = 0 \tag{7.229}$$

$$w(\ell, t) = 0 \tag{7.230}$$

and initial conditions

$$w(x,0) = w_0 \cos\left(\frac{\pi x}{2\ell}\right) \tag{7.231}$$

$$\left[\frac{\partial w}{\partial t}\right]_{t=0} = 0 \tag{7.232}$$

To satisfy the boundary conditions (7.229) and (7.230), we require

$$A_n = 0 \tag{7.233}$$

$$B_n \cos(\lambda_n \ell) = 0; \quad \text{or } \lambda_n = \frac{n\pi}{2\ell}; \quad (n = 1, 3, 5, ...) \tag{7.234}$$

The displacement of the stretched string can now be reduced to the form

$$w(x,t) = \sum_{n=1,3,}^{\infty} \cos\left(\frac{n\pi x}{2\ell}\right) \left\{C_n^* \sin\left(c_0 \frac{n\pi t}{2\ell}\right)\right.$$

$$+ D_n^* \cos\left(c_0 \frac{n\pi t}{2\ell}\right)\Bigg\} \tag{7.235}$$

where C_n^* and D_n^* are 'revised' arbitrary constants. To satisfy the initial condition (7.232) we require

$$C_n^* = 0 \tag{7.236}$$

and the initial condition (7.231) gives

$$w(x,0) = \sum_{n=1,3,}^{\infty} \cos\left(\frac{n\pi x}{2\ell}\right)[D_n^*] = w_0 \cos\left(\frac{\pi x}{2\ell}\right) \tag{7.237}$$

The initial condition can be *exactly satisfied* by setting

$$D_1^* = w_0 \tag{7.238}$$

and all other $D_n^* \equiv 0$. Hence the final solution for the time-dependent displacement of the stretched string is given by

$$w(x,t) = w_0 \cos\left(\frac{\pi x}{2\ell}\right) \sin\left(\frac{c_0 \pi t}{2\ell}\right) \tag{7.239}$$

• • •

(b) *An elastically constrained boundary*

A further type of end condition applicable to a stretched string is the *elastically constrained boundary*, which could be visualized as an *elastic spring* which, during deflection, exerts a force proportional to the deflection at the elastically constrained location. There are no constraints on the derivative of the displacement at an elastically restrained end. An example of an elastically restrained end at $x = 0$ is shown in Figure 7.21. By considering force equilibrium at the elastically restrained end $x = 0$, we can show that the boundary condition at $x = 0$ is

$$k[w(x,t)]_{x=0} = T_0 \left[\frac{\partial w}{\partial x}\right]_{x=0} \tag{7.240}$$

where k is the *elastic stiffness* of the spring with dimensions of force/unit length.

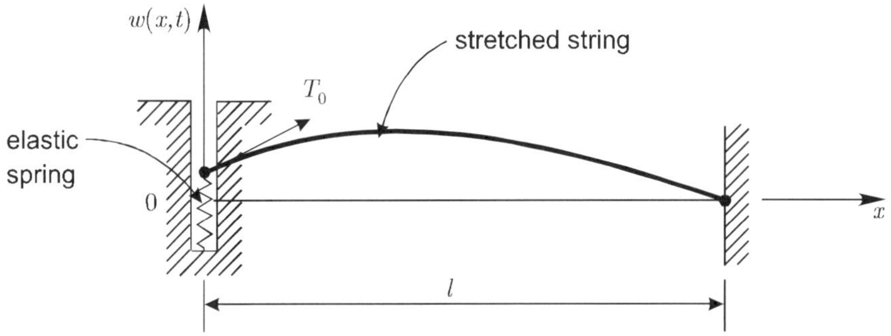

Figure 7.21: Elastically restrained boundary for a stretched string

---

Example 7.9

Consider the elastically restrained stretched finite string problem illustrated in Figure 7.21. Assuming that the generalized solution for the time-dependent deflection of the string can be written as

$$w(x,t) = \sum_{n=1,2,}^{\infty} \{A_n \sin(\lambda_n x) + B_n \cos(\lambda_n x)\}\{C_n \sin(c_0 \lambda_n t) + D_n \cos(c_0 \lambda_n t)\} \tag{7.241}$$

determine the characteristic equation that needs to be solved to determine λ_n, if the string is released from an initially displaced position with zero velocity.

Solution

In order to satisfy the zero initial velocity condition we require

$$C_n = 0 \tag{7.242}$$

The solution (7.241) can now be written as

$$w(x,t) = \sum_{n=1,2} \{A_n^* \sin(\lambda_n x) + B_n^* \cos(\lambda_n x)\} \cos(c_0 \lambda_n t) \tag{7.243}$$

The boundary conditions applicable to the problem are

$$k\left[w(x,t)\right]_{x=0} = T_0 \left[\frac{\partial w}{\partial x}\right]_{x=0} \tag{7.244}$$

and

$$w(\ell, t) = 0 \tag{7.245}$$

Considering (7.243) and the boundary conditions (7.244) and (7.245) we obtain, respectively,

$$\begin{gathered} \lambda_n T_0 A_n^* - k B_n^* = 0 \\ \sin(\lambda_n \ell) A_n^* + \cos(\lambda_n \ell) B_n^* = 0 \end{gathered} \tag{7.246}$$

For a non-trivial solution we require

$$\left(\frac{T_0}{k\ell}\right)(\lambda_n\ell)\cos(\lambda_n\ell) + \sin(\lambda_n\ell) = 0 \tag{7.247}$$

This is the *characteristic equation* which needs to be solved to obtain the eigenvalues λ_n. The time-dependent deflections of the stretched finite string with an elastically restrained end can be obtained once the remaining initial condition for the displacement is prescribed.

7.2.5 Forced vibration of a stretched finite string

In previous sections, we have concentrated largely on the examination of free vibrations of stretched strings of both finite and infinite extent, described by the homogeneous form of the wave equation (7.10). The analysis of the free vibration problem is particularly useful from the point of view of identification of basic processes and solution techniques associated with the dynamics of the stretched string. Forced vibrations of a stretched string can occur when external forces induced by actions resulting from gravity loads, wind forces, hydrodynamic forces and other mechanical effects are in operation. The derivation of the partial differential equation which governs forced motions of the stretched string is given in section 7.1.

For a forcing term $f^*(x,t)$ (with dimensions of force/unit length), the governing partial differential equation is given by

$$\frac{\partial^2 w}{\partial x^2} - \frac{1}{c_0^2}\frac{\partial^2 w}{\partial t^2} = -\frac{f^*(x,t)}{T_0} \tag{7.248}$$

We have already discussed the solution of this type of equation in connection with the development of the *Green's function* for the stretched string of infinite extent (see e.g. section 7.1.6), where both Laplace and Fourier transforms were used to develop the required solution. In the case of a stretched string of finite length, however, appropriate *boundary conditions* need to be specified in addition to the initial conditions. Consider, for example, the initial conditions

$$w(x,0) - N(x) \tag{7.249}$$

$$\left[\frac{\partial w}{\partial t}\right]_{t=0} = M(x) \tag{7.250}$$

where $M(x)$ and $N(x)$ are arbitrary functions. The boundary conditions are selected to be restrained end conditions which give

$$w(0,t) = 0 \quad ; \quad w(\ell,t) = 0 \tag{7.251}$$

If we apply the Laplace transform to (7.248) we obtain (see also (1.92))

$$c_0^2 \frac{d^2\ \widehat{w}(x,s)}{dx^2} - s^2 \widehat{w}(x,s) = -\frac{\widehat{f^*}}{\rho^*}(x,s) - sN(x) - M(x) \tag{7.252}$$

where $\widehat{w}(x,s)$ denotes the Laplace transformed variable.

If we apply the finite Fourier sine transform to $\widehat{w}(x,s)$ defined by (see e.g. (1.203))

$$\widetilde{\widehat{w}}(n,s) = \frac{1}{\ell}\int_0^\ell \widehat{w}(x,s) \sin\left(\frac{n\pi x}{\ell}\right) dx \tag{7.253}$$

to (7.252) we obtain

$$\left\{-c_0^2 \frac{n^2\pi^2}{\ell^2} - s^2\right\} \widetilde{\widehat{w}}(n,s) = -\frac{\widetilde{\widehat{f^*}}(n,s)}{\rho^*} - s\widetilde{N}(n) - \widetilde{M}(n) \tag{7.254}$$

and the solution of the problem in the Laplace-Fourier transformed domain is given by

$$\widetilde{\overline{w}}(n,s) = \frac{\widetilde{\widehat{f}}^*(n,s) + \rho^*\{s\widetilde{N}(n) + \widetilde{M}(n)\}}{\rho^*\left\{\dfrac{c_0^2 n^2 \pi^2}{\ell^2} + s^2\right\}} \tag{7.255}$$

Inverting the Fourier sine transform (see e.g. (1.200) and (1.216)) we obtain

$$\begin{aligned}\overline{w}(x,s) = \frac{2}{\ell}\sum_{n=1,2,} \frac{\sin(n\pi x/\ell)}{\left[\frac{c_0^2 n^2\pi^2}{\ell^2} + s^2\right]} &\left\{\frac{1}{\rho^*}\int_0^\ell \widehat{f}^*(\xi,s)\sin\left(\frac{n\pi\xi}{\ell}\right)d\xi\right. \\ &\left. + s\int_0^\ell N(\xi)\sin\left(\frac{n\pi\xi}{\ell}\right)d\xi + \int_0^\ell M(\xi)\sin\left(\frac{n\pi\xi}{\ell}\right)d\xi\right\}\end{aligned} \tag{7.256}$$

The Laplace inversion of (7.256) can be achieved by taking advantage of the convolution theorem (see e.g. (1.133) and (1.135)).

We note that

$$\mathcal{L}^{-1}\left\{\frac{1}{s^2 + (c_0^2 n^2\pi^2/\ell^2)}\right\} = \frac{1}{\zeta_n}\sin(\zeta_n t) \tag{7.257}$$

$$\mathcal{L}^{-1}\left\{\frac{s}{s^2 + (c_0^2 n^2\pi^2/\ell^2)}\right\} = \cos(\zeta_n t) \tag{7.258}$$

where

$$\zeta_n = \frac{n c_0 \pi}{\ell} \tag{7.259}$$

Using these results and (1.135) we have

$$\mathcal{L}^{-1}\left\{\frac{\widehat{f}^*(\xi,s)}{(s^2+\zeta_n^2)}\right\} = \frac{1}{\zeta_n}\int_0^t f^*(\xi,\tau)\sin\{\zeta_n(t-\tau)\}d\tau \tag{7.260}$$

If we carry out the inversion of the three integral terms of (7.256) separately, we obtain

$$\begin{aligned} w(x,t) = & \frac{2}{\rho^* c_0 \ell} \sum_{n=1,2,}^{\infty} \frac{\sin(n\pi x/\ell)}{(n\pi/\ell)} \int_0^\ell \sin\left(\frac{n\pi\xi}{\ell}\right) d\xi \\ & \cdot \int_0^t f^*(\xi,\tau) \sin\zeta_n(t-\tau) d\tau \\ & + \frac{2}{\ell} \sum_{n=1,2,}^{\infty} \sin\left(\frac{n\pi x}{\ell}\right) \cos(\zeta_n t) \int_0^\ell N(\xi) \sin\left(\frac{n\pi\xi}{\ell}\right) d\xi \\ & + \frac{2}{\ell} \sum_{n=1,2,}^{\infty} \sin\left(\frac{n\pi x}{\ell}\right) \frac{\sin(\zeta_n t)}{\zeta_n} \int_0^\ell M(\xi) \sin\left(\frac{n\pi\xi}{\ell}\right) d\xi \end{aligned} \tag{7.261}$$

The result (7.261) can be evaluated either exactly or numerically depending upon the form of $f^*(x,t)$, $N(x)$ and $M(x)$. The analysis of the *forced vibration problem* is somewhat simpler when $f^*(x,t)$ is either a function of x or a function of t.

Example 7.10

A stretched string of finite length ℓ is acted by a force of constant magnitude p_0 /unit length (Figure 7.22).The string is held stationary at both ends and subjected to an initial displacement which is given by

$$w(x,0) = w_0 \sin\left(\frac{\pi x}{\ell}\right) \tag{7.262}$$

If the string is released from this initial position with zero velocity, determine the resulting time-dependent deflections of the stretched string.

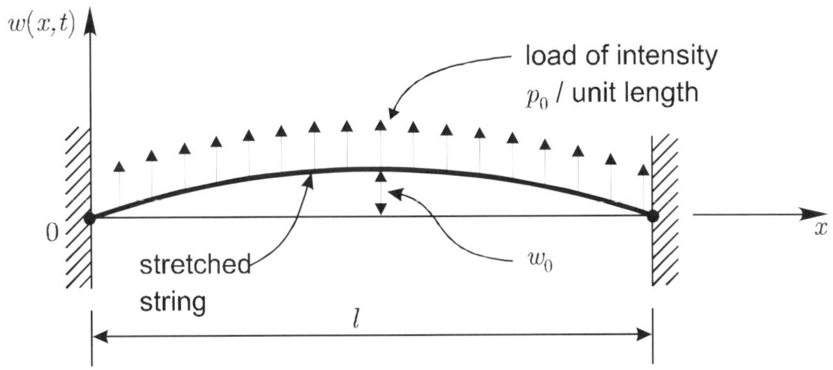

Figure 7.22: Forced vibration of a stretched string

Solution

The partial differential equation governing the deflection of the stretched string is given by

$$\frac{\partial^2 w}{\partial x^2} - \frac{1}{c_0^2}\frac{\partial^2 w}{\partial t^2} = -\frac{p_0}{T_0} = \quad \text{const.} \tag{7.263}$$

and the initial conditions are

$$w(x,0) = w_0 \sin\left(\frac{\pi x}{\ell}\right) \tag{7.264}$$

$$\left[\frac{\partial w}{\partial t}\right]_{t=0} = 0 \tag{7.265}$$

with boundary conditions

$$w(0,t) = 0 \quad ; \quad w(\ell,t) = 0 \tag{7.266}$$

The procedure for investigating the forced vibration problem is to assume that the solution can be represented in the form

$$w(x,t) = W(x,t) + U(x) \tag{7.267}$$

and to select the function $U(x)$ such that the problem governing $W(x,t)$ corresponds to a free vibration problem with $p_0 = 0$. Substituting (7.267) into (7.263) we obtain

$$\frac{\partial^2 W}{\partial x^2} + \frac{d^2 U}{dx^2} - \frac{1}{c_0^2}\frac{\partial^2 W}{\partial t^2} = -\frac{p_0}{T_0} \tag{7.268}$$

The partial differential equation governing $W(x,t)$ will be homogeneous if

$$\frac{d^2 U}{dx^2} + \frac{p_0}{T_0} = 0 \tag{7.269}$$

This ordinary differential equation can be integrated to give

$$U(x) = -\frac{p_0 x^2}{2T_0} + Ex + G \tag{7.270}$$

where E and G are arbitrary constants. The boundary conditions governing the problem (equation (7.266)) are

$$w(0,t) = W(0,t) + U(0) = 0 \tag{7.271}$$

$$w(\ell,t) = W(\ell,t) + U(\ell) = 0 \tag{7.272}$$

The boundary conditions governing $W(x,t)$ will be homogeneous if

$$U(0) = 0 \quad ; \quad U(\ell) = 0 \tag{7.273}$$

Using (7.273), (7.270) can be evaluated in the form

$$U(x) = -\frac{p_0 x(x-\ell)}{2T_0} \tag{7.274}$$

The initial conditions (7.264) and (7.265) can be written as

$$w(x,0) = W(x,0) - \frac{p_0 x(x-\ell)}{2T_0} = w_0 \sin\left(\frac{\pi x}{\ell}\right) \tag{7.275}$$

$$\left[\frac{\partial w}{\partial t}\right]_{t=0} = \left[\frac{\partial W}{\partial t}\right]_{t=0} = 0 \tag{7.276}$$

We can now formulate the *auxiliary initial boundary value problem* governing $W(x,t)$ as follows: the partial differential equation takes the form

$$\frac{\partial^2 W}{\partial x^2} - \frac{1}{c_0^2}\frac{\partial^2 W}{\partial t^2} = 0 \tag{7.277}$$

with boundary conditions

$$W(0,t) = 0 \quad ; \quad W(\ell,t) = 0 \tag{7.278}$$

and initial conditions

$$W(x,0) = w_0 \sin\left(\frac{\pi x}{\ell}\right) + \frac{p_0 x(x-\ell)}{2T_0} \tag{7.279}$$

$$\left[\frac{\partial W}{\partial t}\right]_{t=0} = 0 \tag{7.280}$$

This initial boundary value problem has the same general form as the problems we have discussed in previous sections.

Consider an *eigenfunction expansion* solution of the form

$$W(x,t) = \sum_{n=1,2}^{\infty} \{A_n \sin(\lambda_n x) + B_n \cos(\lambda_n x)\}\{C_n \sin(c_0 \lambda_n t) + D_n \cos(c_0 \lambda_n t)\} \tag{7.281}$$

To satisfy the boundary conditions (7.278) we require

$$B_n \equiv 0 \tag{7.282}$$

and

$$A_n \sin(\lambda_n \ell) = 0 \tag{7.283}$$

$$\text{i.e.}\quad \lambda_n = \frac{n\pi}{\ell} \quad ; \quad n = 1, 2, ... \tag{7.284}$$

Similarly, to satisfy the initial condition (7.280) we require

$$C_n \equiv 0 \tag{7.285}$$

The eigenfunction expansion solution can now be written as

$$W(x,t) = \sum_{n=1,2,} A_n^* \sin\left(\frac{n\pi x}{\ell}\right) \cos\left(c_0 \frac{n\pi t}{\ell}\right) \tag{7.286}$$

where A_n^* is a new constant which combines A_n and D_n. The initial condition (7.279) gives

$$\sum_{n=1,2,}^{\infty} A_n^* \sin\left(\frac{n\pi x}{\ell}\right) = w_0 \sin\left(\frac{\pi x}{\ell}\right) + \frac{p_0 x(\ell - x)}{2T_0} \tag{7.287}$$

The constants A_n^* can be obtained by using a Fourier sine series representation of $p_0 x(\ell - x)/2T_0$. It can be shown that

$$A_n^* = \frac{p_0}{2T_0}\left(\frac{4\ell^2}{n^3\pi^3}\right)\{1 - (-1)^n\} \quad ; \quad n = 1, 3, ... \tag{7.288}$$

The additional solution

$$A_1^* = w_0 \tag{7.289}$$

corresponds to the first term in the right hand side of (7.287). The time-dependent deflection of the stretched string due to the forced motion is given by

$$\begin{aligned} w(x,t) = & -\frac{p_0 x(x-\ell)}{2T_0} + w_0 \sin\left(\frac{\pi x}{\ell}\right) \cos\left(c_0 \frac{\pi t}{\ell}\right) \\ & + \frac{p_0}{2T_0} \sum_{n=1,3,} \frac{4\ell^2}{n^3\pi^3}[1 - (-1)^n] \times \\ & \left\{\sin\left(\frac{n\pi x}{\ell}\right) \cos\left(c_0 \frac{n\pi t}{\ell}\right)\right\} \end{aligned} \tag{7.290}$$

It can be verified that the solution (7.290) satisfies the inhomogeneous partial differential equation (7.263), the initial conditions (7.264) and (7.265) and the boundary conditions (7.266).

7.3 Wave motion in stretched strings: non-classical effects

In this section we will briefly review certain modifications to the stretched string problem which needs to be considered in the application of the classical analysis to practical problems. A detailed consideration of these effects are beyond the scope of this volume. The basic methods of analyses outlined earlier can, however, be adopted for the study of these non-classical effects.

7.3.1 Elastically supported string

The stretched string problems discussed in the previous sections focus primarily on waves and vibrations which are governed by the classical wave equation for either free vibrations or forced vibrations. An important aspect of these solutions is that the waves propagate without distortion . The classical solution by d'Alembert illustrates this aspect of the wave equation. This particular attribute will not be evident in other problems which are not governed by variations of the classical wave equation. An example of such a situation is the problem of an *elastically supported string.* The elastic support can be visualized as a system of closely spaced independent springs with constant stiffness which can exert a reactive force proportional to the deflection of the stretched string at a particular point. An illustration of the elastically supported string is shown in Figure 7.23.

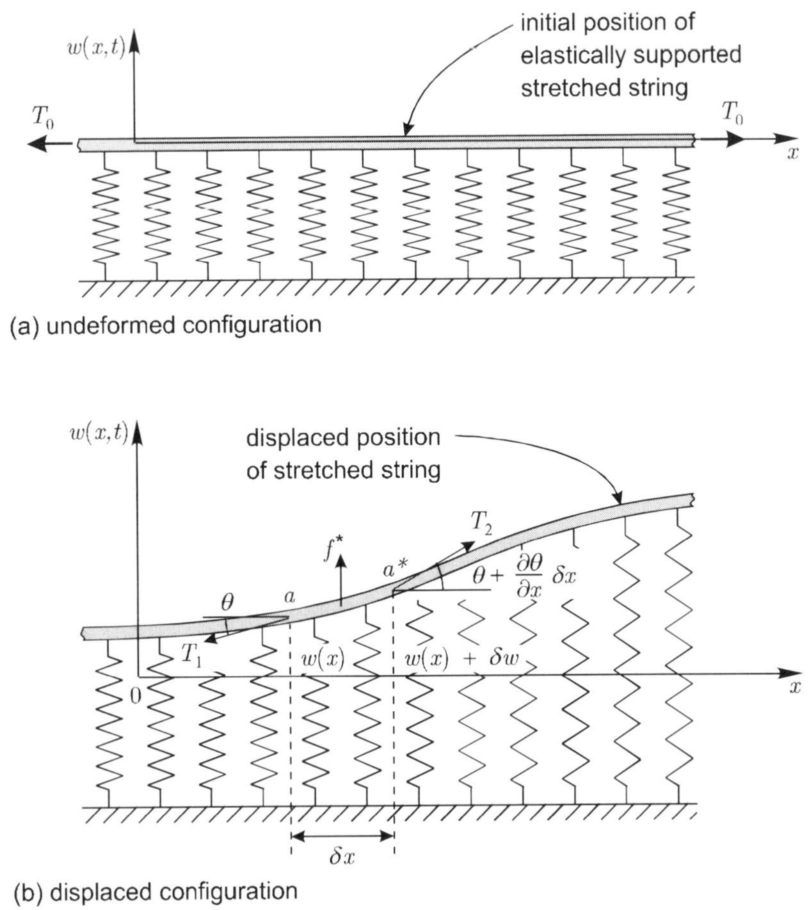

Figure 7.23: Displacement of the elastically supported stretched string.

Following the procedure outlined in Section 7.1, in equations (7.1) to (7.6) we can modify the equation of motion by incorporating a term

$$k\ w(x,t)(\delta x)$$

on the right hand side of (7.8) to account for the restraining force exerted by the springs which are connected to the stretched string. Here k represents the stiffness of the elastic support and has dimensions of force / (length)^2. The resulting equation of the forced vibration of the stretched string can be written in the form

$$T_0 \frac{\partial^2 w}{\partial x^2} - kw + f^* = \rho^* \frac{\partial^2 w}{\partial t^2} \tag{7.291}$$

In the case of free vibrations of the elastically supported stretched string (7.291) reduces to

$$\frac{\partial^2 w}{\partial x^2} - \left(\frac{k}{T_0}\right) w - \frac{1}{c_0^2} \frac{\partial^2 w}{\partial t^2} = 0 \tag{7.292}$$

It is clear that the addition of the elastic support makes solutions of the form $f(x \pm c_0 t)$ inadmissible. As a result, a propagating pulse is expected to *distort.* We can observe the character of the *wave distortion* by examining the propagation of a harmonic wave such as

$$w(x,t) = Ae^{i(\gamma x - \omega t)} \tag{7.293}$$

Substituting (7.293) in (7.292) we obtain

$$A\left[-\gamma^2 - \frac{k}{T_0} + \frac{\omega^2}{c_0^2}\right] e^{i(\gamma x - \omega t)} = 0 \tag{7.294}$$

For a non-trivial solution we require either

$$\omega^2 = c_0^2 \left(\gamma^2 + \frac{k}{T_0}\right) \tag{7.295}$$

or

$$\gamma^2 = \frac{\omega^2}{c_0^2} - \frac{k}{T_0} \tag{7.296}$$

These imply

$$\omega = \omega(\gamma) \quad ; \quad \gamma = \gamma(\omega) \tag{7.297}$$

Hence, we can present the results for the deflection in the form

$$w(x,t) = Ae^{i\gamma(x-ct)} \tag{7.298}$$

where $c(=\omega/\gamma)$ is the *phase velocity* of the wave defined by

$$c = c_0 \left[1 + \frac{k}{T_0\gamma^2}\right]^{\frac{1}{2}} \tag{7.299}$$

or

$$\gamma^2 = \frac{k/T_0}{[(c/c_0)^2 - 1]} \tag{7.300}$$

The results (7.299) and (7.300) imply that

$$c = c(\gamma) \quad ; \quad \gamma = \gamma(c) \tag{7.301}$$

and when the elastic support is absent $c \equiv c_0$. A further set of representations can be obtained by using the relationship for the phase velocity $\omega = \gamma c$ and eliminating either γ from (7.295) or ω from (7.296). We obtain

$$\omega^2 = \frac{kT_0 c^2}{[(c/c_0)^2 - 1]} \quad ; \quad \gamma^2 = \frac{\omega^2 c_0^2}{[\omega^2 - (kc_0^2/T_0)]} \tag{7.302}$$

i.e.

$$\omega = \omega(c) \quad ; \quad \gamma = \gamma(c) \tag{7.303}$$

These combinations of results (7.297), (7.302) and (7.303) indicate that a harmonic frequency ω can propagate only at a specific velocity c governed by $\omega = \omega(c)$. If we consider any pulse to be composed of *Fourier superposition of harmonic waves*, then each Fourier component will propagate with its own individual velocity. These separate components will become increasingly out of phase as they propagate from the source. This will lead to the *distortion* of the pulse.

Conditions governing wave propagation in an elastically supported string can also be examined in the light of these results.

(i) *Propagating waves*

An important inference of the propagation of a wave in an elastically supported string can also be obtained by considering (7.296). We have

$$\gamma = \pm \left(\frac{\omega^2}{c_0^2} - \frac{k}{T_0} \right)^{\frac{1}{2}} \tag{7.304}$$

The above roots are real if $(\omega/c_0)^2 > (k/T_0)$. The form of the harmonic wave (7.293) can be written as

$$w(x,t) = Ae^{-i(\pm\gamma x + \omega t)} \tag{7.305}$$

The real roots therefore will give rise to *leftward* or *rightward* propagating waves depending upon the sign selected.

(ii) *Non-propagating waves*

If $(\omega/c_0)^2 < (k/T_0)$, then the wave number γ is *imaginary*. If we define $\overline{\gamma}^2 = -\gamma^2$, then the form of the wave (7.293) can be written as

$$w(x,t) = Ae^{-\overline{\gamma}x}e^{i\omega t} \tag{7.306}$$

This corresponds to a spatially varying but non-propagating disturbance. Hence imaginary wave numbers are not acceptable since they are non-propagating waves. Imaginary wave numbers are, however, important from the point of view of the study of transient effects.

(iii) *Transition conditions*

The condition, $(\omega/c_0)^2 = (k/T_0)$, represents the point of transition from a propagating to a non-propagating wave. From (7.296) we have $\gamma = 0$ and (7.295) gives $\omega_c = c_0(k/T_0)^{\frac{1}{2}}$. The motion of the stretched string takes the form

$$w(x,t) = Ae^{-i\omega_c t} \tag{7.307}$$

where ω_c is referred to as the *cut-off* frequency. There is no spatial variation in the motion and the stretched string vibrates as a single degree of freedom system with mass and stiffness.

7.3.2 Energy dissipation and damping in a stretched string

The problems dealing with the mechanics of a stretched string discussed in this chapter have solely focused on linear systems in which there are no provisions for dissipation of energy. This results in an idealized model of wave propagation and vibration problems where disturbances created will either propagate over unrestricted distances or systems will continue to vibrate without any evidence of attenuation of the initial conditions. In reality no such idealized system exists. The dynamics of the stretched string will be influenced by a variety of dissipative mechanisms including internal friction or material hysteresis of the string, air resistance, hydrodynamic resistance (in strings immersed in fluids) and other hysteretic phenomena. The processes contributing to dissipation can be broadly classified as viscous damping, and such damping effects can significantly alter the response of the stretched string.

As discussed in Section 6.1.6, Newtonian viscous effects are characterized by a linear relationship between the *shear stress* and the spatial gradient of velocity. Following such an argument,we assume that the viscous energy dissipation in the system can be represented by a *resistive force* to the motion of the string which takes the form

$$q(x,t) = \eta_D \frac{\partial w}{\partial t} \tag{7.308}$$

where η_D is the dynamic viscosity of the resistance with dimensions of (Force)(Time) / (Length)^2. This resistance can be incorporated in the basic equation of forced motion of the stretched string by assuming that $q(x,t)$ acts in a direction opposite to $f^*(x,t)$ (Figure 7.1) and is distributed in exactly the same way as f^*. The resulting equation of forced motion of the stretched string is given by

$$T_0 \frac{\partial^2 w}{\partial x^2} - \eta_D \frac{\partial w}{\partial t} + f^* = \rho^* \frac{\partial^2 w}{\partial t^2} \tag{7.309}$$

In the absence of a forcing function f^*, the "free vibrations" of a stretched string with viscous damping takes the form

$$\frac{\partial^2 w}{\partial x^2} - \frac{1}{c_0^2} \frac{\partial^2 w}{\partial t^2} = \frac{\eta_D}{T_0} \frac{\partial w}{\partial t} \tag{7.310}$$

when the "*damping*" term disappears $\eta_D \to 0$, the partial differential equation reduces to the *classical one-dimensional wave equation.* On the other hand when the mass per unit length $\rho^* \to 0$, the partial differential equation reduces to the *one-dimensional diffusion equation.* The character of the partial differential equation changes from a *hyperbolic* (wave equation) to a *parabolic* (diffusion equation) form as either $\eta_D \to 0$ or $\rho^* \to 0$. This suggests that there are competing influences of damping which will influence the wave propagation aspects.

7.4 Waves and vibrations in stretched membranes

The study of waves and vibrations in stretched flat membranes is a natural extension of the stretched string problem to two dimensions. The study of the dynamics of stretched membranes has a long history starting with the investigations of Leonhard Euler (1707-1783) and Joseph Lagrange (1736-1813) who introduced mathematical models for vibrating drumheads and other slender structures. The study of the dynamics of membrane-like structures is important to the modelling of vibrations in acoustical devices such as loudspeakers and receivers. In biomedical applications, the nasal passages and ear drums represent membranes with vibrational characteristics. Vibrational characteristics of membranes are also important to the study of pressure regulators used in flow monitoring. In civil engineering applications, vibrations of membranes is important to studies related to wind-induced oscillations of membrane roofs for large enclosures. Other flow-induced vibrations of membrane structures are encountered in mass transfer applications involving semi-permeable membranes. Similar membrane vibration problems are encountered in the study of sails and hot air balloons.

For the purposes of the presentation, a membrane is defined as a flat material region which has no flexural stiffness but possesses extensional stiffness. The membrane region is assumed to be subjected to a state of in-plane tension which is isotropic. Referring to Figure 7.24 it is assumed that, prior to deflection, every point in the membrane has an *isotropic state* of tension (i.e. the tension T_m is *oricntation independent*).

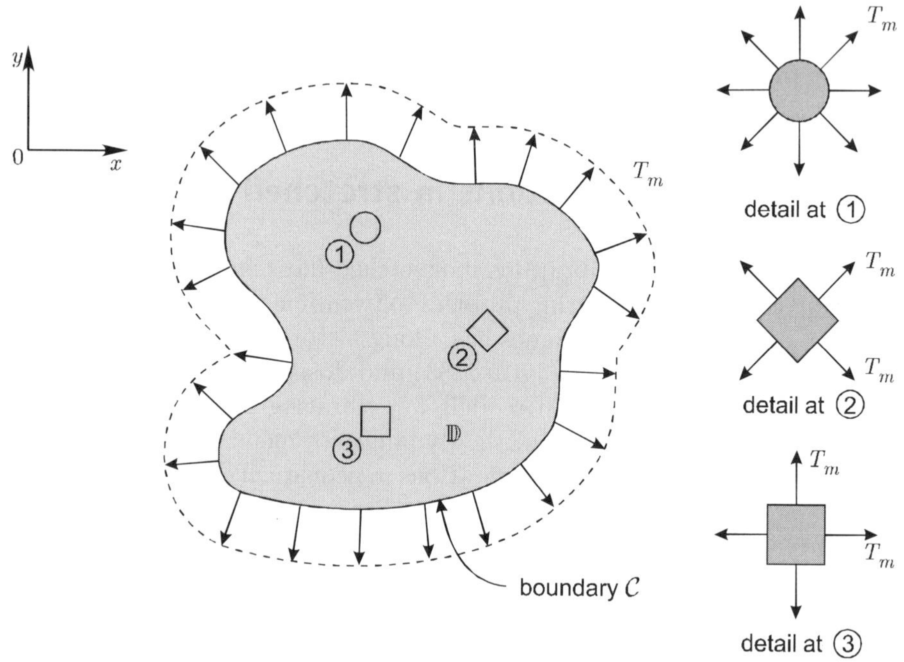

Figure 7.24: Isotropy of tension in a stretched membrane.

At the boundary of the membrane the tension T_m acts normal to the boundary $\mathcal{C}$ of the region $\mathbb{D}$. The configuration of the outer boundary is assumed to be such that the application of an all round tension does not lead to *wrinkling* of the deformed surface.

7.4.1 Equation of motion for a stretched membrane

We consider the problem of a membrane which is subjected to an isotropic tension T_m (with dimensions of force/unit length) in its flat, undisplaced state. During the dynamic behaviour, the membrane is assumed to experience *small amplitude* displacement in a direction normal to its undisplaced plane (transverse direction). The term *small amplitude* is intended to signify displacements which are small in relation to a characteristic length parameter in the problem. The deflected configuration of a segment of the stretched membrane is shown in Figure 7.25.

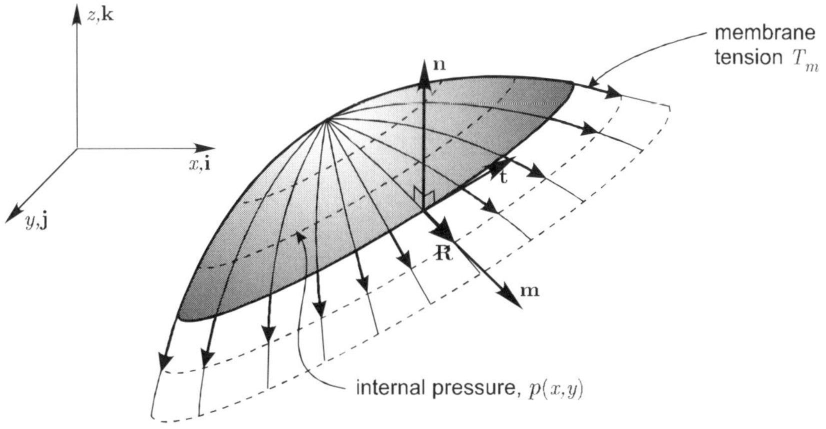

Figure 7.25: Segment of stretched membrane

It is assumed that although the membrane experiences a transverse deflection, the magnitude of the tension in the membrane remains unaltered. The tension T_m will act along the direction $\mathbf{m}$ in Figure 7.25, which is normal to the boundary of the segment and acts along the entire edge. The normal to the deflected surface of the membrane at the point under consideration is denoted by $\mathbf{n}$ and the vectors $\mathbf{n}, \mathbf{m}$ and $\mathbf{t}$ are unit tangent vectors which form an orthogonal set. The coordinate directions associated with the reference or undisplaced configuration of the membrane are x and y with unit vectors $\mathbf{i}$ and $\mathbf{j}$, respectively, and the coordinate for the vertical displacement of the membrane is denoted by $w(x, y)$ with unit vector $\mathbf{k}$. Referring to Figure 7.25, the direction of the tensile force can be obtained by taking the cross product of the unit tangent vector $\mathbf{t}$ with the unit normal vector $\mathbf{n}$. The tension T_m is assumed to be constant along the boundary of the segment. From vector algebra we have

$$\mathbf{m} = \mathbf{t} \times \mathbf{n} \tag{7.311}$$

Hence the vector of membrane forces $\mathbf{R}$ (see Figure 7.25) is given by

$$\mathbf{R} = T_m \mathbf{m} = T_m(\mathbf{t} \times \mathbf{n}) \tag{7.312}$$

For consideration of the vertical equilibrium of the membrane element we require the vertical component of $\mathbf{R}$, aligned in the z-direction. This is given by

$$\mathbf{R}_v = \mathbf{R} \cdot \mathbf{k} = T_m(\mathbf{t} \times \mathbf{n}) \cdot \mathbf{k} \tag{7.313}$$

where $\mathbf{k}$ is the unit vector in the z-direction.

If we consider an arbitrary region of the membrane segment with a plan projected area $\mathbb{D}$ in the x,y plane and closed contour $\mathcal{C}$, we can consider the global equilibrium of the segment. Let the vertical acceleration of a point in the membrane be $\partial^2 w/\partial t^2$ and $\widetilde{\rho}$ its mass per unit area. Considering the equation of motion for the entire segment of the membrane occupying the region $\mathbb{D}$ we have

$$\int\int_{\mathbb{D}} \widetilde{\rho}\frac{\partial^2 w}{\partial t^2}\, dS = \int\int_{\mathbb{D}} p(x,y,t)\, dS + T_m \oint_{\mathcal{C}} (\mathbf{t} \times \mathbf{n}) \cdot \mathbf{k}\, d\mathcal{C} \tag{7.314}$$

where dS is the infinitesimal area element in the x, y plane, $d\mathcal{C}$ is a differential arc element along $\mathcal{C}$ and $p(x,y,t)$ is a pressure acting normal to the displaced membrane. Considering the results of vector products (section 1.3) we can write (7.314) as

$$\int\int_{\mathbb{D}} \left[\widetilde{\rho}\frac{\partial^2 w}{\partial t^2} - p(x,y,t)\right] dS = T_m \oint_{\mathcal{C}} (\mathbf{n} \times \mathbf{k}) \cdot \mathbf{t}\, d\mathcal{C} \tag{7.315}$$

We can now convert the line integral in (7.315) to an area integral by invoking Stokes' theorem (Section 1.9) i.e.

$$\oint_{\mathcal{C}} [\mathbf{F}] \cdot \mathbf{t} \, d\mathcal{C} = \int\int_{\mathbb{D}} (\nabla \times [\mathbf{F}]) \cdot \mathbf{n} \, dS \tag{7.316}$$

where ∇ is the gradient operator. If we identify $[\mathbf{F}]$ as $(\mathbf{n} \times \mathbf{k})$ we can write (7.315) as

$$\int\int_{\mathbb{D}} \left[\tilde{\rho} \frac{\partial^2 w}{\partial t^2} - p(x, y, t) \right] dS = T_m \int\int_{\mathbb{D}} [\nabla \times (\mathbf{n} \times \mathbf{k})] \cdot \mathbf{n} \, dS \tag{7.317}$$

or

$$\int\int_{\mathbb{D}} \left\{ \tilde{\rho} \frac{\partial^2 w}{\partial t^2} - p(x, y, t) - T_m [\nabla \times (\mathbf{n} \times \mathbf{k})] \cdot \mathbf{n} \right\} dS = 0 \tag{7.318}$$

since the region $\mathbb{D}$ is arbitrary, by virtue of the *Dubois-Reymond Lemma*, the integral can vanish only if the integrand is zero. This gives

$$\tilde{\rho} \frac{\partial^2 w}{\partial t^2} - p(x, y) = T_m [\nabla \times \mathbf{n} \times \mathbf{k}] \cdot \mathbf{n} \tag{7.319}$$

Since $w(x, y, t)$ characterizes the displaced surface of the membrane, the unit normal $\mathbf{n}$ to the deflected surface can be calculated by the following way. The normal to the surface $z = w(x, y, t)$ is parallel to

$$\text{grad}\{z - w(x, y, t)\} = \left(-\frac{\partial w}{\partial x} \right) \mathbf{i} + \left(-\frac{\partial w}{\partial y} \right) \mathbf{j} + \mathbf{k} \tag{7.320}$$

Hence, the unit normal to the deflected surface is

$$\mathbf{n} = \left(-\frac{\partial w}{\partial x} \right) \mathbf{i} + \left(-\frac{\partial w}{\partial y} \right) \mathbf{j} + \mathbf{k} \tag{7.321}$$

and

$$\mathbf{n} \times \mathbf{k} = \begin{vmatrix} \mathbf{i} & \mathbf{j} & \mathbf{k} \\ -\dfrac{\partial w}{\partial x} & -\dfrac{\partial w}{\partial y} & 1 \\ 0 & 0 & 1 \end{vmatrix} = \left(-\frac{\partial w}{\partial y}\right)\mathbf{i} + \left(\frac{\partial w}{\partial x}\right)\mathbf{j} + (0)\mathbf{k} \quad (7.322)$$

Also from (1.26) we have

$$\nabla \times (\mathbf{n} \times \mathbf{k}) = \begin{vmatrix} \mathbf{i} & \mathbf{j} & \mathbf{k} \\ \dfrac{\partial}{\partial x} & \dfrac{\partial}{\partial y} & \dfrac{\partial}{\partial z} \\ -\dfrac{\partial w}{\partial y} & \dfrac{\partial w}{\partial x} & 0 \end{vmatrix} = \left[0\mathbf{i} + 0\mathbf{j} + \left(\nabla^2 w\right)\mathbf{k}\right] \quad (7.323)$$

where

$$\nabla^2 = \frac{\partial^2}{\partial x^2} + \frac{\partial^2}{\partial y^2} \quad (7.324)$$

is Laplace's operator referred to the rectangular Cartesian coordinate system *(x,y)*. Using (7.321) and (7.323) we can show that

$$\{\nabla \times \mathbf{n} \times \mathbf{k}\} \cdot \mathbf{n} = (\nabla^2 w)\mathbf{k} \cdot \left[\left(-\frac{\partial w}{\partial x}\right)\mathbf{i} + \left(-\frac{\partial w}{\partial y}\right)\mathbf{i} + \mathbf{k}\right] \quad (7.325)$$

or

$$\{\nabla \times \mathbf{n} \times \mathbf{k}\} \cdot \mathbf{n} = \nabla^2 w \quad (7.326)$$

The equation of motion for the stretched membrane undergoing small amplitude motions is thus given by

$$T_m \nabla^2 w(x, y, t) + p(x, y, t) = \widetilde{\rho}\, \frac{\partial^2 w}{\partial t^2} \tag{7.327}$$

Since the result is in an invariant form, we can simply change the form of the operator ∇^2 to suit other two-dimensional coordinate systems. For example,

(i) in *plane polar coordinates* (r, θ) we have

$$x = r\cos\theta \quad ; \quad y = r\sin\theta \quad ; \quad z = z \tag{7.328}$$

and

$$\nabla^2 = \frac{\partial^2}{\partial r^2} + \frac{1}{r}\frac{\partial}{\partial r} + \frac{1}{r^2}\frac{\partial^2}{\partial \theta^2} \tag{7.329}$$

(ii) in *elliptical coordinates* u,v we have

$$x = a\cosh u \cos v \quad ; \quad y = a\sinh u \sin v \quad ; \quad z = z \tag{7.330}$$

and

$$\nabla^2 = \frac{1}{a^2(\sinh^2 u + \sin^2 v)} \left(\frac{\partial^2}{\partial u^2} + \frac{\partial^2}{\partial v^2} \right) \tag{7.331}$$

(iii) in *bi-polar coordinatcs* u, v wc have

$$x = \frac{a \sinh v}{\cosh v - \cos u} \quad ; \quad y = \frac{a \sin u}{\cosh v - \cos u} \quad ; \quad z = z \tag{7.332}$$

$$\nabla^2 = \frac{(\cosh v - \cos u)^2}{a^2} \left(\frac{\partial^2}{\partial u^2} + \frac{\partial^2}{\partial v^2} \right) \tag{7.333}$$

The coordinate system can thus be chosen to facilitate the description of the boundary conditions associated with a particular membrane problem. It is also evident that the equation of motion of the stretched string, (7.9) can be recovered as a special case of the result for the stretched membrane (7.327).

7.4.2 Plane wave motion in a stretched infinite membrane

In the study of stretched strings we examined the propagation of pulses, the solution for which was given by d' Alembert's general result (7.51). In the case of a stretched membrane, the two dimensional analogue of the pulse is a plane wave. Plane waves are defined as propagating waves where the motion of each point in the planes perpendicular to the direction of propagation is the same. The Figure 7.26 illustrates a propagating plane wave with an arbitrary orientation.

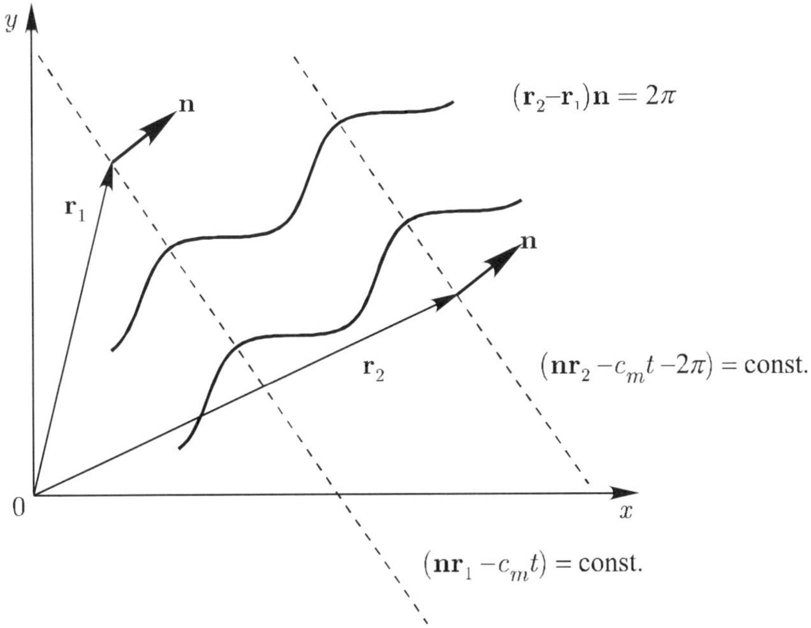

Figure 7.26: Propagation of plane waves.

The motion of every point along the wavefront is characterized by

$$\mathbf{n} \cdot \mathbf{r} - c_m t \quad = \quad \text{const} \tag{7.334}$$

where

$$c_m^2 = \frac{T_m}{\tilde{\rho}} \tag{7.335}$$

where c_m is the propagation velocity and $\mathbf{n}$ is the normal to the wave front. Since in two dimensions

$$\mathbf{n} = \ell\mathbf{i} + m\mathbf{j} \qquad ; \qquad \mathbf{r} = x\mathbf{i} + y\mathbf{j} \tag{7.336}$$

we havo

$$(\mathbf{n}\cdot\mathbf{r} - c_m t) = (\ell x + my - c_m t) \quad = \quad \text{const.} \tag{7.337}$$

We will now assume the existence of plane waves such that

$$w(x, y, t) = F(\ell x + my - c_m t) \tag{7.338}$$

where F is an arbitrary differentiable function. Substituting (7.338) into the equation for free vibration of a stretched membrane

$$\nabla^2 w(x, y, t) = \frac{1}{c_m^2} \frac{\partial^2 w}{\partial t^2} \tag{7.339}$$

we can show that (7.338) constitutes a solution of (7.339) provided

$$\ell^2 + m^2 = 1 \tag{7.340}$$

which, of course, is satisfied by virtue of the fact that ℓ and m are direction cosines. By assuming a harmonic wave of the form

$$w(x, y, t) = Ae^{i\gamma(\ell x + my - c_m t)} \tag{7.341}$$

where γ is the frequency and A is the amplitude of the wave it can shown that the *propagation velocity* of plane waves of *any* frequency in the stretched membrane is constant. Since the propagation of the wave is one-dimensional it is intuitively clear that propagation will occur without distortion of the wave.

It is natural to enquire whether an equivalent analogue with a circular wave form is possible in a stretched membrane. If we consider a *polar symmetric motion* of the stretched membrane we have

$$\nabla^2 w(r,t) = \frac{\partial^2 w}{\partial r^2} + \frac{1}{r}\frac{\partial w}{\partial r} = \frac{1}{c_m^2}\frac{\partial^2 w}{\partial t^2} \tag{7.342}$$

and ∇^2 is the polar symmetric form of Laplace's-operator in two dimensions. If we assume radially symmetric waves of the form

$$w(r,t) = AF(r - c_m t) \tag{7.343}$$

where A is a constant and F is an arbitrary function it is evident that (7.343) does not satisfy the wave equation (7.342). This is not an unexpected result since simple observations that can be made from the wave pattern on the still surface of a pond when a pebble is thrown in, indicate that, the amplitude of the wave decreases as the wave *radiates* from the source. Indeed from purely considerations of energy conservation, the amplitude of the wave is expected to decrease as the radially symmetric wave propagates outward.

7.4.3 Free vibrations of a stretched membrane of infinite extent

We consider the problem of free vibrations of a stretched membrane of infinite extent which is subjected to certain initial conditions. This is the analogue of d'Alembert's solution which deals with the general result for the propagation of waves in a stretched string of infinite extent which is subjected to certain initial conditions. The initial boundary value problem can be formulated as follows: a stretched membrane of infinite extent (Figure 7.27) satisfies the equation of motion

$$\nabla^2 w(x,y,t) = \frac{1}{c_m^2}\frac{\partial^2 w}{\partial t^2} \tag{7.344}$$

for the transverse displacement $w(x, y, t)$ where ∇^2 is Laplace's operator in two- dimensions x and y.

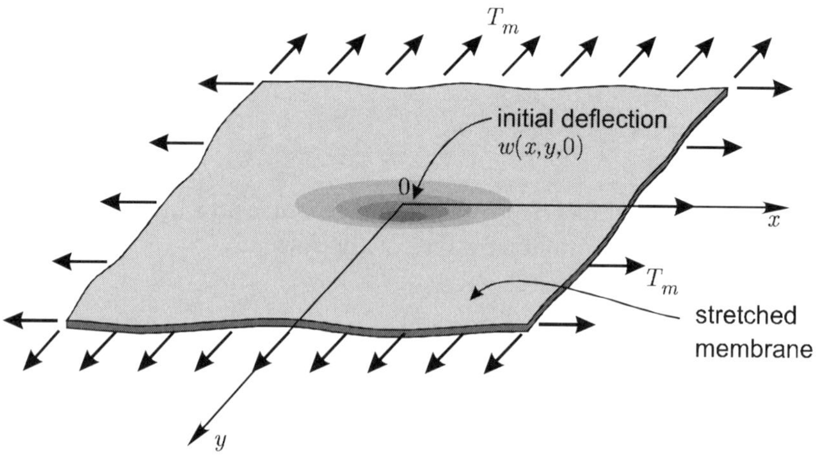

Figure 7.27: Initial displacement of stretched membrane

The membrane is subjected to an initial transverse displacement

$$w(x, y, 0) = F(x, y) \tag{7.345}$$

and an initial transverse velocity

$$\left[\frac{\partial w}{\partial t}\right]_{t=0} = G(x, y) \tag{7.346}$$

Since the domain of the membrane is infinite, there are no boundary conditions as such; however, we will impose the conditions that

$$\left\{w(x, y, t), \frac{\partial w(x, y, t)}{\partial x}, \frac{\partial w(x, y, t)}{\partial y}\right\} \to 0 \quad \text{as} \quad (x^2 + y^2) \to \infty \tag{7.347}$$

Owing to these *regularity conditions*, the problem can be examined by adopting a Fourier transform approach. We define the exponential Fourier transform of $w(x, y, t)$ with respect to x as follows:

$$\mathcal{F}\{w(x,y,t);\xi\} = \tilde{W}(\xi,y,t) = \frac{1}{\sqrt{2\pi}}\int_{-\infty}^{\infty} e^{i\xi x} w(x,y,t)\, dx \tag{7.348}$$

The exponential Fourier transform of $\tilde{W}(\xi, y, t)$ with respect to y is defined as

$$\mathcal{F}\{\tilde{W}(\xi,y,t);\eta\} = W(\xi,\eta,t) = \frac{1}{\sqrt{2\pi}}\int_{-\infty}^{\infty} e^{i\eta y}\tilde{W}(\xi,y,t)dy \tag{7.349}$$

Hence the exponential *double Fourier transform* of $w(x, y, t)$ is given by

$$W(\xi,\eta,t) = \frac{1}{2\pi}\int_{-\infty}^{\infty}\int_{-\infty}^{\infty} w(x,y,t)e^{i(\xi x+\eta y)}dxdy \tag{7.350}$$

It can be shown that if the regularity conditions given by (7.347) are applicable, the *double Fourier transforms of the derivatives* of $w(x, y, t)$ with respect to x and y are given by

$$\begin{aligned}\mathcal{F}\left\{\frac{\partial^2 w}{\partial x^2};\xi,\eta\right\} &= \frac{1}{2\pi}\int_{-\infty}^{\infty}\int_{-\infty}^{\infty}\frac{\partial^2 w}{\partial x^2}e^{i(\xi x+\eta y)}dxdy \\ &= -\xi^2 W(\xi,\eta,t)\end{aligned} \tag{7.351}$$

$$\begin{aligned}\mathcal{F}\left\{\frac{\partial^2 w}{\partial y^2};\xi,\eta\right\} &= \frac{1}{2\pi}\int_{-\infty}^{\infty}\int_{-\infty}^{\infty}\frac{\partial^2 w}{\partial y^2}e^{i(\xi x+\eta y)}dxdy \\ &= -\eta^2 W(\xi,\eta,t)\end{aligned} \tag{7.352}$$

or

$$\mathcal{F}\{\nabla^2 w(x,y,t);\xi,\eta\} = -(\xi^2+\eta^2)W(\xi,\eta,t) \tag{7.353}$$

Using the result (7.353), the partial differential equation governing free vibrations of the stretched membrane can be written as

$$\mathcal{F}\left\{\nabla^2 w(x,y,t);\xi,\eta\right\} = \frac{1}{c_m^2}\mathcal{F}\left\{\frac{\partial^2 w}{\partial t^2}(x,y,t);\xi,\eta\right\} \tag{7.354}$$

This gives the ordinary differential equation

$$\frac{d^2W(\xi,\eta,t)}{dt^2} + c_m^2(\xi^2+\eta^2)W(\xi,\eta,t) = 0 \tag{7.355}$$

The solution of (7.355) gives

$$W(\xi,\eta,t) = A\cos(\mu t) + B\sin(\mu t) \tag{7.356}$$

where A and B are constants and

$$\mu^2 = c_m^2(\xi^2+\eta^2) \tag{7.357}$$

To determine the arbitrary constants A and B we represent the initial conditions (7.345) and (7.346) in terms of their respective double Fourier transforms, i.e.

$$\begin{aligned} f(\xi,\eta) &= \frac{1}{2\pi}\int_{-\infty}^{\infty}\int_{-\infty}^{\infty} w(x,y,0)e^{i(\xi x+\eta y)}dxdy \\ &= \frac{1}{2\pi}\int_{-\infty}^{\infty}\int_{-\infty}^{\infty} F(x,y)e^{i(\xi x+\eta y)}dxdy \end{aligned} \tag{7.358}$$

and

$$g(\xi,\eta) = \frac{1}{2\pi}\int_{-\infty}^{\infty}\int_{-\infty}^{\infty}\left[\frac{\partial w}{\partial t}\right]_{t=0} e^{i(\xi x+\eta y)}dxdy$$
$$= \frac{1}{2\pi}\int_{-\infty}^{\infty}\int_{-\infty}^{\infty} G(x,y)e^{i(\xi x+\eta y)}dxdy \tag{7.359}$$

Using the results (7.356), (7.358) and (7.359) we have

$$W(\xi,\eta,0) = A = f(\xi,\eta) \tag{7.360}$$

$$\left[\frac{dW}{dt}\right]_{t=0} = \frac{B}{\mu} = g(\xi,\eta) \tag{7.361}$$

The solution of (7.355) in the double Fourier transform domain is

$$W(\xi,\eta,t) = f(\xi,\eta)\cos(\mu t) + \frac{g(\xi,\eta)}{\mu}\sin(\mu t) \tag{7.362}$$

The solution for $w(x,y,t)$ can be obtained by the *inversion theorem* for the double Fourier transform i.e.

$$w(x,y,t) = \frac{1}{2\pi}\int_{-\infty}^{\infty}\int_{-\infty}^{\infty} W(\xi,\eta,t)e^{-i(\xi x+\eta y)}d\xi d\eta \tag{7.363}$$

From the (7.362) and (7.363) we obtain

$$w(x,y,t) = \frac{1}{2\pi}\int_{-\infty}^{\infty}\int_{-\infty}^{\infty} f(\xi,\eta)\cos(c_m t\sqrt{\xi^2+\eta^2})e^{-i(\xi x+\eta y)}d\xi d\eta$$
$$+\frac{1}{2\pi}\int_{-\infty}^{\infty}\int_{-\infty}^{\infty}\frac{g(\xi,\eta)}{c_m(\xi^2+\eta^2)^{\frac{1}{2}}}\sin(c_m t\sqrt{\xi^2+\eta^2})e^{-i(\xi x+\eta y)}d\xi d\eta \tag{7.364}$$

Tho result (7.364) represents the formal solution to the free vibration problem related to a *stretched membrane of infinite extent* which is subjected to an *initial displacement* of $F(x, y)$ and an *initial velocity* of $G(x, y)$. This solution can be evaluated either analytically or numerically depending on the form of $F(x, y)$ and $G(x, y)$. It should, however, be noted that if the transforms for $\cos(c_m t\sqrt{\xi^2 + \eta^2})$ and $\sin(c_m t\sqrt{\xi^2 + \eta^2})/(c_m\sqrt{\xi^2 + \eta^2})$ are evaluated via a relationship of the type (7.363), it will result in *divergent integrals*. It is, however, possible to introduce transformations such as

$$\Omega_\lambda(\xi, \eta) = e^{-\lambda\mu}\cos(\mu t) \quad ; \quad \Phi_\lambda(\xi, \eta) = \frac{e^{-\lambda\mu}\sin(\mu t)}{\mu} \tag{7.365}$$

which leads to *convergent integrals* which makes the inversion of (7.362) possible, through an application of the Faltung theorem for Fourier transforms. Omitting details of the analysis it can be shown that the inverse transforms of the functions $\Omega_\lambda(\xi, \eta)$ and $\Phi_\lambda(\xi, \eta)$ as $\lambda \to 0$ can be written as, respectively,

$$\begin{aligned}\mathcal{F}^{-1}\left\{\cos(c_m t\sqrt{\xi^2 + \eta^2}); x, y\right\} &= \omega_0(x, y)\\ &= \frac{-1}{c_m}\frac{\partial}{\partial t}\left\{\frac{1}{(c_m^2 t^2 - x^2 - y^2)^{\frac{1}{2}}}\right\}\end{aligned} \tag{7.366}$$

$$\begin{aligned}\mathcal{F}^{-1}\left\{\frac{\sin(c_m t\sqrt{\xi^2 + \eta^2})}{c_m\sqrt{\xi^2 + \eta^2}}; x, y\right\} &= \varphi_0(x, y)\\ &= \frac{-1}{c_m(c_m^2 t^2 - x^2 - y^2)^{\frac{1}{2}}}\end{aligned} \tag{7.367}$$

Utilizing these results in (7.364) we obtain

$$\int_{-\infty}^{\infty}\int_{-\infty}^{\infty} f(\xi, \eta)\cos(\mu t)e^{-i(\xi x + \eta y)}d\xi d\eta$$

$$= -\frac{1}{c_m}\frac{\partial}{\partial t}\int_{-\infty}^{\infty}\int_{-\infty}^{\infty}\frac{F(\alpha,\beta)d\alpha d\beta}{[c_m^2 t^2-(x-\alpha)^2-(y-\beta)^2]^{\frac{1}{2}}} \tag{7.368}$$

and

$$\int_{-\infty}^{\infty}\int_{-\infty}^{\infty} g(\xi,\eta)\frac{\sin(\mu t)}{\mu}e^{-i(\xi x+\eta y)}d\xi d\eta$$
$$= -\frac{1}{c_m}\int_{-\infty}^{\infty}\int_{-\infty}^{\infty}\frac{G(\alpha,\beta)d\alpha d\beta}{[c_m^2 t^2-(x-\alpha)^2-(y-\beta)^2]^{\frac{1}{2}}} \tag{7.369}$$

Using the results (7.368) and (7.369) in (7.365) we obtain the following expression for the time-dependent displacement of the stretched membrane:

$$w(x,y,t) = -\frac{1}{2\pi c_m}\int_{-\infty}^{\infty}\int_{-\infty}^{\infty}\frac{\partial}{\partial t}\left\{\frac{F(\alpha,\beta)d\alpha d\beta}{[c_m^2 t^2-(x-\alpha)^2-(y-\beta)^2]^{\frac{1}{2}}}\right\}$$
$$-\frac{1}{2\pi c_m}\int_{-\infty}^{\infty}\int_{-\infty}^{\infty}\left\{\frac{G(\alpha,\beta)d\alpha d\beta}{[c_m^2 t^2-(x-\alpha)^2-(y-\beta)^2]^{\frac{1}{2}}}\right\} \tag{7.370}$$

From this solution we can recover results for the cases where the stretched membrane can be subjected separately to the initial conditions of the type (7.345) and (7.346).

7.4.4 Symmetric free vibrations of the stretched membrane

In the case when the free vibrations of the stretched infinite membrane are symmetric about an axis (Figure 7.28) the problem can be formulated in relation to the axisymmetric form of the equation of motion governing the free vibration problem. The partial differential equation governing the initial boundary value problem is

$$\nabla^2 w(r,t) = \frac{\partial^2 w}{\partial r^2} + \frac{1}{r}\frac{\partial w}{\partial r} = \frac{1}{c_m^2}\frac{\partial^2 w}{\partial t^2} \tag{7.371}$$

The initial conditions are specified on the displacement and velocity of the stretched membrane, i.e.

$$w(r,0) = f_0(r) \tag{7.372}$$

$$\left[\frac{\partial w}{\partial t}\right]_{t=0} = g_0(r) \tag{7.373}$$

In addition, we assume that the deflections of the membrane are bounded as $r \to 0$ and as $r \to \infty$: i.e.

$$\left\{r\frac{\partial w}{\partial r}\right\} \to 0 \quad \text{at} \quad r = 0 \quad ; \quad r = \infty \tag{7.374}$$

This condition ensures that the deflections of the stretched membrane are bounded irrespective of the choices of $f_0(r)$ and $g_0(r)$.

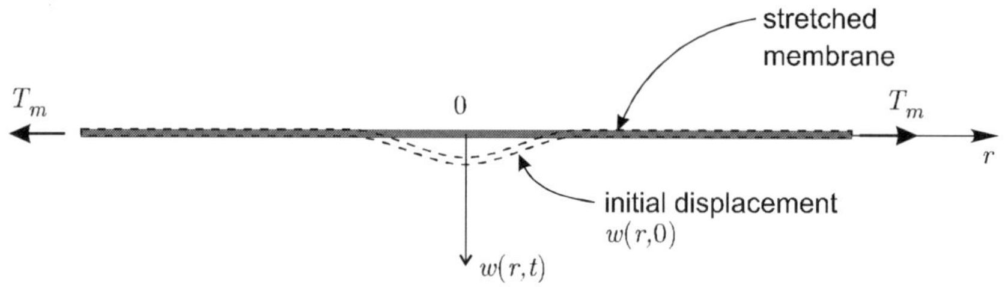

Figure 7.28: Initial axisymmetric displacement of stretched membrane

Owing to the axial symmetry of the problem, the formulation of the problem can be most conveniently achieved by employing a Hankel transform technique. We define the Hankel transform, of zero-order, of the membrane deflection $w(r,t)$ as follows:

$$\mathcal{H}_0\{w(r,t);\xi\} = \overline{w}(\xi,t) = \int_0^\infty rw(r,t)J_0(\xi r)dr \tag{7.375}$$

Using results given in Chapter 1, we have

$$\mathcal{H}_0\left\{\left[\frac{\partial^2 w}{\partial r^2} + \frac{1}{r}\frac{\partial w}{\partial r}\right];\xi\right\} = -\,\xi^2\overline{w}(\xi,t) \tag{7.376}$$

Using (7.375) and (7.376), the governing partial differential equation (7.371) can be reduced to the solution of the following ordinary differential equation for $\overline{w}(\xi,t)$; i.e.

$$\frac{d^2\overline{w}(\xi,t)}{dt^2} + c_m^2\xi^2\overline{w}(\xi,t) = 0 \tag{7.377}$$

The solution of (7.377) is

$$\overline{w}(\xi,t) = A(\xi)\cos(c_m\xi t) + B(\xi)\sin(c_m\xi t) \tag{7.378}$$

where $A(\xi)$ and $B(\xi)$ are arbitrary functions. To determine these functions, we utilize the initial conditions. The initial conditions (7.372) and (7.373) can be expressed in terms of their Hankel transforms as follows:

$$\overline{w}(\xi,0) = \int_0^\infty rf_0(r)J_0(\xi r)dr = \overline{f}_0(\xi) \tag{7.379}$$

$$\left[\frac{\partial\overline{w}}{\partial t}\right]_{t=0} = \int_0^\infty rg_0(r)J_0(\xi r)dr = \overline{g}_0(\xi) \tag{7.380}$$

where $\overline{f}_0(\xi)$ and $\overline{g}_0(\xi)$ are the zeroth-order Hankel transforms of $f_0(r)$ and $g_0(r)$ respectively. Using (7.379), (7.380) and (7.378) we can determine an explicit solution for the transformed equivalent of the membrane deflection:

i.e.

$$\overline{w}(\xi, t) = \overline{f}_0(\xi)\cos(c_m\xi t) + \frac{\overline{g}(\xi)}{c_m\xi}\sin(c_m\xi t) \tag{7.381}$$

The time-dependent deflection of the stretched membrane, $w(r,t)$ can be obtained by making use of the Hankel inversion theorem. We have

$$\begin{aligned} w(r,t) &= \int_0^\infty \xi\overline{f}_0(\xi)\cos(c_m\xi t)J_0(\xi r)d\xi \\ &+ \frac{1}{c_m}\int_0^\infty \overline{g}_0(\xi)\sin(c_m\xi t)J_0(\xi r)d\xi \end{aligned} \tag{7.382}$$

The result (7.382) represents the formal solution to the time-dependent axisymmetric motion of a stretched membrane which is subjected to the initial displacement and initial velocity defined by (7.372) and (7.373) respectively. Although the integrands of the integrals occurring in (7.382) are oscillatory in nature, the integrals themselves are convergent and can be evaluated quite conveniently by making use of *quadrature techniques* or by making use of *symbolic computational schemes* offered by mathematical software such as MAPLE® and MATHEMATICA ®. An alternative representation for (7.382) takes the form

$$w(r,t) = -\frac{1}{2\pi c_m}\int_0^\infty \zeta\left\{f_0(\zeta)\frac{\partial}{\partial t} + g_0(\zeta)\right\}\frac{F(k,\pi/2)d\zeta}{\{c_m^2t^2 - r^2 - \zeta^2\}^{\frac{1}{2}}} \tag{7.383}$$

where

$$k^2 = \frac{4\zeta r}{[c_m^2t^2 - (r-\zeta)^2]} \tag{7.384}$$

and $F(k,\pi/2)$ is the complete elliptic integral defined by

$$F(k, \pi/2) = \int_0^{\pi/2} \frac{d\theta}{\sqrt{1 - k^2 \sin^2 \theta}} \tag{7.385}$$

7.4.5 Green's function for the vibration of a stretched membrane

The results derived in the previous section for the free vibrations of a stretched membrane of infinite extent can be conveniently adopted to develop a Green's function for the stretched membrane (Figure 7.29). We consider the problem where the membrane is subjected to a localized initial displacement of the form

$$w(r, 0) = \begin{cases} V_0/\pi a_0^2 & ; \quad 0 \le r \le a_0 \\ 0 & ; \quad r \ge a_0 \end{cases} \tag{7.386}$$

and a zero initial velocity. The initial deflection profile can be visualized as a cylindrical indentation of radius a_0 and volume V_0.

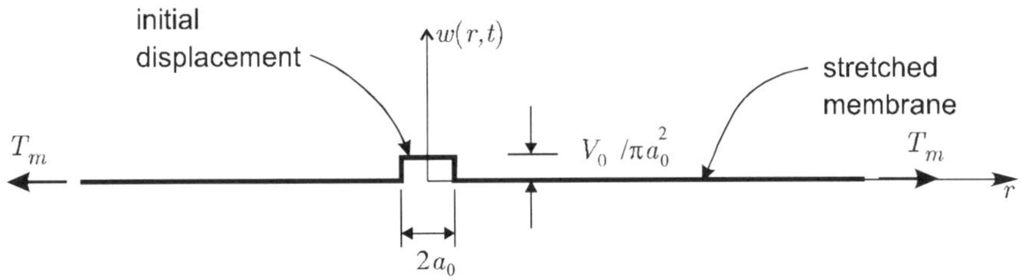

Figure 7.29: Initial displacement of stretched membrane: Green's function.

The Hankel transform of this displacement is

$$\overline{w}(\xi, 0) = \frac{V_0}{\pi \xi a_0} J_1(\xi a_0) \tag{7.387}$$

In the limit as $a_0 \to 0$, with V_0 remaining constant, $J_1(\xi a_0) \to (\xi a_0/2)$ such that

$$[\overline{w}(\xi, 0)]_{a_0 \to 0} = \frac{V_0}{2\pi} \tag{7.388}$$

We can identify (7.388) as the $\overline{f}_0(\xi)$ term occurring in (7.382) and since the initial velocity of the stretched membrane is zero, $\overline{g}_0(\xi) \equiv 0$. The reduced form of the solution for the dynamic deflections of the stretched membrane is given by

$$w(r,t) = \frac{V_0}{2\pi} \int_0^\infty \xi \cos(c_m \xi t) J_0(\xi r) d\xi \tag{7.389}$$

Rewriting (7.389) in the form

$$w(r,t) = \frac{V_0}{2\pi c_m} \frac{\partial}{\partial t} \int_0^\infty \sin(c_m \xi t) J_0(\xi r) d\xi \tag{7.390}$$

and using *Tables of Hankel Transforms*, we can evaluate (7.390) in the form

$$w(r,t) = \begin{cases} \dfrac{V_0}{2\pi c_m} \dfrac{\partial}{\partial t} \left\{ \dfrac{1}{[(c_m t)^2 - r^2]^{\frac{1}{2}}} \right\} & ; 0 \le r \le a \\ 0 & ; \quad r \ge a \end{cases} \tag{7.391}$$

Performing the differentiation with respect to t, we have

$$w(r,t) = \begin{cases} -\dfrac{V_0}{2\pi} \dfrac{c_m t}{[(c_m t)^2 - r^2]^{\frac{3}{2}}} & ; 0 \le r \le a \\ 0 & ; \quad r \ge a \end{cases} \tag{7.392}$$

7.5 Vibrations of stretched finite membranes

The class of problems which deal with the determination of vibrational characteristics of stretched membranes of finite extent is of particular interest to the analysis and design of acoustical devices. Such studies are also important to applications of membranes in biomedical engineering and aeronautical engineering where the vibrations are induced by fluid-membrane interactions. Similar applications can be found in studies involving fluid-induced vibrations of reinforced membrane roofs for large enclosures. The detailed analysis of stretched membranes with an arbitrary planform invariably requires recourse to computational and numerical methods for the solution of the dynamic equations of motion of the stretched membrane. The objective of this section is to illustrate, by appeal to some specific examples, the essential features of the solution associated with the vibration of stretched membranes with square, rectangular and circular planforms. The analysis will be performed primarily by appeal to a separation of variables technique; procedures based on integral transform techniques can be readily adopted to investigate membrane vibration problems with the appropriate geometrical and loading configurations.

7.5.1 Vibrations of a stretched square membrane

We consider the problem of a stretched square membrane with an edge length a, which is held fixed along the edges (Figure 7.30). The partial differential equation governing the free vibrations of the stretched membrane is

$$\left(\frac{\partial^2 w}{\partial x^2} + \frac{\partial^2 w}{\partial y^2}\right) = \frac{1}{c_m^2}\frac{\partial^2 w}{\partial t^2} \tag{7.393}$$

where $c_m^2 = T_m/\widetilde{\rho}$, and $w(x, y, t)$ is the transverse displacement. The boundary conditions governing the problem are

$$w(x, 0, t) = w(0, y, t) = 0 \tag{7.394}$$

$$w(a, y, t) = w(x, a, t) = 0 \tag{7.395}$$

The membrane is subjected to an initial displacement and released with zero velocity at $t = 0$. The initial conditions governing the problem are

$$w(x, y, 0) = f(x, y) \tag{7.396}$$

$$\left[\frac{\partial w}{\partial t}\right]_{t=0} = 0 \tag{7.397}$$

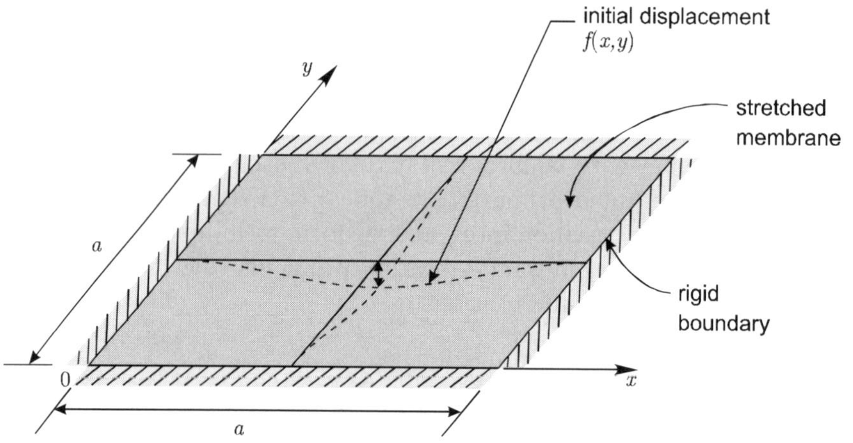

Figure 7.30: Initial displacement of a stretched square membrane.

We employ a separation of variables technique to solve the initial boundary value problem defined by (7.393) to (7.397). We assume that $w(x, y, t)$ admits a solution of the form

$$w(x, y, t) = X(x)Y(y)T(t) \tag{7.398}$$

Substituting (7.398) in (7.393) we obtain

$$\frac{d^2X/dx^2}{X} + \frac{d^2Y/dy^2}{Y} = \frac{d^2T/dt^2}{c_m^2 T} = -\lambda^2 \tag{7.399}$$

where λ^2 is a constant. The ordinary differential equations resulting from (7.399) are as follows:

$$\frac{d^2X(x)}{dx^2} + \mu^2 X(x) = 0 \tag{7.400}$$

$$\frac{d^2Y(y)}{dy^2} + (\lambda^2 - \mu^2)Y(y) = 0 \tag{7.401}$$

$$\frac{d^2T(t)}{dt^2} + \lambda^2 c_m^2 T(t) = 0 \tag{7.402}$$

The general solution of (7.400), (7.401) and (7.402) are respectively

$$X(x) = A_1 \sin(\mu x) + B_1 \cos(\mu x) \tag{7.403}$$

$$Y(y) = A_2 \sin(\sqrt{\lambda^2 - \mu^2}\; y) + B_2 \cos(\sqrt{\lambda^2 - \mu^2}\; y) \tag{7.404}$$

$$T(t) = A_3 \sin(\lambda c_m t) + B_3 \cos(\lambda c_m t) \tag{7.405}$$

The general solution for the time-dependent deflections of the stretched membrane is given by

$$\begin{aligned} w(x, y, t) = \{A_1 \sin(\mu x) + B_1 \cos(\mu x)\}\{A_2 \sin(\sqrt{\lambda^2 - \mu^2}\; y) \\ + B_2 \cos(\sqrt{\lambda^2 - \mu^2}\; y)\}\{A_3 \sin(\lambda c_m t) + B_3 \cos(\lambda c_m t)\} \end{aligned} \tag{7.406}$$

To determine the arbitrary constants A_1, B_1, A_2, B_2 ...etc., we make the use of the first set of boundary conditions (7.394) and the initial condition (7.397). To satisfy these conditions we require

$$B_1 = B_2 = A_3 = 0 \tag{7.407}$$

The remaining solutions give

$$w(x, y, t) = A \sin(\mu x) \sin(\sqrt{\lambda^2 - \mu^2}\; y) \cos(\lambda c_m t) \tag{7.408}$$

where A is an arbitrary constant. To satisfy the second set of displacement boundary conditions (7.395) we require

$$\sin(\mu a) = 0, \qquad \text{or} \qquad \mu = \frac{\alpha\pi}{a}; \qquad (\alpha = 1, 2, 3, ...) \tag{7.409}$$

$$\sin(\sqrt{\lambda^2 - \mu^2}\; a) = 0, \qquad \text{or} \qquad (\lambda^2 - \mu^2)^{\frac{1}{2}} = \frac{\beta\pi}{a}; \qquad (\beta = 1, 2, 3, ...) \tag{7.410}$$

Alternatively we have

$$\lambda = \frac{\pi}{a}(\alpha^2 + \beta^2)^{\frac{1}{2}} \tag{7.411}$$

Hence, a solution which satisfies all the boundary conditions (7.394) and (7.395) and the initial condition (7.397) is given by

$$w(x, y, t) = A \sin\left(\frac{\alpha\pi x}{a}\right) \sin\left(\frac{\beta\pi y}{a}\right) \cos\left(\sqrt{\alpha^2 + \beta^2}\; \frac{c_w \pi t}{a}\right) \\ ;\ (\alpha = 1, 2, 3, ...; \beta = 1, 2, 3, ...) \tag{7.412}$$

The only remaining condition pertaining to the stretched square membrane problem is the initial condition (7.396). The solution (7.412) represents the eigen-function solutions associated with the vibration of a stretched square membrane. We can now construct a generalized solution, which consists of a *superposition of the eigen-function solutions* for $m \to \infty$ and $n \to \infty$. The resulting series solution is

$$w(x,y,t) = \sum_{\alpha=1,2,}^{\infty} \sum_{\beta=1,2,}^{\infty} A_{\alpha\beta} \sin\left(\frac{\alpha\pi x}{a}\right) \sin\left(\frac{\beta\pi y}{a}\right) \cdot \cos\left(\sqrt{\alpha^2+\beta^2}\frac{c_w \pi t}{a}\right) \tag{7.413}$$

The constants $A_{\alpha\beta}$ can now be determined by making use of the initial condition (7.392), which gives

$$w(x,y,0) = f(x,y) = \sum_{\alpha=1,2,}^{\infty} \sum_{\beta=1,2,}^{\infty} A_{\alpha\beta} \sin\left(\frac{\alpha\pi x}{a}\right) \sin\left(\frac{\beta\pi y}{a}\right) \tag{7.414}$$

We now assume that the function $f(x,y)$ also admits a double Fourier series representation such that by making use of the orthogonality properties of the sine functions we obtain

$$A_{\alpha\beta} = \frac{4}{a^2} \int_0^a \int_0^a f(x,y) \sin\left(\frac{\alpha\pi x}{a}\right) \sin\left(\frac{\beta\pi y}{a}\right) dxdy \tag{7.415}$$

This formally completes the solution of the problem of the vibration of a stretched square membrane. Once the form of the initial displacement $f(x,y)$ is specified the constants $A_{\alpha\beta}$ can be evaluated through (7.415) and the resulting expressions can be substituted in (7.413) to determine $w(x,y,t)$.

It is instructive to examine the *frequency characteristics* associated with the vibration of the square membrane. The *fundamental frequencies* of the natural modes are given by

$$\zeta_{\alpha\beta} = \frac{1}{2}(\alpha^2+\beta^2)^{\frac{1}{2}} \frac{c_m}{a} = \frac{1}{2}\frac{(\alpha^2+\beta^2)^{\frac{1}{2}}}{a}\sqrt{\frac{T_m}{\tilde{\rho}}} \tag{7.416}$$

The frequency of the *lowest mode* ($\alpha = 1; \beta = 0$ or *vice versa*) is

$$\zeta_{\alpha\beta}^{10} = \frac{1}{2a}\sqrt{\frac{T_m}{\widetilde{\rho}}} \tag{7.417}$$

The frequency of the next highest mode (i.e. with $\alpha = 1; \beta = 1$) is

$$\zeta_{\alpha\beta}^{11} = \frac{1}{2a}\sqrt{\frac{2T_m}{\widetilde{\rho}}} = \sqrt{2}\zeta_{\alpha\beta}^{10} \tag{7.418}$$

which is not an *integer multiple* of $\zeta_{\alpha\beta}^{10}$. The same comment applies to other higher modes which invariably *do not have frequencies that are integer multiples of the lowest mode.* The absence of such integer multiples of the lowest fundamental mode means that the stretched square membrane is void of the *musical overtones.* Therefore, the end result of the excitation of the stretched membrane is *noise* rather than *music.* If on the other hand the tension in the membrane T_m is increased substantially, the lowest mode will be the dominant mode which would result in a membrane having a *marginally musical tone.*

Example 7.11

A stretched square membrane of edge length a (Figure 7.31) is subjected to an initial velocity of

$$\left[\frac{\partial w}{\partial t}\right]_{t=0} = v_0 \sin\left(\frac{\pi x}{a}\right) \sin\left(\frac{\pi y}{a}\right) \tag{7.419}$$

If the membrane is in an undeflected equilibrium position prior to the application of the initial velocity, derive an expression for the time-dependent dynamic displacement of the membrane.

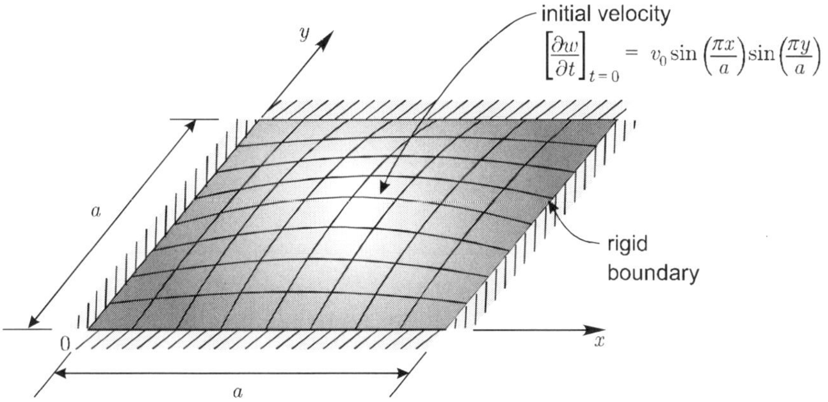

Figure 7.31: Initial velocity distribution in the stretched membrane.

Solution

Following the separation of variables technique presented earlier, we can show that the general result for the time-dependent displacement of the membrane is given by (7.406). The boundary conditions along the fixed edges will be satisfied by the solution

$$w(x,y,t) = \sum_{\alpha=1,2,}^{\infty} \sum_{\beta=1,2,}^{\infty} A'_{\alpha\beta} \sin\left(\frac{\alpha\pi x}{a}\right) \sin\left(\frac{\beta\pi y}{a}\right) \{A_3 \sin(\lambda c_m t) + B_3 \cos(\lambda c_m t)\} \tag{7.420}$$

where $A'_{\alpha\beta}$, A_3 and B_3 are constants and

$$\lambda = \frac{\pi}{a}(\alpha^2 + \beta^2)^{\frac{1}{2}} \tag{7.421}$$

In order to satisfy the initial condition

$$w(x,y,0) = 0 \tag{7.422}$$

we require $B_3 \equiv 0$. Applying the initial condition (7.419) we have

$$v_0 \sin\left(\frac{\pi x}{a}\right) \sin\left(\frac{\pi y}{a}\right) = \sum_{\alpha=1,2,}^{\infty} \sum_{\beta=1,2,}^{\infty} A^*_{\alpha\beta}\, \lambda c_m \sin\left(\frac{\alpha \pi x}{a}\right) \sin\left(\frac{\beta \pi y}{a}\right) \tag{7.423}$$

where $A^*_{\alpha\beta}$ are constants. It is evident that a single term of the double series will satisfy the initial condition explicitly: i.e.,

$$A^*_{11} = \frac{v_0}{\lambda c_m} \quad ; \quad \alpha = 1,\ \beta = 1:\ \lambda = \frac{\sqrt{2}\pi}{a} \tag{7.424}$$

The resulting expression for the time-dependent displacement of the stretched membrane is given by

$$w(x,y,t) = \frac{v_0 a}{\sqrt{2}\pi c_m} \sin\left(\frac{\pi x}{a}\right) \sin\left(\frac{\pi y}{a}\right) \sin\left(\sqrt{2}\frac{\pi c_m t}{a}\right) \tag{7.425}$$

7.5.2 Free vibrations of a stretched rectangular membrane

In this section we generalize the solution developed in section 7.5.1 for the problem of a stretched *square membrane* to the case of a stretched *rectangular membrane* which is rigidly constrained along the boundaries (Figure 7.32). The basic approach for the analysis of the problem can, of course, follow the separation of variables techniques outlined previously for the square membrane. In this section, however, we will illustrate the application of a *finite Fourier transform*, technique for the solution of the problem.

The partial differential equation governing free vibrations of the stretched membrane is

$$\nabla^2 w(x,y,t) = \frac{1}{c_m^2} \frac{\partial^2 w}{\partial t^2} \tag{7.426}$$

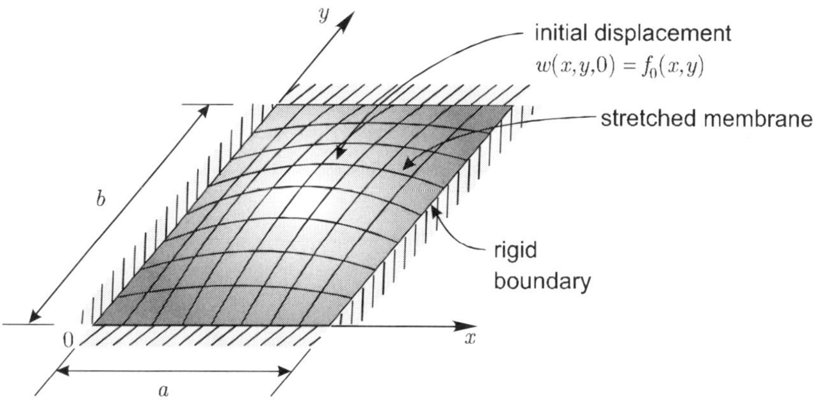

Figure 7.32: Stretched rectangular membrane subjected to initial displacement and initial velocity.

The boundary conditions governing the problem are

$$w(0, y, t) = w(x, 0, t) = 0 \tag{7.427}$$

$$w(a, y, t) = w(x, b, t) = 0 \tag{7.428}$$

We assume that the stretched membrane is subjected to the initial displacement

$$w(x, y, 0) = f_0(x, y) \tag{7.429}$$

and the membrane is released from this displaced position with an initial velocity of

$$\left[\frac{\partial w}{\partial t}\right]_{t=0} = g_0(x,y) \tag{7.430}$$

Since the deflections of the membrane are zero at $x = 0$ and $x = a$, using the properties of finite Fourier transforms we have

$$\int_0^a \frac{\partial^2 w}{\partial x^2} \sin\left(\frac{\alpha\pi x}{a}\right) dx = -\frac{\alpha^2\pi^2}{a^2} \int_0^a w(x,y,t) \sin\left(\frac{\alpha\pi x}{a}\right) dx \tag{7.431}$$

If we define the *finite double Fourier sine transform* of $w(x,y,t)$ as

$$\overline{W}_s(\alpha,\beta,t) = \int_0^a \int_0^b w(x,y,t) \sin\left(\frac{\alpha\pi x}{a}\right) \sin\left(\frac{\beta\pi y}{b}\right) dxdy \tag{7.432}$$

Then

$$\int_0^a \int_0^b \frac{\partial^2 w}{\partial x^2} \sin\left(\frac{\alpha\pi x}{a}\right) \sin\left(\frac{\beta\pi y}{b}\right) dxdy = -\frac{\alpha^2\pi^2}{a^2} \overline{W}_s(\alpha,\beta,t) \tag{7.433}$$

and

$$\int_0^a \int_0^b \frac{\partial^2 w}{\partial y^2} \sin\left(\frac{\alpha\pi x}{a}\right) \sin\left(\frac{\beta\pi y}{b}\right) dxdy = -\frac{\beta^2\pi^2}{b^2} \overline{W}_s(\alpha,\beta,t) \tag{7.434}$$

Considering the governing partial differential equation (7.426), we can multiply both sides of the equation by $\{\sin(\alpha\pi x/a)\sin(\beta\pi y/b)\}$ and integrate over the region of the rectangular membrane; this gives

$$\begin{aligned}&\int_0^a \int_0^b \nabla^2 w(x,y,t) \sin\left(\frac{\alpha\pi x}{a}\right) \sin\left(\frac{\beta\pi y}{b}\right) dxdy \\ &= \frac{1}{c_m^2} \int_0^a \int_0^b \frac{\partial^2 w}{\partial t^2} \sin\left(\frac{\alpha\pi x}{a}\right) \sin\left(\frac{\beta\pi y}{b}\right) dxdy\end{aligned} \tag{7.435}$$

Using the results (7.432) to (7.434) in (7.435) we obtain

$$-\pi^2\left(\frac{\alpha^2}{a^2}+\frac{\beta^2}{b^2}\right)\overline{W}_s(\alpha,\beta,t)=\frac{1}{c_m^2}\frac{d^2\overline{W}_s}{dt^2} \tag{7.436}$$

The general solution of the ordinary differential equation for $\overline{W}_s(\alpha,\beta,t)$ can be written as

$$\overline{W}_s(\alpha,\beta,t)=A\sin(c_m\pi\Omega_{\alpha\beta}t)+B\cos(c_m\pi\Omega_{\alpha\beta}t) \tag{7.437}$$

where A and B are constants and

$$\Omega_{\alpha\beta}=\left(\frac{\alpha^2}{a^2}+\frac{\beta^2}{b^2}\right)^{\frac{1}{2}} \tag{7.438}$$

Using double Fourier sine transforms, we can express the initial conditions (7.429) and (7.430) as follows:

$$\overline{F}_s(\alpha,\beta)=\int_0^a\int_0^b w(x,y,0)\sin\left(\frac{\alpha\pi x}{a}\right)\sin\left(\frac{\beta\pi y}{b}\right)dxdy \tag{7.439}$$

and

$$\overline{G}_s(\alpha,\beta)=\int_0^a\int_0^b\left[\frac{\partial w}{\partial t}\right]_{t=0}\sin\left(\frac{\alpha\pi x}{a}\right)\sin\left(\frac{\beta\pi y}{b}\right)dxdy \tag{7.440}$$

Using (7.439) and (7.440) in (7.437) we have

$$\overline{W}_s(\alpha,\beta)=B=\overline{F}_s(\alpha,\beta) \tag{7.441}$$

$$\left[\frac{d\overline{W}_o}{dt}\right]_{t=0} = Ac_m\pi\Omega_{\alpha\beta} = G_s(\alpha,\beta) \tag{7.442}$$

The explicit solution for the double Fourier sine transform of the deflection of the stretched membrane is given by

$$\begin{aligned}\overline{W}_s(\alpha,\beta,t) = \cos(c_m\pi\Omega_{\alpha\beta}t)&\left[\int_0^a\int_0^b f_0(\xi,\eta)\sin\left(\frac{\alpha\pi\xi}{a}\right)\right.\\ &\left.\sin\left(\frac{\beta\pi\eta}{b}\right)d\xi d\eta\right]\\ &+\frac{\sin(c_m\pi\Omega_{\alpha\beta}t)}{c_m\pi\Omega_{\alpha\beta}}\left[\int_0^a\int_0^b g_0(\xi,\eta)\sin\left(\frac{\alpha\pi\xi}{a}\right)\right.\\ &\left.\sin\left(\frac{\beta\pi\eta}{b}\right)d\xi d\eta\right]\end{aligned} \tag{7.443}$$

Using the inversion theorem for double Fourier sine transforms we have

$$\begin{aligned}\frac{w(x,y,t)}{4/ab} = \sum_{\alpha=1,2,}^{\infty}\sum_{\beta=1,2,}^{\infty}&\left\{\sin\left(\frac{\alpha\pi x}{a}\right)\sin\left(\frac{\beta\pi y}{b}\right)\cos(c_m\pi\Omega_{\alpha\beta}t)\right\}\\ &\cdot\left[\int_0^a\int_0^b f_0(\xi,\eta)\sin\left(\frac{\alpha\pi\xi}{a}\right)\sin\left(\frac{\beta\pi\eta}{b}\right)d\xi d\eta\right.\\ &+\frac{ab\tan(c_m\pi\Omega_{\alpha\beta}t)}{c_m\pi(\alpha^2b^2+\beta^2a^2)^{\frac{1}{2}}}\int_0^a\int_0^b g_0(\xi,\eta)\sin\left(\frac{\alpha\pi\xi}{a}\right)\\ &\left.\sin\left(\frac{\beta\pi\eta}{b}\right)d\xi d\eta\right]\end{aligned} \tag{7.444}$$

This represents the complete solution for the time-dependent transverse response of a stretched rectangular membrane which is subjected to arbitrary initial conditions (7.429) and (7.430) pertaining, respectively, to an initial displacement $f_0(x,y)$ and an initial velocity $g_0(x,y)$. Depending upon the nature of the functions $f_0(x,y)$ and $g_0(x,y)$ the integrals occurring in (7.444) can be evaluated either in an explicit form or alternatively, numerical techniques can be used to evaluate both the integrals and to generate the expression for $w(x, y, t)$.

Example 7.12

The rectangular membrane shown in Figure 7.33 is constrained along the edges. It is subjected to the discontinuous initial displacement

$$w(x, y, 0) = w_0\, \delta\left(x - \frac{a}{2}\right) \delta\left(y - \frac{b}{2}\right) \sin\left(\frac{\pi x}{a}\right) \sin\left(\frac{\pi y}{b}\right) \tag{7.445}$$

where $\delta(x - a/2)$ and $\delta(y - b/2)$ are Dirac delta functions. Formulate the initial boundary value problem and develop a series expansion for the dynamic displacements of the stretched membrane if the membrane is released from this position with zero initial velocity.

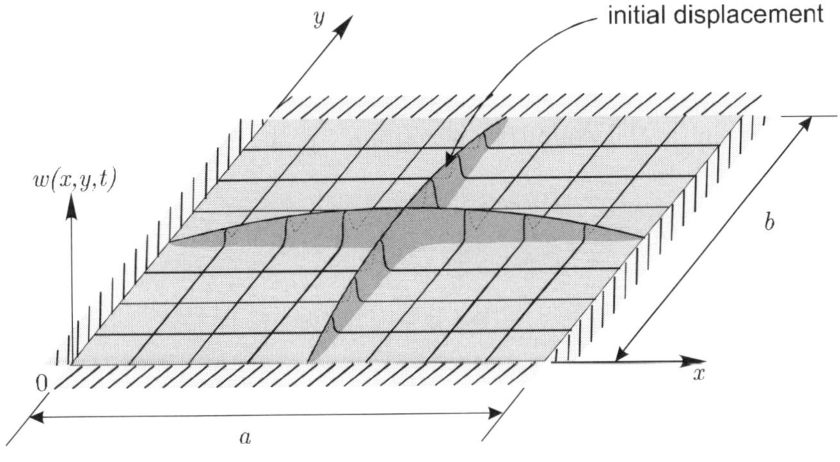

Figure 7.33: Discontinuous displacement of a stretched rectangular membrane.

Solution

We can apply either the separation of variables procedure or an integral transform technique to solve this problem. Using the separation of variables technique outlined for the square membrane problem we can show that a displacement field of the type

$$w(x,y,t) = \sum_{\alpha=1,2,}^{\infty} \sum_{\beta=1,2,}^{\infty} A_{\alpha\beta} \sin\left(\frac{\alpha\pi x}{a}\right) \sin\left(\frac{\beta\pi y}{b}\right) \cos\left(\sqrt{\frac{\alpha^2}{a^2}+\frac{\beta^2}{b^2}}\, c_m \pi t\right) \tag{7.446}$$

satisfies the displacement boundary conditions

$$w(0,y,t) = 0 \quad ; \quad w(a,y,t) = 0 \tag{7.447}$$

$$w(x,0,t) = 0 \quad ; \quad w(x,b,t) = 0 \tag{7.448}$$

and satisfies the initial condition pertaining to the initial velocity of the stretched membrane i.e.

$$\left[\frac{\partial w}{\partial t}\right]_{t=0} = 0 \tag{7.449}$$

The arbitrary constants $A_{\alpha\beta}$ can be obtained by making use of the double Fourier series representation of (7.445) and the orthogonality properties: i.e.

$$A_{\alpha\beta} = \frac{4}{ab}\int_0^a \int_0^b w(x,y,0) \sin\left(\frac{\alpha\pi x}{a}\right) \sin\left(\frac{\beta\pi y}{b}\right) dx dy \tag{7.450}$$

Substituting (7.445) in (7.450) we obtain

$$A_{\alpha\beta} = \frac{4w_0}{ab} \int_0^a \delta\left(x - \frac{a}{2}\right) \sin\left(\frac{\alpha\pi x}{a}\right) \sin\left(\frac{\pi x}{a}\right) dx \cdot \int_0^b \delta\left(y - \frac{b}{2}\right) \sin\left(\frac{\beta\pi y}{b}\right) \sin\left(\frac{\pi y}{b}\right) dy \tag{7.451}$$

Evaluating (7.451) we have

$$A_{\alpha\beta} = 4w_0 \sin\left(\frac{\alpha\pi}{2}\right) \sin\left(\frac{\beta\pi}{2}\right) \tag{7.452}$$

The time-dependent deflection of the stretched rectangular membrane is given by

$$\frac{w(x,y,t)}{w_0} = 4 \sum_{\alpha=1,3,5}^{\infty} \sum_{\beta=1,3,5}^{\infty} \sin\left(\frac{\alpha\pi}{2}\right) \sin\left(\frac{\beta\pi y}{2}\right) \sin\left(\frac{\alpha\pi x}{a}\right) \sin\left(\frac{\beta\pi y}{b}\right) \cos\left(\sqrt{\frac{\alpha^2}{a^2} + \frac{\beta^2}{b^2}}\, c_m \pi t\right) \tag{7.453}$$

7.5.3 Forced vibrations of a stretched rectangular membrane

In this section we consider the problem of the forced vibration of a stretched rectangular membrane which is governed by the partial differential equation

$$\nabla^2 w(x,y,t) - \frac{1}{c_m^2} \frac{\partial^2 w(x,y,t)}{\partial t^2} = -\frac{p(x,y,t)}{T_m} \tag{7.454}$$

The forcing function $p(x,y,t)$ in general can have any form a distribution both spatially and temporally. In this discussion, however, we consider the special case where the spatial distribution is arbitrary but the temporal variation is of a harmonic form. Following this assumption, the forcing function

$p(x, y, t)$ is represented in the form

$$p(x, y, t) = p_0 \; P(x, y)e^{-i\omega t} \tag{7.455}$$

where p_0 and ω are constants and $P(x, y)$ is an arbitrary function of x and y. (p_0 has dimensions of pressure)

We assume that the forcing function acts over the entire region of stretched rectangular membrane (Figure 7.34). We also assume that for the forcing function defined by (7.455), the solution for $w(x, y, t)$ admits a representation of the form

$$w(x, y, t) = W(x, y)e^{-i\omega t} \tag{7.456}$$

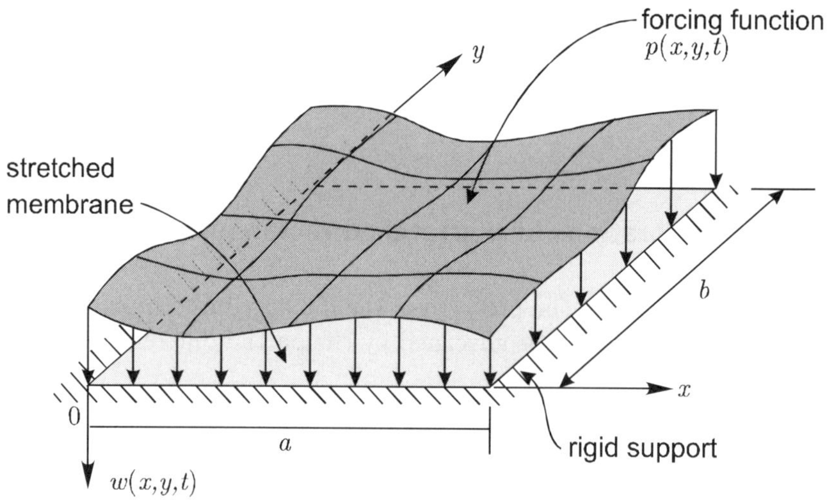

Figure 7.34: Forced vibration of a stretched rectangular membrane.

Substituting (7.455) and (7.456) in (7.454) we obtain the following partial differential equation for $W(x, y)$;

$$\nabla^2 W(x,y) - \frac{\omega^2}{c_m^2} W(x,y) = -\frac{p_0}{T_m} P(x,y) \tag{7.457}$$

This equation is an inhomogeneous form of Helmholtz's equation discussed in Section 6.4.1, which can be solved by using the double Fourier transform technique presented in Section 7.5.2. Multiplying both sides of (7.457) by $\{\sin(\alpha\pi x/a)\sin(\beta\pi y/b)\}$ and integrating over the region of the rectangular membrane we obtain

$$\int_0^a \int_0^b \left[\nabla^2 W(x,y) - \frac{\omega^2}{c_m^2} W(x,y)\right] \sin\left(\frac{\alpha\pi x}{a}\right) \sin\left(\frac{\beta\pi y}{b}\right) dx dy$$
$$= -\frac{p_0}{T_m} \int_0^a \int_0^b P(x,y) \sin\left(\frac{\alpha\pi x}{a}\right) \sin\left(\frac{\beta\pi y}{b}\right) dx dy \tag{7.458}$$

Using the results (7.432) to (7.434) we obtain

$$\overline{W}_s(\alpha,\beta) = \frac{p_0 \overline{P}_s(\alpha,\beta)}{T_m\left[\pi^2\left(\frac{\alpha^2}{a^2}+\frac{\beta^2}{b^2}\right)+\frac{\omega^2}{c_m^2}\right]} \tag{7.459}$$

The inversion of (7.459) gives an expression for $W(x,y)$; the time-dependent displacements of the stretched membrane can be obtained in the form

$$w(x,y,t) = \frac{4p_0 e^{-i\omega t}}{abT_m} \sum_{\alpha=1,2,}^{\infty} \sum_{\beta=1,2,}^{\infty} \frac{\sin\left(\frac{\alpha\pi x}{a}\right)\sin\left(\frac{\beta\pi y}{b}\right)}{\left[\pi^2\left(\frac{\alpha^2}{a^2}+\frac{\beta^2}{b^2}\right)+\frac{\omega^2}{c_m^2}\right]}$$
$$\cdot \int_0^a \int_0^b P(\xi,\eta) \sin\left(\frac{\alpha\pi\xi}{a}\right) \sin\left(\frac{\beta\pi\eta}{b}\right) d\xi d\eta \tag{7.460}$$

where $P(\xi,\eta)$ is defined by (7.455).

In the special case when the harmonic loading $p(x,y,t)$ has a constant intensity p_0 and occupies the entire region of the stretched membrane $P(x,y)=1$ and

$$\int_0^a \int_0^b \sin\left(\frac{\alpha\pi\xi}{a}\right) \sin\left(\frac{\beta\pi\eta}{b}\right) d\xi d\eta = \frac{4ab}{\pi^2\alpha\beta} \begin{cases} \alpha = 1,3,5,... \\ \beta = 1,3,5,... \end{cases} \tag{7.461}$$

and (7.460) reduces to

$$w(x,y,t) = \frac{16p_0 e^{-i\omega t}}{\pi^2 T_m} \sum_{\alpha=1,3,}^{\infty} \sum_{\beta=1,3,}^{\infty} \frac{\sin\left(\frac{\alpha\pi x}{a}\right) \sin\left(\frac{\beta\pi y}{b}\right)}{\alpha\beta\left[\pi^2\left(\frac{\alpha^2}{a^2}+\frac{\beta^2}{b^2}\right)+\frac{\omega^2}{c_m^2}\right]} \tag{7.462}$$

Other forms of spatial and temporal distributions can be examined by suitable applications of Laplace and finite double Fourier transform techniques.

7.5.4 Free vibrations of a stretched circular membrane

The problem of the vibration of a stretched circular membrane serves as a seminal model for the study of the vibrational characteristics of drums, other musical instruments and acoustical devices in general. The models of stretched circular and annular membranes are also utilized to monitor sensitive pressure regulators and as sensors for monitoring fluid pressure transients. Owing to the circular configuration usually associated with such membranes, it is convenient to formulate problems in relation to the plane polar coordinate system. Similar procedures were adopted in Sections 7.4.4 and 7.4.5, respectively, in connection with the study of the symmetric free vibrations of a stretched membrane of infinite extent and for the development of a Green's function for a stretched membrane subjected to an initial displacement. Prior to considering the more general case where the deflections of the membrane depend on r, θ and t, we shall examine the problem where the circular membrane experiences motions which are symmetric about the central axis (Figure 7.35).

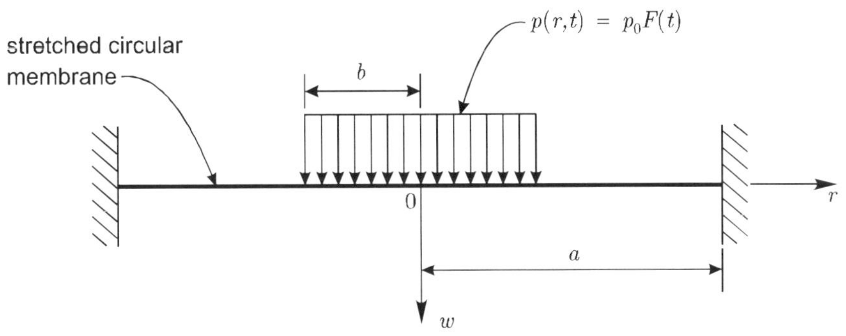

Figure 7.35: Symmetric motion of stretched circular membrane.

During free vibrations, the pressure $p(r,t) = 0$ and the partial differential equation governing the dynamic deflection of the membrane can be expressed in the polar symmetric form

$$\nabla^2 w(r,t) = \frac{1}{r}\frac{\partial}{\partial r}\left(r\frac{\partial w}{\partial r}\right) = \frac{1}{c_m^2}\frac{\partial^2 w}{\partial t^2} \tag{7.463}$$

The boundary condition applicable to a rigid boundary support is

$$w(a,t) = 0 \tag{7.464}$$

In addition to this boundary condition, the displacements within the membrane region should be *finite* and *bounded*. This requirement will exclude certain classes of solutions which can exhibit singular behaviour as $r \to 0$. The initial boundary value will be complete and well-posed with the specification of initial conditions pertaining to the displacement and velocity of the stretched membrane. For the polar symmetric case, these can take the forms

$$w(r,0) = f_0(r) \tag{7.465}$$

$$\left[\frac{\partial w}{\partial t}\right]_{t=0} = g_0(r) \tag{7.466}$$

This completes the formulation of the initial boundary value problem. We now consider a variables separable solution of the form

$$w(r,t) = R(r)T(t) \tag{7.467}$$

Substituting (7.467) into (7.463) we obtain

$$\frac{1}{R}\left\{\frac{d^2R}{dr^2} + \frac{1}{r}\frac{dR}{dr}\right\} = \frac{1}{c_m^2}\frac{1}{T}\frac{d^2T}{dt^2} = -k^2 \tag{7.468}$$

where for periodic vibrations, the constant k^2 must be negative. From (7.468) we obtain the following ordinary differential equations for *R(r)* and *T(t)*:

$$\frac{d^2R}{dr^2} + \frac{1}{r}\frac{dR}{dr} + k^2R = 0 \tag{7.469}$$

and

$$\frac{d^2T}{dt^2} + k^2c_m^2T = 0 \tag{7.470}$$

The solution of (7.469) was discussed in Chapter 6 in connection with Bessel's equation of order zero (see e.g. (6.209)). The solution for *R(r)* is

$$R(r) = A_1J_0(kr) + B_1Y_0(kr) \tag{7.471}$$

and

$$T(t) = A_2\sin(kc_mt) + B_2\cos(kc_mt) \tag{7.472}$$

Since the membrane region includes the location $r = 0$, the deflections of the membrane must be finite within $0 \leq r \leq a$. Hence the term $B_1 Y_0(kr)$ should be excluded from the solution, since, $Y_0(kr)$ tends to infinity as $r \to 0$. The solution of (7.463) can now be written as

$$w(r,t) = J_0(kr)\{A\sin(kc_m t) + B\cos(kc_m t)\} \tag{7.473}$$

where A and B are arbitrary constants. To satisfy the zero displacement boundary condition we require

$$w(a,t) = J_0(ka)\{A\sin(kc_m t) + B\cos(kc_m t)\} \tag{7.474}$$

For a non-trivial result we require

$$J_0(ka) = 0 \tag{7.475}$$

This is the *eigen-value equation*, the roots of which are

$$k_n a = 2.404;\ 5.520;\ 8.653;\ 11.791;\ 14.93;\ etc. \tag{7.476}$$

corresponding to $n = 1,2,3,\ldots$. We can construct an eigen-function expansion solution of the form

$$w(r,t) = \sum_{n=1,2,}^{\infty} J_0(k_n r)\{A_n \sin(k_n c_m t) + B_n \cos(k_n c_m t)\} \tag{7.477}$$

The arbitrary constants occurring in (7.477) can be obtained by making use of the initial conditions (7.465) and (7.466). We now assume that the initial conditions can be expanded as series in terms of $J_0(k_n r)$ (see e.g. (6.233)) i.e.

$$w(r,0) = f_0(r) = \sum_{n=1,2,}^{\infty} F_n J_0(k_n r) \tag{7.478}$$

$$\left[\frac{\partial w}{\partial t}\right]_{t=0} = g_0(r) = \sum_{n=1,2,}^{\infty} G_n J_0(k_n r) \tag{7.479}$$

Using the orthogonality properties of Bessel functions we can express the constants F_n and G_n as follows (see e.g. section 1.12.3):

$$F_n = \frac{2}{a^2\{J_1(k_n a)\}^2} \int_0^a r f_0(r) J_0(k_n r) dr \tag{7.480}$$

$$G_n = \frac{2}{a^2\{J_1(k_n a)\}^2} \int_0^a r g_0(r) J_0(k_n r) dr \tag{7.481}$$

Considering (7.477), the initial condition (7.465) and representation (7.478) we obtain

$$\sum_{n=1,2,} B_n J_0(k_n r) = \sum_{n=1,2,} F_n J_0(k_n r) \tag{7.482}$$

or

$$B_n = F_n \tag{7.483}$$

Similarly using (7.477), the initial condition (7.466) and the representation (7.479) we obtain

$$\sum_{n=1,2,}^{\infty} A_n k_n c_m J_0(k_n r) = \sum_{n=1,2,} G_n J_0(k_n r) \tag{7.484}$$

or

$$A_n = \frac{G_n}{k_n c_m} \tag{7.485}$$

Hence, the result for the time-dependent deflection of the stretched circular membrane can be written as

$$w(r,t) = \sum_{n=1,2,}^{\infty} \frac{2J_0(k_n r)}{a^2\{J_1(k_n a)\}^2} \left[\left\{ \int_0^a \lambda g_0(\lambda) J_0(k_n \lambda) d\lambda \right\} \frac{\sin(k_n c_m t)}{k_n c_m} \right.$$
$$\left. + \left\{ \int_0^a \lambda f_0(\lambda) J_0(k_n \xi) d\xi \right\} \cos(k_n c_m t) \right] \tag{7.486}$$

The circular membrane problem with polar symmetry can also be solved by employing integral transform techniques. In view of the axisymmetric nature of the problem it is most conveniently formulated in relation to a Hankel transform approach.

Example 7.13

A stretched circular membrane of radius a is subjected to a harmonic motion of the form

$$w(r,t) = W(r)e^{i\omega t} \tag{7.487}$$

where $W(r)$ is an arbitrary function and ω is a constant. Determine an appropriate solution to the problem given the maximum amplitude of motion is W_0.

Solution

Substituting the assumed form of the deflection (7.487) in the governing partial differential equation for *free* vibrations, (7.463), we have

$$\frac{d^2W}{dr^2} + \frac{1}{r}\frac{dW}{dr} + \frac{\omega^2}{c_m^2}W = 0 \tag{7.488}$$

The complete solution of Bessel's equation (7.488) is

$$W(r) = AJ_0\left(\frac{\omega r}{c_m}\right) + BY_0\left(\frac{\omega r}{c_m}\right) \tag{7.489}$$

where A and B are constants. As observed previously, since the deflections of the stretched membrane are finite within $0 \leq r \leq a$, we require $B \equiv 0$. To satisfy the boundary conditions at the *fixed end* we require

$$w(a,t) = W(a)e^{i\omega t} = AJ_0\left(\frac{\omega a}{c_m}\right) = 0 \tag{7.490}$$

which gives the frequency equation

$$J_0\left(\frac{\omega a}{c_m}\right) = 0; \tag{7.491}$$

Since $J_0(0) = 1$, we have

$$W(0) = W_0 = A. \tag{7.492}$$

The deflection of the stretched membrane is given by

$$w(r,t) = W_0 J_0 \left(\frac{\omega r}{c_m}\right) e^{i\omega t} \tag{7.493}$$

and the admissible frequencies are determined from (7.491).

7.5.5 Hankel transform analysis of free vibrations of a stretched circular membrane

The *symmetric* free vibrations of a stretched circular membrane can also be conveniently formulated by considering a Hankel transform approach. Consider the symmetric free vibrations of a stretched circular membrane governed by the partial differential equation (7.463). We consider the finite$^{(f)}$ Hankel transform of order zero of the deflection of the stretched membrane defined by (see e.g. section 1.12.3)

$$\widetilde{w}(\xi_i,t) = \mathcal{H}^f\{w(r,);\xi_i\} = \int_0^a rw(r,t)J_0(\xi_i r)dr \tag{7.494}$$

where ξ_i are zeros of $J_0(\xi_i a)$. The partial differential equation (7.463) can be transformed by multiplying each term by $\{rJ_0(\xi_i r)\}$ and integrating between the limits 0 to a. The resulting transformed equation is

$$\left(\frac{d^2}{dt^2} + c_m^2\xi_i^2\right)\widetilde{w}(\xi_i,t) = 0 \tag{7.495}$$

The solution of (7.495) takes the form

$$\widetilde{w}(\xi_i,t) = A(\xi_i)\sin(c_m\xi_i t) + B(\xi_i)\cos(c_m\xi_i t) \tag{7.496}$$

where $A(\xi_i)$ and $B(\xi_i)$ are arbitrary functions of ξ_i. To determine these functions, we use of the initial conditions (7.465) and (7.466) which can be transformed according to

$$\widetilde{w}(\xi_i, 0) = \mathcal{H}^f\{f_0(r); \xi_i\} = \int_0^a r g_0(r) J_0(\xi_i r) dr = \widetilde{f}_0(\xi_i) \tag{7.497}$$

and

$$\left[\frac{\partial \widetilde{w}}{\partial t}\right]_{t=0} = \mathcal{H}^f\{g_0(r); \xi_i\} = \int_0^a r g_0(r) J_0(\xi_i r) dr = \widetilde{g}_0(\xi_i) \tag{7.498}$$

Using the results (7.496), (7.497) and (7.498) we obtain

$$B(\xi_i) = \widetilde{f}_0(\xi_i) \quad ; \quad A(\xi_i) = \frac{\widetilde{g}_0(\xi_i)}{c_m \xi_i} \tag{7.499}$$

The finite Hankel transform solution for the deflection of the stretched membrane can be written as

$$\widetilde{w}(\xi_i, t) = \frac{\widetilde{g}_0(\xi_i)}{c_m \xi_i} \sin(c_m \xi_i t) + \widetilde{f}_0(\xi_i) \cos(c_m \xi_i t) \tag{7.500}$$

Inverting the finite Hankel transform, we obtain

$$\begin{aligned} w(r, t) = {} & \frac{2}{a^2} \sum_{n=1,2,}^{\infty} \frac{J_0(\xi_i r) \cos(c_m \xi_i t)}{[J_1(\xi_i a)]^2} \int_0^a \lambda f_0(\lambda) J_0(\lambda \xi_i) d\lambda \\ & + \frac{2}{c_m a^2} \sum_{n=1,2,}^{\infty} \frac{J_0(\xi_i r)}{[J_1(\xi_i a)]^2} \frac{\sin(c_m \xi_i t)}{\xi_i} \int_0^a \lambda g_0(\lambda) J_0(\lambda \xi_i) d\lambda \end{aligned} \tag{7.501}$$

where the summation extends over the positive roots of the transcendental equation

$$J_0(\xi_i a) = 0 \tag{7.502}$$

The membrane problem discussed here can also be examined by adopting a Laplace transform approach where the transform is taken with respect to the *time variable*. The resulting analysis is, however, much more complicated and involves transform inversion via contour integrals.

7.5.6 Hankel transform analysis of forced vibrations of a stretched circular membrane

We consider the *polar symmetric* problem of the forced vibrations of a circular membrane starting from its static position (i.e. zero initial displacement and zero initial velocity). The vibrations are induced by a time-dependent pressure of uniform intensity (Figure 7.36).

$$p(r,t) = p_0 F(t) \tag{7.503}$$

The partial differential equation governing the deflections of the stretched membrane is

$$\nabla^2 w(r) = \frac{1}{c_m^2}\frac{\partial^2 w}{\partial t^2} - \frac{p(r,t)}{T_m} \tag{7.504}$$

where $p(r,t)$ is assumed to act in the direction of the membrane deflection. The boundary condition governing the problem is

$$w(a,t) = 0 \tag{7.505}$$

along with the requirement that $w(r,t)$ be finite in the region $0 \leq r \leq a$. Since the free vibrations are induced from a static position we have

$$w(r,0) = \left[\frac{dw}{dt}\right]_{t=0} = 0 \tag{7.506}$$

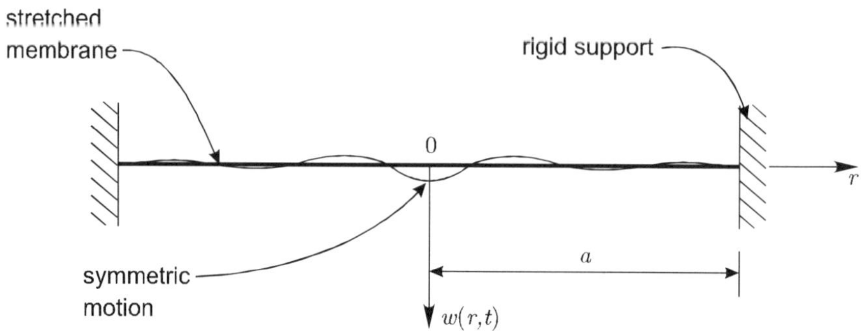

Figure 7.36: Forced vibration of a stretched circular membrane.

We consider the finite Hankel transform development of the governing partial differential equation. Applying the finite Hankel transform to (7.504) we have

$$\left(\frac{d^2}{dt^2} + c_m^2 \xi_i^2\right) \widetilde{w}(\xi_i, t) = \frac{\widetilde{p}(\xi_i, t)}{T_m} \tag{7.507}$$

where $\widetilde{p}(\xi_i, t)$ is the finite Hankel transform of $p(r,t)$ defined by

$$\widetilde{p}(\xi_i, t) = \mathcal{H}^f\{p(r,t); \xi_i\} = \int_0^a r p(r,t) J_0(\xi_i r) dr \tag{7.508}$$

Since

$$p(r,t) = \begin{cases} p_0 F(t) \; ; 0 \leq r < b \\ \quad 0 \quad \; ; b < r < a \end{cases} \tag{7.509}$$

we can write (7.508) as

$$\widetilde{p}(\xi_i, t) = p_0 F(t) \int_0^b r J_0(\xi_i r) dr = \frac{b p_0 F(t) J_1(\xi_i b)}{\xi_i} \tag{7.510}$$

and ξ_i are interpreted as the roots of the equation (7.502) which ensures that the boundary condition (7.505) is automatically satisfied for all choices of ξ_i. Since the stretched membrane is set in motion by the pressure $p(r,t)$ with zero initial displacement and zero initial velocity (i.e.(7.506)), we have

$$\widetilde{w}(\xi_i, 0) = \left[\frac{\partial \widetilde{w}}{\partial t}\right]_{t=0} = 0 \tag{7.511}$$

Applying Laplace transform to the ordinary differential equation (7.507) and noting the conditions (7.511) we obtain

$$\overline{\widetilde{w}}(\xi_i, s) = \frac{p_0 b J_1(\xi_i b)}{T_m \xi_i}\left\{\frac{\overline{F}(s)}{(s^2 + c_m^2 \xi_i^2)}\right\} \tag{7.512}$$

where $\overline{F}(s)$ is the Laplace transform of the time variation of $p(r,t)$. Inverting the Laplace transform and making use of the convolution theorem, we obtain

$$\widetilde{w}(\xi_i, t) = \frac{p_0 b J_1(\xi_i b)}{c_m T_m \xi_i^2} \int_0^t F(u) \sin(c_m \xi_i [t-u]) du \tag{7.513}$$

Inverting the finite Hankel transform, (7.513) gives

$$w(r,t) = \frac{2 p_0 b c_m}{T_m a^2} \sum_{i=1,2,}^{\infty} \frac{J_1(\xi_i b) J_0(\xi_i r)}{\xi_i^2 [J_1(\xi_i a)]^2} \int_0^t F(u) \sin(c_m \xi_i [t-u]) du \tag{7.514}$$

and the sum is interpreted as being over all the positive roots of $J_0(\xi_i a) = 0$. Alternatively, since mathematical software and tabulated values for the evaluation of roots of

$$J_0(\lambda_i) = 0 \tag{7.515}$$

are readily available, we can express (7.514) in the alternative form by replacing ξ_i by λ_i/a; we have

$$w(r,t) = \frac{2p_0bc_m}{T_m}\sum_{i=1,2,}^{\infty}\frac{J_1(\lambda_i b/a)J_1(\lambda_i r/a)}{\lambda_i^2[J_1(\lambda_i)]^2}$$
$$\int_o^t F(u)\sin\left(c_m\frac{\lambda_i}{a}[t-u]\right)du \tag{7.516}$$

We can recover from these general expressions (7.514) and (7.516) results of practical interest. Consider the case when

$$F(t) = F_0 = \text{ const. }, \quad t > 0 \tag{7.517}$$

When $F_0 = 1$, the result (7.516) now reduces to

$$w(r,t) = \frac{2p_0ab}{T_m}\sum_{i=1,2,}^{\infty}\frac{J_1(\lambda_i b/a)J_0(\lambda_i r/a)}{\lambda_i^3[J_1(\lambda_i)]^2}\left\{1-\cos\left(\frac{\lambda_i c_m t}{a}\right)\right\} \tag{7.518}$$

When $b = a$, (7.518) reduces to

$$w(r,t) = \frac{p_0}{T_m}\left[2a^2\sum_{i=1,2,}^{\infty}\frac{J_0(\lambda_i r/a)}{\lambda_i^3 J_1(\lambda_i)}\right.$$
$$\left.-2a^2\sum_{i=1,2,}^{\infty}\frac{J_0(\lambda_i r/a)}{\lambda_i^3 J_1(\lambda_i)}\cos(\lambda_1 c_m t/a)\right] \tag{7.519}$$

The first term on the right hand side of (7.519) can be expressed as

$$2a^2\sum_{i=1,2,}^{\infty}\frac{J_0(\lambda_i r/a)}{\lambda_i^3 J_1(\lambda_i)} = \frac{1}{4}(a^2-r^2) \tag{7.520}$$

which reduces (7.519) to the form

$$w(r,t) = \frac{p_0}{T_m}\left[\frac{(a^2-r^2)}{4} - 2a^2\sum_{i-1,2,}^{\infty}\frac{J_0(\lambda_i r/a)}{\lambda_i^3 J_1(\lambda_i)}\cos(\lambda_i c_m t/a)\right] \quad (7.521)$$

The second-term is convergent and can be calculated using mathematical software such as MAPLE®and MATHEMATICA®to evaluate the time-dependent displacement profile of the stretched membrane.

7.5.7 Vibrations of a circular membrane-general formulation

In previous sections, we have focussed on the analysis of membrane vibration problems where the stretched membrane is circular and the deflections induced during the dynamic behaviour exhibit polar symmetry. In the more general case, the deflections of the membrane depend on r, θ and t and the partial differential equation governing free vibrations is

$$\nabla^2 w(r,\theta,t) = \frac{\partial^2 w}{\partial r^2} + \frac{1}{r}\frac{\partial w}{\partial r} + \frac{1}{r^2}\frac{\partial^2 w}{\partial \theta^2} = \frac{1}{c_m^2}\frac{\partial^2 w}{\partial t^2} \quad (7.522)$$

In keeping with the procedure for the development of a variables separable solution we assume that the $w(r,\theta,t)$ admits a solution of the form

$$w(r,\theta,t) = R(r)\Theta(\theta)T(t) \quad (7.523)$$

Substituting (7.523) in (7.522) we obtain

$$c_m^2\left\{\Theta T\frac{d^2R}{dr^2} + \frac{\Theta T}{r}\frac{dR}{dr} + \frac{RT}{r^2}\frac{d^2\Theta}{d\theta^2}\right\} = R\Theta\frac{d^2T}{dt^2} \quad (7.524)$$

Dividing throughout by $R\Theta T$ we obtain

$$\frac{d^2T/dt^2}{c_m^2 T} = \frac{1}{R}\left(\frac{d^2R}{dr^2} + \frac{1}{r}\frac{dR}{dr}\right) + \frac{1}{r^2}\frac{d^2\Theta}{d\theta^2} = -\lambda^2 \tag{7.525}$$

where λ is a positive constant. This gives the following

$$\frac{d^2T}{dt^2} + \lambda^2 c_m^2 T = 0 \tag{7.526}$$

$$\frac{1}{R}\left\{\frac{d^2R}{dr^2} + \frac{1}{r}\frac{dR}{dr}\right\} + \frac{1}{r^2\Theta}\frac{d^2\Theta}{d\theta^2} = -\lambda^2 \tag{7.527}$$

Multiplying (7.527) by r^2 we can separate the variables again to obtain

$$\frac{r^2}{R}\frac{d^2R}{dr^2} + \frac{r}{R}\frac{dR}{dR} + \lambda^2 r^2 = -\frac{1}{\Theta}\frac{d^2\Theta}{d\theta^2} = \mu^2 \tag{7.528}$$

where μ is a positive constant. The equation (7.528) gives the following

$$\frac{d^2\Theta}{d\theta^2} + \mu^2\Theta = 0 \tag{7.529}$$

and

$$r^2\frac{d^2R}{dr^2} + r\frac{dR}{dr} + (\lambda^2 r^2 - \mu^2)R = 0 \tag{7.530}$$

The general solutions of (7.526), (7.529) and (7.530) take the forms

$$T(t) = A_1 \sin(\lambda c_m t) + B_1 \cos(\lambda c_m t) \tag{7.531}$$

$$\Theta(\theta) = A_2 \sin(\mu\theta) + B_2 \cos(\mu\theta) \tag{7.532}$$

$$R(r) = A_3 J_\mu(\lambda r) + B_3 Y_\mu(\lambda r) \tag{7.533}$$

The solution for $w(r, \theta, t)$ is therefore a product of (7.531), (7.532) and (7.533). We can, however place restrictions on the form of the product solution especially as it pertains to a circular region with $0 \leq r \leq a$; $0 \leq \theta \leq 2\pi$.

Since the deflected shape of the stretched membrane must have a period 2π in the variable θ we require

$$\mu = \alpha, \qquad \text{where} \qquad \alpha = 0, 1, 2, 3, \tag{7.534}$$

Also since the deflections are finite as $r \to 0$,

$$B_3 \equiv 0 \tag{7.535}$$

The resulting general solution applicable to a stretched circular membrane occupying the membrane region $0 \leq r \leq a$, $0 \leq \theta \leq 2\pi$, takes the form

$$\begin{aligned} w(r, \theta, t) = &\{A_1 \sin(\lambda c_m t) + B_1 \cos(\lambda c_m t)\} \\ &\{A_2^* \sin(\alpha\theta) + B_2^* \cos(\alpha\theta)\} \, J_\alpha(\lambda r) \end{aligned} \tag{7.536}$$

In particular, if the circular membrane is held fixed at the boundary, $r = a$, we require, for a non-trivial solution

$$J_\alpha(\lambda a) = 0 \tag{7.537}$$

The positive roots of (7.537) are denoted by

$$\lambda = \lambda_{\alpha\beta} \quad ; \quad \beta = 1, 2, 3, ... \tag{7.538}$$

We can now develop a general solution which consists of the superposition of solutions which involve summations over both α and β, such that

$$w(r, \theta, t) = \sum_{\alpha=0,1,}^{\infty} \sum_{\beta=1,2,}^{\infty} \{A_{\alpha\beta} \sin(\lambda_{\alpha\beta} c_m t) + B_{\alpha\beta} \cos(\lambda_{\alpha\beta} c_m t)\} \tag{7.539}$$
$$\cdot \left\{A^*_{\alpha\beta} \sin(\alpha\theta) + B^*_{\alpha\beta} \cos(\alpha\theta)\right\} J_\alpha(\lambda_{\alpha\beta} r)$$

The solution (7.539) satisfies the displacement boundary conditions at $r = a$ and gives finite membrane deflections within the region $0 \leq r \leq a$. In order to determine the arbitrary constants $A_{\alpha\beta}, B_{\alpha\beta}, A^*_{\alpha\beta}$ and $B^*_{\alpha\beta}$ it is necessary to specify initial conditions applicable to a particular problem.

Example 7.14

Considering the general solution for the vibration of a stretched circular membrane (radius a) given by (7.539), determine the result for the time-dependent deflections of the membrane if it is subjected to an initial displacement

$$w(r, \theta, 0) = f_0(r, \theta) \tag{7.540}$$

and released from this position with zero velocity.

Solution

Considering the general solution (7.539), the initial velocity is given by

$$\left[\frac{\partial w}{\partial t}\right]_{t=0} = \left[\sum_{\alpha=0,1,}^{\infty}\sum_{\beta=1,2,}^{\infty} \lambda_{\alpha\beta}c_m \left\{A_{\alpha\beta}\cos(\lambda_{\alpha\beta}c_m t) - B_{\alpha\beta}\sin(\lambda_{\alpha\beta}c_m t)\right\} \right.$$
$$\left. \cdot \left\{A^*_{\alpha\beta}\sin(\alpha\theta) + B^*_{\alpha\beta}\cos(\alpha\theta)\right\} J_\alpha(\lambda_{\alpha\beta} r)\right]_{t=0} = 0 \quad (7.541)$$

For the initial velocity to be zero for all r, θ, we require

$$A_{\alpha\beta} \equiv 0. \quad (7.542)$$

The resulting solution takes the form

$$w(r,\theta,t) = \sum_{\alpha=0,1,}^{\infty} \left\{ \left[\sum_{\beta=1,2,}^{\infty} \widetilde{A}_{\alpha\beta} J_\alpha\left(\lambda_{\alpha\beta} r\right)\right] \sin(\alpha\theta) \right.$$
$$\left. + \left[\sum_{\beta=1,2,}^{\infty} \widetilde{B}_{\alpha\beta} J_\alpha\left(\lambda_{\alpha\beta} r\right)\right] \cos(\alpha\theta) \right\} \cos(\lambda_{\alpha\beta} c_m t) \quad (7.543)$$

Using the initial condition (7.540) we can write

$$w(r,\theta,0) = f_0(r,\theta) = \sum_{\alpha=0,1,}^{\infty} \left\{C_\alpha \sin(\alpha\theta) + D_\alpha\cos(\alpha\theta)\right\} \quad (7.544)$$

where

$$C_\alpha = \sum_{\beta=1,2,}^{\infty} \left\{\widetilde{A}_{\alpha\beta} J_\alpha(\lambda_{\alpha\beta} r)\right\} \quad (7.545)$$

$$D_\alpha = \sum_{\beta=1,2,}^{\infty} \left\{\widetilde{B}_{\alpha\beta} J_\alpha(\lambda_{\alpha\beta} r)\right\} \quad (7.546)$$

Since (7.544) is essentially a Fourier series representation for an arbitrary function; we can determine C_α and D_α by making use of the Fourier series representations of the type

$$C_\alpha = \begin{cases} \dfrac{1}{\pi}\displaystyle\int_0^{2\pi} f_0(r,\theta)\cos(\alpha\theta)d\theta & ; \quad \alpha = 1,2,3, \\ \dfrac{1}{2\pi}\displaystyle\int_0^{2\pi} f_0(r,\theta)d\theta & ; \quad \alpha = 0 \end{cases} \tag{7.547}$$

and

$$D_\alpha = \frac{1}{\pi}\int_0^{2\pi} f_0(r,\theta)\sin(\alpha\theta)d\theta \quad ; \quad \alpha = 1,2,3, \tag{7.548}$$

Using (7.547) and (7.548) and the results of Bessel series expansions we obtain

$$\widetilde{A}_{\alpha\beta} = \frac{2}{a^2[J_{\alpha+1}(\lambda_{\alpha\beta}a)]^2}\int_0^a rJ_\alpha(\lambda_{\alpha\beta}r)C_\alpha\, dr \tag{7.549}$$

or alternatively

$$\begin{aligned} \widetilde{A}_{\alpha\beta} &= \frac{2}{\pi a^2[J_{\alpha+1}(\lambda_{\alpha\beta}a)]^2}\int_0^a\int_0^{2\pi} rf_0(r,\theta)J_\alpha(\lambda_{\alpha\beta}r)\cos(\alpha\theta)drd\theta \\ &\qquad ; \quad \alpha = 1,2, \\ \widetilde{A}_{\alpha\beta} &= \frac{1}{\pi a^2[J_{\alpha+1}(\lambda_{\alpha\beta}a)]^2}\int_0^a\int_0^{2\pi} rf_0(r,\theta)J_0(\lambda_{0\beta}r)drd\theta \\ &\qquad ; \quad \alpha = 0 \end{aligned} \tag{7.550}$$

Similarly

$$\widetilde{B}_{\alpha\beta} = \frac{2}{a^2[J_{\alpha+1}(\lambda_{\alpha\beta}a)]^2} \int_0^a rJ_\alpha(\lambda_{\alpha\beta}r)D_\alpha \, dr \tag{7.551}$$

or, alternatively

$$\widetilde{B}_{\alpha\beta} = \frac{2}{\pi a^2[J_{\alpha+1}(\lambda_{\alpha\beta}a)]^2} \int_0^a \int_0^{2\pi} rf_0(r,\theta)J_\alpha(\lambda_{\alpha\beta}r)\sin(\alpha\theta)drd\theta$$
$$; \quad \alpha = 0, 1, 2, \tag{7.552}$$

The solution of the problem is formally complete and explicit expressions for $w(r, \theta, t)$ consistent with the initial condition (7.540) can be obtained in series form by making use of general solution (7.543) and the expressions for $\tilde{A}_{\alpha\beta}$ and $\tilde{B}_{\alpha\beta}$ defined by (7.549) to (7.552) and $\lambda_{\alpha\beta}$ are determined by (7.537).

7.6 Wave motion and vibrations in membranes: non-classical effects

The membrane vibration problems considered in the preceding sections have dealt with highly idealized situations where there are no provisions made for phenomena which can attenuate the vibrations and waves generated by the initial disturbances. As indicated in section 7.3, in connection with non-classical effects associated with the stretched string problem, to realistically model the dynamic behaviour of stretched membranes it is necessary to incorporate the influence of dissipative phenomena. In particular, with stretched membrane type structures, the incorporation of dissipative phenomena occurring within the material of the membrane and dissipative phenomena associated with interaction between the membrane and the ambient (fluid) environment can lead to a significant alteration in the vibrational and wave propagation characteristics of the membrane.

The global representation of all dissipative phenomena by an *effective viscous damping* model is a convenient approach for examining such non-classical effects. This approach, however, has its drawbacks in the sense that the accurate determination of the "*viscosity*" contributing to the dissipative processes is a difficult exercise. Nonetheless, the "*viscous damping*" approach is a useful first approximation for the study of dissipative phenomena. We can assume that the viscous damping processes follow "Newtonian viscous" characteristics and provides a resistive force to the motion of the membrane which takes the form

$$q^*(x,y,t) = \eta_D \frac{\partial w}{\partial t}(x,y,t) \tag{7.553}$$

where η_D is the dynamic viscosity of the resistance, with dimensions of (force)(time)/(length). This resistive force can be incorporated in the basic equation of motion for the stretched membrane by assuming that $q^*(x,y,t)$ acts in a direction opposite to $p(x,y,t)$, and is distributed in the same manner as $p(x,y,t)$.

As a further non-classical effect we can also consider the influence of an *elastic foundation* which can be visualized as a support consisting of a system of a closely spaced independent springs of constant stiffness which can exert a reactive force proportional to the deflection of the membrane. The equation of motion can be modified by taking into consideration the reactive force term $kw(x,y,t)$ per unit area of the membrane, where k is the stiffness of the spring which has dimensions of force/(length)^3. If these non-classical effects are taken into account, the equation of motion of the membrane can be expressed in the form

$$\{T_m \nabla^2 - k\} w(x,y,t) + p(x,y,t) = \left\{ \tilde{\rho}\frac{\partial^2}{\partial t^2} + \eta_D \frac{\partial}{\partial t} \right\} w(x,y,t) \tag{7.554}$$

A detailed examination of these non-classical effects are beyond the scope of this introductory volume. The basic analytical procedures outlined in the preceding sections can, however, be extended to develop solutions to these non-classical initial boundary value problems involving stretched membranes with circular and rectangular planforms.

7.7 Wave equation for problems in solid mechanics

The primary emphasis of this chapter has been on the exposition of the classical wave equation as applied to stretched strings and membranes with a view to illustrating the fundamental aspects of the solutions and the various methods available for obtaining such solutions. The literature in solid mechanics also contains an extensive range of problems which deal with wave and vibration phenomena. These include longitudinal waves in bar elements, torsional waves in circular shafts, flexural waves in beams, plates and shells and waves in elastic continua. The detailed formulation of these wave propagation and vibration problems associated with either discrete or continuum regions requires some exposition of the concepts of *stress* and *strain* in continua, constitutive relationships applicable to elastic media and the dynamics of such solids. While such topics will be examined in more detail in a subsequent chapter it is a general interest to formulate, for completeness, the basic equations governing one-dimensional wave phenomena in structural elements such as slender rods which are subjected to longitudinal and rotational motion. In both instances the partial differential equations are of the *second-order* in both the spatial and time variables which enables the application of many of the solution schemes developed for the study of the stretched string problem.

7.7.1 Longitudinal wave motion in a slender elastic rod

We consider a slender elastic rod of uniform cross section, A_0, the longitudinal displacements in which are characterized by the displacement $u(x,t)$ (Figure 7.37). In general, if the slender rod is composed of a material which has *continuum characteristics*, we can expect a displacement in the longitudinal direction to be accompanied by a transverse displacements. In the definition of a *slender rod* we make the implicit assumption that the transverse displacements can be neglected in comparison with the longitudinal displacement. Accordingly, during the longitudinal motion of a slender rod, the cross section of the rod is assumed to remain constant.

We consider the longitudinal motion of a slender rod in which an element of length dx in the undeformed configuration experiences an extension du in longitudinal direction. During this motion we assume that cross sections of the bar which were originally normal to the axis of the bar will remain normal. The forces that are exerted on the planes $a-a$ and $b-b$ of the element are denoted by F and $F+(\partial F/\partial x)dx$. The latter term accounts for the variation of F (i.e. dF) over the length dx.

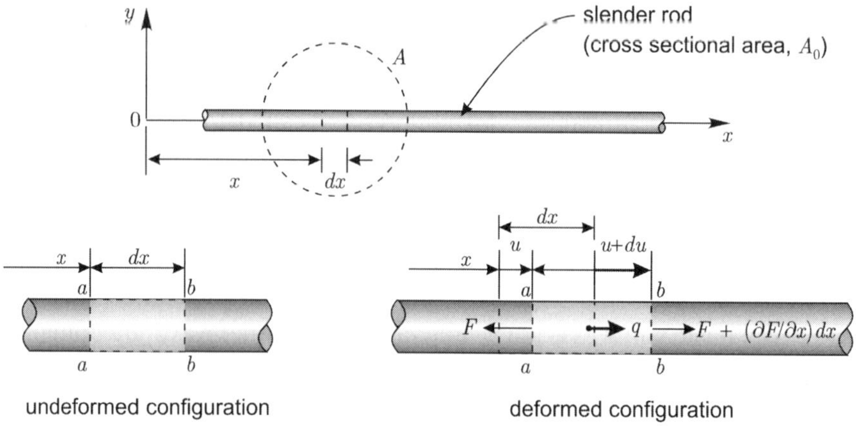

Figure 7.37: Longitudinal wave motion in a slender rod

Considering the dynamic equilibrium of the element undergoing motion we have

$$\left\{F + \frac{\partial F}{\partial x} dx\right\} - \{F\} + qA_0 dx = \{\rho_0 A_0 dx\} \frac{\partial^2 u}{\partial t^2} \tag{7.555}$$

where ρ_0 is the mass density, q is a body force term which can include externally applied forces and $\partial^2 u/\partial t^2$ represents the acceleration experiences by the element during motion. In order to complete the formulation we need to satisfy a *constitutive relationship* which links the kinematic variable $u(x,t)$ with the mechanical variable $F(x,t)$. We assume that the material of the slender rod exhibits elastic behaviour which is characterized by a linear relationship between the *axial stress* (σ_{xx}) and the *axial strain* (ϵ_{xx}). The proposal of such a linear relationship between stress and strain is generally attributed to Robert Hooke (1635-1703). For the one dimensional state of stress in the rod characterized by

$$\sigma_{xx}(x,t) = \frac{F(x,t)}{A_0} \quad ; \quad \epsilon_{xx}(x,t) = \frac{\partial u(x,t)}{\partial x} \tag{7.556}$$

we obtain, from Hooke's Law,

$$\sigma_{xx}(x,t) = E\epsilon_{xx}(x,t) \tag{7.557}$$

where E is referred to as the *elastic modulus* or *Young's modulus.* Using (7.556) and (7.557) we can reduce the equation of dynamic equilibrium (7.555) to the form

$$\frac{\partial^2 u}{\partial x^2} + q(x,t) = \frac{1}{c_L^2}\frac{\partial^2 u}{\partial t^2} \tag{7.558}$$

which is the partial differential equation governing longitudinal forced wave motion in a slender rod or vibration of a slender rod; and

$$c_L = \sqrt{\frac{E}{\rho_0}} \tag{7.559}$$

is the *longitudinal wave velocity* in the rod.

(i) *Boundary Conditions*

It is evident from (7.558) that the free longitudinal vibration of the slender rod is governed by a second-order partial differential equation which has a form identical to that governing the stretched string (see e.g. (7.10)). As such, many of the techniques and procedures described for the analysis of the stretched string problem can be adopted for the analysis of the longitudinal wave motion in the slender rod.

For example, considering a slender rod; i.e.

$$u(x,t) = f(x - c_L t) + g(x + c_L t) \tag{7.560}$$

which indicates that longitudinal waves can propagate at the velocity c_L *without distortion.*

Subtle differences, however, exist between the stretched string and slender rod when it comes to interpreting the boundary conditions applicable to various problems. In the case of a stretched string we have noted that it is necessary to provide fixity conditions for semi-infinite strings in order to maintain the tension T_0. In the case of slender rods, semi-infinite regions can be defined with either *fixed* or *free* ends. The boundary condition applicable to a fixed end requires that at the point of fixity the slender rod will not experience longitudinal displacements, i.e.

$$u(0,t) = 0 \tag{7.561}$$

Once again an image wave form as indicated in Figure 7.38 can be used to visualize the longitudinal motion in the bar with a fixed end and the associated influence of the interaction of the wave and its image on the amplification of stresses at the fixed boundary. It can be observed that the stresses will be superimposed to give an *amplification factor* of two. The sign of stress, however, remains unchanged, i.e. compressional waves which approach the fixed boundary will be reflected back as compressional waves and tensional waves will be reflected back as tensional waves.

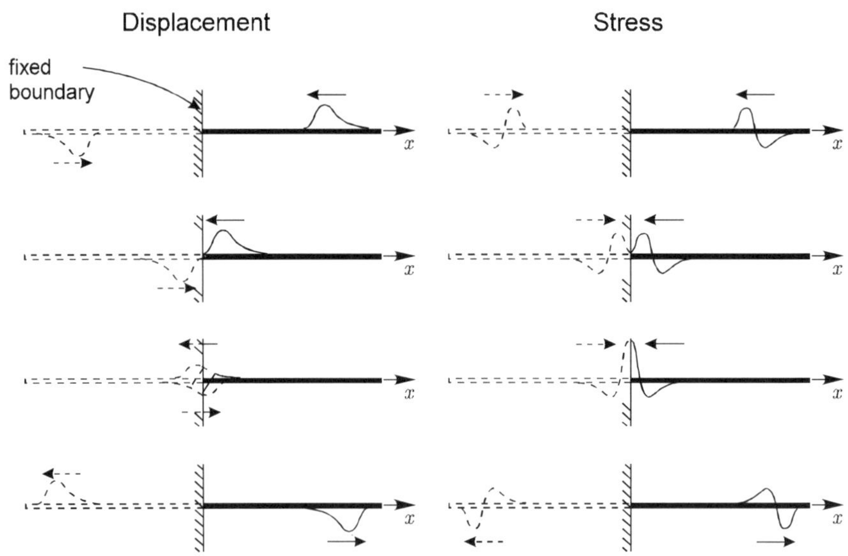

Figure 7.38: Reflection of a longitudinal wave at a fixed end

In the case of an end of a slender rod which is maintained free, the displacement is not defined but the force on the free end should be zero at all times. If the longitudinal force or alternatively the longitudinal stress is to be zero, we require

$$\left[\frac{\partial u}{\partial x}\right]_{x=0} = 0 \tag{7.562}$$

Again, we can apply the image pulse techniques to observe the process of interaction of a longitudinal wave with a free end (see e.g. (Figure 7.39)). An important observation is that the displacement at the free end is doubled as the incident pulse is reflected from the free end. Also the stress pulse is reflected with an opposite sense. Hence, compressional longitudinal waves are reflected from a free end as tensional longitudinal waves (and vice versa) without any amplification or attenuation.

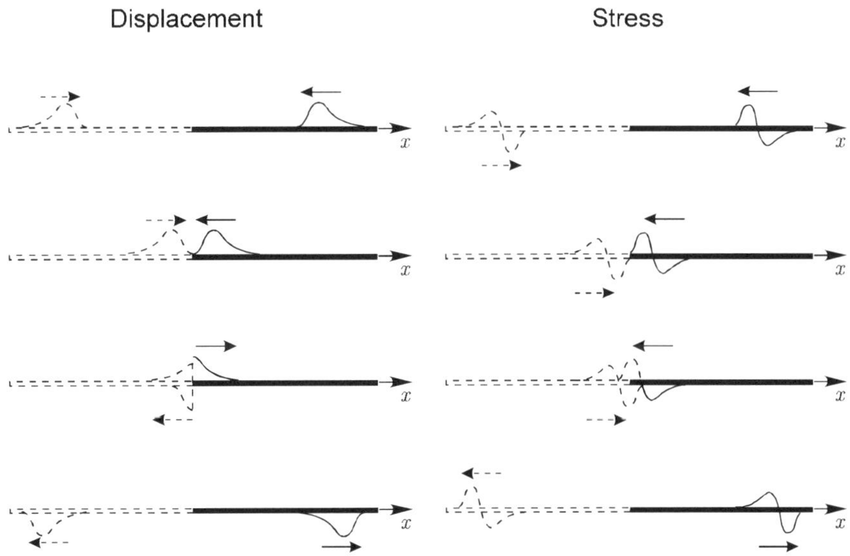

Figure 7.39: Reflection of a longitudinal wave at a free end

(ii) *Initial conditions*

As in the case of the formulation of the dynamic problem related to a stretched string, initial conditions can be specified in relation to the initial longitudinal displacement and the initial longitudinal velocity in the slender rod. These will include

$$u(x,0) = f_0(x) \tag{7.563}$$

$$\left[\frac{\partial u}{\partial t}\right]_{t=0} = g_0(x) \tag{7.564}$$

(iii) *Longitudinal free vibrations*

From a practical point of view it is not possible to clamp the ends of a slender rod so that standing longitudinal waves could occur with *nodes* at the clamped ends as would be the case for a stretched string. Standing longitudinal waves can be induced in the slender rod by allowing the ends to be free and assuming that clamped conditions are provided along the rod at some intermediate point, corresponding to a node. For example, it can be shown that (when $q(x,t) = 0$) the equation for longitudinal wave motion has a solution similar to

$$\begin{aligned} u(x,t) = \sum_{n=1,2,}^{\infty} & \{A_n \sin(\lambda_n x) + B_n \cos(\lambda_n x)\} \\ & \{C_n \sin(c_L \lambda_n t) + D_n \cos(c_L \lambda_n t)\} \end{aligned} \tag{7.565}$$

where $A_n, B_n, ...$ are arbitrary constants which can be determined by satisfying the boundary conditions and the initial conditions applicable to a particular problem. If the slender rod, of length ℓ, has both ends free we require

$$\left[\frac{\partial u}{\partial x}\right]_{x=0} = 0 \quad ; \quad \left[\frac{\partial u}{\partial x}\right]_{x=\ell} = 0 \tag{7.566}$$

The first condition of (7.566) requires $A_n = 0$ and for a non-trivial solution, the second condition requires

$$\sin \lambda_n \ell = 0 \quad ; \quad \text{or} \;\; \lambda_n \ell = n\pi \;\; (n = 0, 1, 2, ...) \tag{7.567}$$

The eigenvalues are

$$\lambda_n = \frac{n\pi}{\ell} \quad ; \quad n = 1, 2, \tag{7.568}$$

where $n = 0$ can be excluded if rigid body terms are excluded. The solution to the longitudinal vibration problem (7.565) can now be written as

$$u(x,t) = \sum_{n=1,2,}^{\infty} \cos\left(\frac{n\pi x}{\ell}\right) \left\{ C_n^* \sin\left(\frac{n\pi c_L t}{\ell}\right) + D^* \cos\left(\frac{n\pi c_L t}{\ell}\right) \right\} \tag{7.569}$$

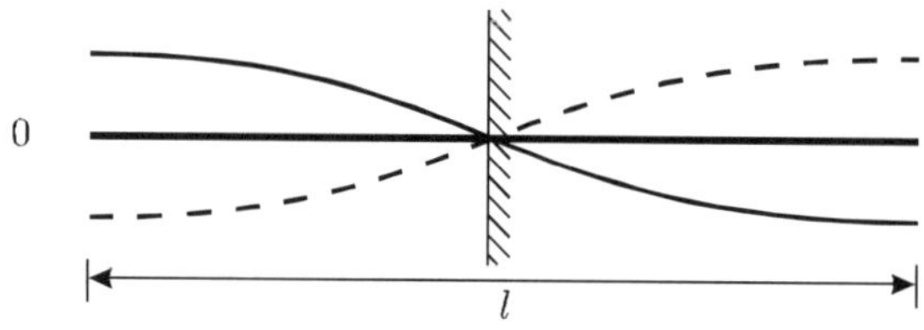

$n = 1$ $$u_1(x,t) = \cos\left(\frac{\pi x}{l}\right)\left\{C_1^* \sin\left(\frac{c_L \pi t}{l}\right) + D_1^* \cos\left(\frac{c_L \pi t}{l}\right)\right\}$$

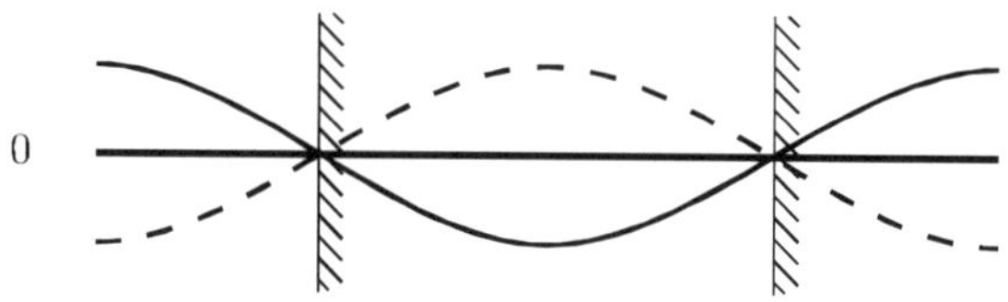

$n = 2$ $$u_2(x,t) = \cos\left(\frac{2\pi x}{l}\right)\left\{C_2^* \sin\left(\frac{2c_L \pi t}{l}\right) + D_2^* \cos\left(\frac{2c_L \pi t}{l}\right)\right\}$$

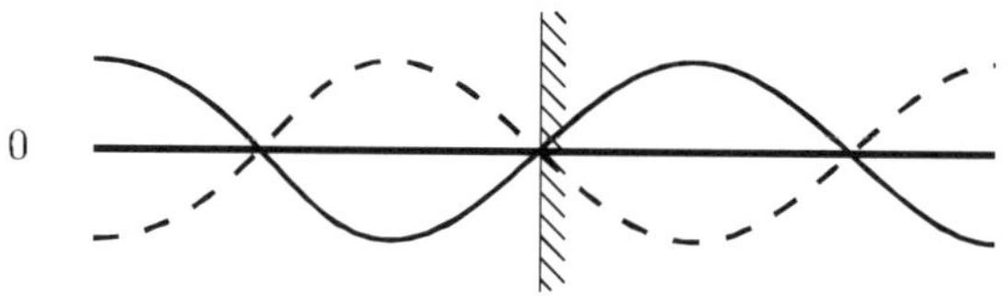

$n = 3$ $$u_3(x,t) = \cos\left(\frac{3\pi x}{l}\right)\left\{C_3^* \sin\left(\frac{3c_L \pi t}{l}\right) + D_3^* \cos\left(\frac{3c_L \pi t}{l}\right)\right\}$$

Figure 7.40: Normal modes of longitudinal vibrations of a slender rod

The normal modes corresponding to the natural frequencies $n = 1, 2$ and 3 are shown in Figure 7.40

(iv) *Longitudinal forced vibrations*

As a simple example involving forced longitudinal vibrations, let us examine the problem of a slender rod which is subjected to a *harmonic forcing function* at one end (Figure 7.41). The partial differential equation governing the motion of the bar is given by

$$\frac{\partial^2 u}{\partial x^2} = \frac{1}{c_L^2}\frac{\partial^2 u}{\partial t^2} \tag{7.570}$$

with boundary conditions

$$\left[\frac{\partial u}{\partial x}\right]_{x=0} = 0 \tag{7.571}$$

$$A_0 E\left[\frac{\partial u}{\partial x}\right]_{x=\ell} = -P_0 e^{i\omega t} \tag{7.572}$$

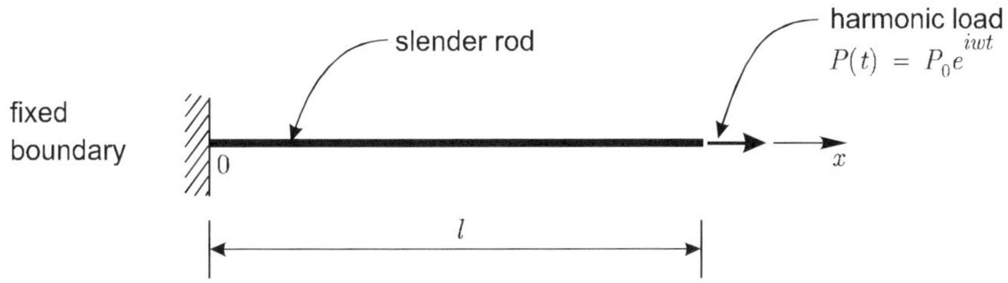

Figure 7.41: Forced longitudinal vibration of a clamped rod by a harmonic load.

The boundary condition (7.572) represents a condition of equilibrium between the *forcing function* and the *internal stresses in the bar* at the location $x = \ell$. We assume that the solution for the longitudinal displacement can be written as

$$u(x,t) = U(x)e^{i\omega t} \tag{7.573}$$

Using (7.573) in (7.570) we obtain

$$\frac{d^2 U}{dx^2} + \frac{\omega^2}{c_L^2}U = 0 \tag{7.574}$$

The solutions of which are

$$U(x) = A^* \sin\left(\frac{\omega x}{c_L}\right) + B^* \cos\left(\frac{\omega x}{c_L}\right) \tag{7.575}$$

The boundary conditions (7.571) and (7.572) can be used to determine the constants A^* and B^*; and the final form of the solution is

$$u(x,t) = \frac{P_0 c_L}{A_0 E \omega}\left[\frac{\cos(\omega x/c_L)}{\sin(\omega \ell/c_L)}\right] e^{i\omega t} \tag{7.576}$$

It is evident from (7.576) that $u(x,t)$ can be unbounded for certain choices of frequencies corresponding to the condition

$$\sin\left(\frac{\omega \ell}{c_L}\right) = 0 \quad ; \quad \text{or} \quad \frac{\omega \ell}{c_L} = n\pi \ ; \ n = 1, 2, \ldots \tag{7.577}$$

Hence the *resonant frequencies* corresponding to longitudinal vibrations of the slender rod are

$$\omega_R = \frac{n\pi}{\ell}\sqrt{\frac{E}{\rho_0}} \quad ; \quad n = 1, 2, \ldots . \tag{7.578}$$

7.7.2 Torsional waves in a slender circular elastic rod

We now briefly examine the problem of torsional waves that are produced in slender rods of circular cross section. The requirement for a circular cross section is essential to the developments which follow, since, by virtue of symmetry bars with circular cross section will not experience any *warping deformations* during relative rotational deformations along the axis. Such a simplification will not be applicable to slender rods with other cross sections

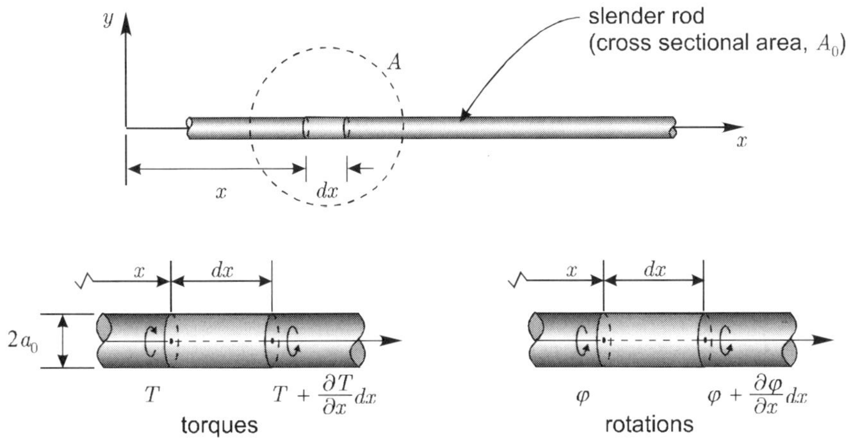

Figure 7.42: Torsional moments and rotations in a circular bar.

including an elliptical shape. This observation was made by B. de Saint Venant (1797-1886) in his classical studies on the theory of torsion of solids (see e.g. Chapter 9). The Figure 7.42 indicates the configuration of the circular bar and the relative rotational deformations experienced by the circular bar during torsional wave motion. We consider the equilibrium of an element of length dx which is subjected to torques T and $T + (\partial T/\partial x)dx$ about the axis of the cylinder. The net unbalanced torque $(\partial T/\partial x)dx$ provide the element with angular acceleration. We assume that the plane at x has rotated an amount φ and the plane at $x+dx$ has rotated an amount $\varphi+(\partial\varphi/\partial x)dx$. The net twist of the cylindrical element is $(\partial\varphi/\partial x)dx$.

The moment of inertia of the element about the axis

$$\Delta I = \frac{1}{2}\Delta m a_0^2 = \frac{\pi\rho_0 a^4}{2}dx \tag{7.579}$$

where Δm is the mass of the element. Considering the rotational equilibrium of the element we have

$$\frac{\partial T}{\partial x} = \frac{\pi\rho_0 a_0^4}{2}\frac{\partial^2\varphi}{\partial t^2} \tag{7.580}$$

where $(\partial^2\varphi/\partial t^2)$ is the rotational acceleration. Again, we have to introduce a constitutive relationship which will relate the *mechanical variable* (T) to the *kinematic variable* (φ). We assume that the torsional effects are produced by the shear stresses which vary linearly over the cross section of the cylindrical element, to a maximum of $a_0\mu(\partial\varphi/\partial x)$ at the location $r = a_0$. The torque at any section is given by

$$T = \int_0^{a_0}\int_0^{2\pi}\left(\mu\frac{\partial\varphi}{\partial x}\right)r^3 dr d\theta = \frac{\pi\mu a_0^4}{2}\frac{\partial\varphi}{\partial x} \tag{7.581}$$

where μ is the *shear modulus* of the elastic material. Combining (7.580) and (7.581) we obtain the partial differential equation governing the torsional wave motion in the slender bar as

$$\frac{\partial^2\varphi}{\partial x^2} = \frac{1}{c_S^2}\frac{\partial^2\varphi}{\partial t^2} \tag{7.582}$$

where c_S is the *shear wave velocity* in the elastic material which is defined by

$$c_S = \sqrt{\frac{\mu}{\rho_0}} \tag{7.583}$$

The boundary conditions which are applicable to either clamped or free ends take the forms (for $x = 0$)

$$\varphi(0,t) = 0 \tag{7.584}$$

$$\left[\frac{\partial\varphi}{\partial x}\right]_{x=0} = 0 \tag{7.585}$$

respectively. These conditions along with appropriate initial conditions governing the initial distributions of $\varphi(x,0)$ and $(\partial\varphi/\partial t)_{t=0}$ along the slender bar constitutes a well-posed torsional vibration problem. Many of the procedures and techniques that were outlined for the study of transverse wave motion in both finite and infinite stretched strings and longitudinal wave motion in both finite and infinite rods can be used to examine problems related propagation of torsional waves and vibrations in elastic bars.

7.8 Shallow water waves

The discussion on any aspect of wave mechanics would not be complete without some examination of wave motions in fluids. The study of wave motions in fluids has had a long historical interest particularly with reference to water waves and developments in this area over the past two centuries are too extensive for discussion in the context of the present section. In this section, however, we shall consider the problem of wave motions in shallow water which will yield partial differential equations which are similar to those discussed in this Chapter. In the shallow water wave problem discussed in this section the depth of water is assumed to be considerably smaller than the horizontal extent of the body of water. The problem discussed here is of interest to the study of tides and bores in rivers and estuaries.

In order to develop the equation governing shallow water waves in a general fashion, it is necessary to introduce some basic notions pertaining to fluid motion applicable to the problem. Let us assume that a material element has a position $\mathbf{x}$ at time t given by

$$\mathbf{x} = \mathbf{x}(t) \tag{7.586}$$

Any property $\mathcal{F}$ for this material element (e.g. velocity, pressure, density, etc.) will vary with time according to

$$\mathcal{F} = \mathcal{F}[x(t), y(t), z(t), t] = \mathcal{F}[\mathbf{x}(t), t] \tag{7.587}$$

The rate of change of $\mathcal{F}$ for the material element is

$$\frac{d\mathcal{F}}{dt} = \frac{\partial \mathcal{F}}{\partial t} + \frac{\partial \mathcal{F}}{\partial x} \cdot \frac{dx}{dt} + \frac{\partial \mathcal{F}}{\partial y} \cdot \frac{dy}{dt} + \frac{\partial \mathcal{F}}{\partial z}\frac{dz}{dt} = \frac{\partial \mathcal{F}}{\partial t} + \frac{d\mathbf{x}}{dt} \text{ grad } \mathcal{F} \quad (7.588)$$

We can interpret $d\mathbf{x}/dt$ as the rate of change of position of the material element; i.e. the fluid velocity

$$\frac{d\mathbf{x}}{dt} = \mathbf{v} = (v_x\mathbf{i} + v_y\mathbf{j} + v_z\mathbf{k}) \quad (7.589)$$

Hence for a material element, $d\mathcal{F}/dt$ is defined as the *material time derivative*

$$\frac{D}{Dt}(\mathcal{F}) = \frac{\partial \mathcal{F}}{\partial t} + \mathbf{v} \cdot \text{ grad } \mathcal{F} \quad (7.590)$$

The concept of a *material time derivative* is useful when dealing with many problems in mechanics and fluid mechanics. We can use this as a conservation equation. For example, if we have a two fluid region with one region containing a chemical at concentration C, and no mass transfer takes place between the two regions, then the concentration in the region containing the chemical must remain constant such that

$$\frac{D}{Dt}(C) = 0 \quad (7.591)$$

We now consider the problem of wave motion in a shallow water depth, by considering a region similar to that shown in Figure 7.43.

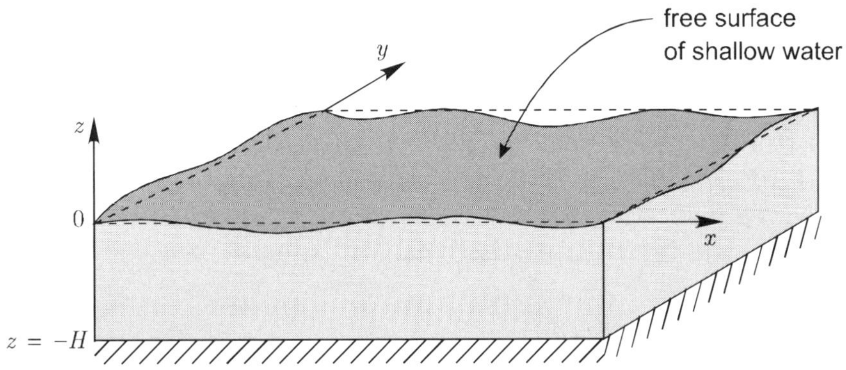

Figure 7.43: Wave motion in shallow water. Displacement of the free surface from the static position at an undisturbed water depth H.

We assume that the equilibrium position of the water level $z = 0$ is perturbed by the arrival of a wave which is defined by a wave height

$$z = h(x, y, t) \tag{7.592}$$

Considering the shallow water wave motion, the condition that all particles of water will remain on the free surface at $z = h(x, y, t)$ requires

$$\frac{D}{Dt}\{z - h(x, y, t)\} = 0 \tag{7.593}$$

or

$$\frac{\partial}{\partial t}\{z - h(x, y, t)\} + \mathbf{v} \cdot \text{ grad } \{z - h(x, y, t)\} = 0 \tag{7.594}$$

Equation (7.594) gives

$$v_z = \frac{\partial h}{\partial t} + v_x \frac{\partial h}{\partial x} + v_y \frac{\partial h}{\partial y} \tag{7.595}$$

If the perturbations are small, $v_x(\partial h/\partial h)$, $v_y(\partial h/\partial y)$ are small in comparison with $(\partial h/\partial t)$ and the vertical velocity induced by the wave motion at the *surface of the wave* is given by

$$v_z = \frac{\partial h}{\partial t} \quad ; \quad \text{at} \quad z = h \tag{7.596}$$

We can now utilize the result (7.596) to develop an alternative expression for the mass conservation equation for the incompressible water; i.e.

$$\text{div}(\mathbf{v}) = 0 \tag{7.597}$$

or

$$\frac{\partial v_x}{\partial x} + \frac{\partial v_y}{\partial y} + \frac{\partial v_z}{\partial z} = 0 \tag{7.598}$$

We can integrate (7.598) over the depth of the water which gives

$$\int_{-H}^{h} \left[\frac{\partial v_x}{\partial x} + \frac{\partial v_y}{\partial y} + \frac{\partial v_z}{\partial z} \right] dz = 0 \tag{7.599}$$

This gives

$$(H + h) \left[\frac{\partial v_x}{\partial x} + \frac{\partial v_y}{\partial y} \right] + \int_0^{\partial h/\partial t} dv_z = 0 \tag{7.600}$$

and we note that the vertical velocity at the base of the water depth is zero.

If we neglect higher-order terms in (7.600) we obtain

$$\frac{\partial h}{\partial t} + H \text{ div } \mathbf{v}^* = 0 \tag{7.601}$$

where

$$\text{div } \mathbf{v}^* = \frac{\partial v_x}{\partial x} + \frac{\partial v_y}{\partial y} \tag{7.602}$$

is referred to the *horizontal divergence* or the divergence of the horizontal components of the velocities. In the preceding, we have used the kinematics of the wave motion to recast the generalized mass conservation equation in a form which is relevant to the study of the shallow water waves. We will now consider the equations governing the dynamics of the wave motion. In its equilibrium state (i.e. without wave motion), the fluid pressure is given by

$$p_0(z) = -\rho g z \tag{7.603}$$

where ρ is the mass density of the fluid and g is the gravitational acceleration. We can define a perturbation pressure $p^*(x, y, t)$ by

$$p(x, y, z, t) = -\rho g z + p^*(x, y, t) \tag{7.604}$$

If we assume that the pressure $p(x, y, t)$ satisfies the hydrostatic equation

$$\frac{\partial p}{\partial z} = -\rho g \tag{7.605}$$

As a result of (7.604) and (7.605) the perturbation pressure $p^*(x, y, t)$ satisfies the equation

$$\frac{\partial p^*}{\partial z} = 0 \tag{7.606}$$

Integrating (7.606) we have

$$p^*(x, y, t) = f(x, y, t) \tag{7.607}$$

Also, the perturbation pressure $p^*(x, y, t)$ should vanish at the surface of the wave. Therefore without loss generality we can write

$$p^*(x, y, t) = \rho g h(x, y, t) \tag{7.608}$$

The momentum equations can be written as

$$\rho \frac{\partial \mathbf{v}}{\partial t} = -\text{ grad } (p^*) \tag{7.609}$$

Using (7.606) and (7.608), this result can be written as

$$\frac{\partial \mathbf{v}^*}{\partial t} = -g \text{ grad } (h) \tag{7.610}$$

Combining equations (7.601) and (7.610) and eliminating $\mathbf{v}^*$ we obtain

$$\frac{\partial^2 h}{\partial t^2} = gH \text{ div grad } (h) = gH\nabla^2 h \tag{7.611}$$

which can be written in the familiar form

$$c_w^2 \nabla^2 h = \frac{\partial^2 h}{\partial t^2} \tag{7.612}$$

where $c_w^2 = gH$ is a positive quantity such that the wave velocities are positive and the characteristics are real. The result (7.611) assumes that the depth of water H is constant. Lagrange in 1781 developed a more general result which is applicable to situations when H is a function of x and y. In this case, the wave equation takes a more complicated form

$$gH\nabla^2 h + g\left\{\frac{\partial h}{\partial x}\frac{\partial H}{\partial x} + \frac{\partial h}{\partial y}\frac{\partial H}{\partial y}\right\} = \frac{\partial^2 h}{\partial t^2} \tag{7.613}$$

7.9 Uniqueness theorem

In the ensuing we shall present a proof of uniqueness of solution for the class of problems which are governed by the wave equation developed for the study of wave motions and vibrations in stretched membranes. As indicated in previous discussions, the proof of uniqueness of a solution to an initial boundary value problem is an essential requirement for the well-posedness of a problem. Although attention is focussed on the proof of uniqueness of solution for the stretched membrane problem, the general procedure can also be employed to prove uniqueness of solution for problems associated with the stretched string and the longitudinal and torsional vibrations of slender rods.

Prior to consideration of a proof it is instructive to establish the domain of applicability of the uniqueness theorem and to outline some parametric representations of the boundary conditions relevant to a generalized formulation of the stretched membrane problem. We consider the forced vibration of a stretched membrane, the transverse deflection of which in the z- direction is governed by

$$T_m \nabla^2 w(x, y, t) + p(x, y, t) = \widetilde{\rho}\frac{\partial^2 w}{\partial t^2} \tag{7.614}$$

This partial differential equation is applicable to situations where the membrane is stretched either over space curve $\mathcal{C}_s$ or over a plane curve $\mathcal{C}$ which is the projection of $\mathcal{C}_s$ on the (x, y)-plane or even if the boundary of the membrane is allowed to move freely along the walls of a right cylinder with a boundary curve $\mathcal{C}$ (Figure 7.44)

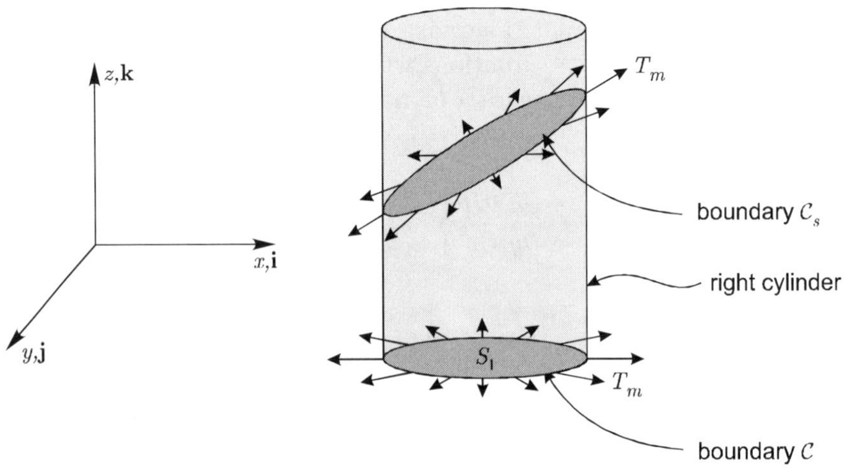

Figure 7.44: Configuration of stretched membrane

When boundary conditions are specified for the stretched membrane, they can be referred to either the plane curve $\mathcal{C}$ or the space curve $\mathcal{C}_s$. In general if Dirichlet boundary conditions are specified in relation to the closed space curve $\mathcal{C}_s$ we can express the boundary condition as

$$w(x(s), y(s), t) = \Omega(s) \tag{7.615}$$

where

$$x = x(s) \quad ; \quad y = y(s) \quad ; \quad w = \Omega(s) \tag{7.616}$$

is the parametric representation of the closed boundary $\mathcal{C}_s$ and s is the arc length referred to the closed boundary $\mathcal{C}$ in the x, y plane.

If the boundary of the membrane is free, the boundary condition will be of the Neumann-type which can be of the form

$$\frac{\partial}{\partial n}\left[w\{x(s), y(s), t\}\right] = 0 \tag{7.617}$$

If the parameter s represents the arc length on $\mathcal{C}$, the tangent vector to $\mathcal{C}$ is given by

$$\mathbf{t}^{\mathcal{C}} = \frac{dx}{ds}\mathbf{i} + \frac{dy}{ds}\mathbf{j} \tag{7.618}$$

The normal to $\mathcal{C}$ is given by

$$\mathbf{n}^{\mathcal{C}} = -\frac{dy}{ds}\mathbf{i} + \frac{dx}{ds}\mathbf{j} \tag{7.619}$$

The derivative of w in the direction normal to $\mathcal{C}$ can be expressed in the form

$$\frac{dw}{dn} = -\frac{\partial w}{\partial x}\frac{dy}{ds} + \frac{\partial w}{\partial y}\frac{dx}{ds} \tag{7.620}$$

A requirement of the Neumann boundary condition

$$\frac{dw}{dn} = 0 \qquad ; \quad (x, y) \in \mathcal{C} \tag{7.621}$$

implies that $w(x, y)$ is orthogonal to the walls of the right cylinder with projection $\mathcal{C}$ along which the membrane can move freely maintaining constant tension T_m. [It may be noted that a similar boundary condition was encountered in the one-dimensional case of the free edge boundary involving a stretched string (see e.g. section 7.2.4 equation (7.225)). The reservations that were expressed in connection with the *zero slope condition* as applied to the free boundary of the stretched string are equally applicable to the stretched membrane problem; i.e. a highly flexible structural element with no flexural resistance is required to maintain a configuration with a slope normal to a boundary (i.e. normal to the boundary of the right cylinder)].

In addition to the Dirichlet and Neumann boundary conditions applicable to the boundary of the stretched membrane, we require two initial conditions to be specified: i.e.

$$w(x,y,0) = \varphi_0(x,y) \tag{7.622}$$

$$\left[\frac{\partial w}{\partial t}\right]_{t=0} = \psi_0(x,y) \tag{7.623}$$

where $\varphi_0(x,y)$ is the initial displacement and $\psi_0(x,y)$ is the initial velocity.

For consistency with the Dirichlet boundary condition (7.615), we require that $\varphi_0(x,y)$ and $\psi_0(x,y)$ satisfy

$$\varphi_0(x(s),y(s)) = \Omega(s) \tag{7.624}$$

$$\psi_0(x(s),y(s)) = 0 \tag{7.625}$$

Similarly, for consistency with the Neumann boundary condition, we require

$$\frac{\partial \varphi_0}{\partial n} = 0 \quad ; \quad (x,y) \in \mathcal{C} \tag{7.626}$$

$$\frac{\partial \psi_0}{\partial n} = 0 \quad ; \quad (x,y) \in \mathcal{C} \tag{7.627}$$

In addition to the generalized formulation of boundary conditions and initial conditions, we need to identify a domain $\mathbb{D}$ in the (x,y,t) space in which the uniqueness theorem is to be formulated and proved. As a convenient representation of $\mathbb{D}$ we can take the region bounded by the plane region S_1 with boundary $\mathcal{C}$ in the x, y plane and extending over $t = 0$ to $t = t_L$ and $0 \leq t \leq t_L$. The geometrical representation of the region is shown in Figure 7.45. In order to prove the uniqueness theorem we once again adopt a *reductio ad absurdum* or proof by contradiction approach.

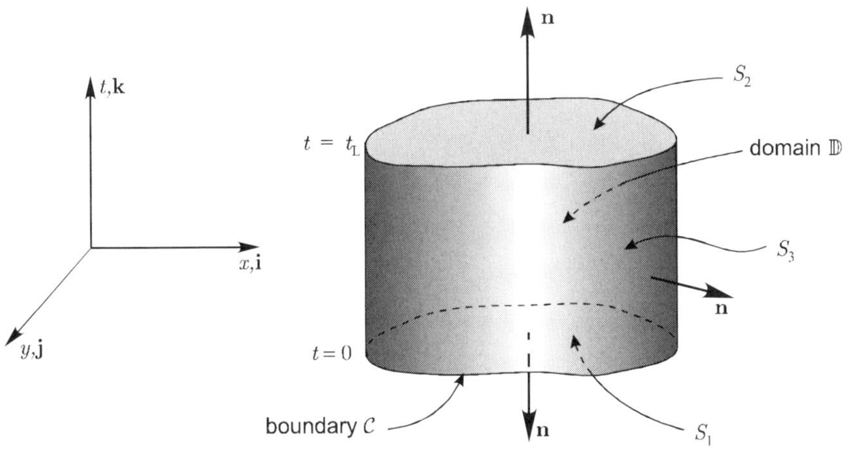

Figure 7.45: Domain for consideration of proof of uniqueness of the wave equation for a stretched membrane.

THEOREM 7.1

Consider the domain $\mathbb{D}$ in the (x, y, t) space defined by $(x, y) \in S_1$ where S_1 is the closed region defined by the projection of $\mathcal{C}_s$, the space curve defined by the boundary of the membrane, in the (x, y) plane and the surface S_2 and S_3 of the right cylinder formed by assigning $t = 0$ and $t = t_L$. The deflection of the membrane is such that it satisfies the wave equation

$$T_m \nabla^2 w(x, y, t) + p(x, y, t) = \tilde{\rho} \frac{\partial^2 w(x, y, t)}{\partial t^2} \quad ; \quad (x, y, t) \in \mathbb{D} \quad (7.628)$$

satisfies either Dirichlet boundary conditions

$$w(x(s), y(s), t) = \Omega(s) \quad \text{in} \quad (x, y, t) \subset S_3 \tag{7.629}$$

or

$$\frac{\partial w(x(s), y(s), t)}{\partial n} = 0 \quad \text{in} \quad (x, y, t) \in S_3 \tag{7.630}$$

and the initial conditions

$$w(x, y, 0) = \varphi_0(x, y) \quad \text{in} \quad (x, y) \in S_1 \tag{7.631}$$

$$\left[\frac{\partial w}{\partial t}\right]_{t=0} = \psi_0(x, y) \quad \text{in} \quad (x, y) \in S_1 \tag{7.632}$$

A solution to the initial boundary value problem obtained by satisfying the governing partial differential equation, (7.628); the appropriate boundary conditions, (7.629) or (7.630); and the initial conditions (7.631) and (7.632) is unique.

PROOF

We assume that there exists two distinct solutions $w_1(x, y, t)$ and $w_2(x, y, t)$ which satisfy the governing partial differential equation, the appropriate boundary condition and the initial conditions defined by (7.628) to (7.632). We can define a third solution $w^*(x, y, t)$ as

$$w^*(x, y, t) = w_1(x, y, t) - w_2(x, y, t) \quad ; \quad (x, y, t) \in \mathbb{D} \tag{7.633}$$

The new solution $w^*(x, y, t)$ will now satisfy the initial boundary value problem defined by

$$T_m \nabla^2 w^*(x, y, t) = \widetilde{\rho} \frac{\partial^2 w(x, y, t)}{\partial t^2} \quad ; \quad (x, y, t) \in \mathbb{D} \tag{7.634}$$

either of the boundary conditions

$$w^*(x(s), y(s), t) = 0 \quad ; \quad (x, y, t) \in S_3 \tag{7.635}$$

or

$$\frac{\partial w^*(x(s), y(s), t)}{\partial n} = 0 \quad ; \quad (x, y, t) \in S_3 \tag{7.636}$$

and the initial conditions

$$w^*(x, y, 0) = 0 \quad ; \quad (x, y) \in S_1 \tag{7.637}$$

$$\left[\frac{\partial w^*}{\partial t}\right]_{t=0} = 0 \quad ; \quad (x, y) \in S_1 \tag{7.638}$$

Since *null initial conditions* are applicable to the initial boundary value problem defined by (7.634) to (7.638) we can state that, on physical grounds, the membrane will be in a static position with $w^*(x, y, t) = 0$ for $(x, y, t) \in \mathbb{D}$. To prove this assertion we consider the integral identity

$$I = \int\int\int_{\mathbb{D}} \frac{\partial w^*}{\partial t}\left[\widetilde{\rho}\frac{\partial^2 w^*}{\partial t^2} - T_m \nabla^2 w^*\right] dxdydt \tag{7.639}$$

Since the term within the square brackets in (7.639) satisfies (7.634) we require $I \equiv 0$. We can transform (7.639) into the form

$$I = \int\int\int_{\mathbb{D}} \left[\frac{1}{2}\frac{\partial}{\partial t}\left\{\widetilde{\rho}\left(\frac{\partial w^*}{\partial t}\right)^2 + T_m\left(\frac{\partial w^*}{\partial x}\right)^2 + T_m\left(\frac{\partial w^*}{\partial y}\right)^2\right\}\right.$$
$$+\frac{\partial}{\partial x}\left[-T_m\left(\frac{\partial w^*}{\partial t}\right)\left(\frac{\partial w^*}{\partial x}\right)\right]$$

$$+\frac{\partial}{\partial y}\left[-T_m\left(\frac{\partial w^*}{\partial t}\right)\left(\frac{\partial w^*}{\partial y}\right)\right]\right] dxdydt \tag{7.640}$$

We can now apply the *divergence theorem* (see e.g. (1.30)) such that

$$\int\int\int_{\mathbb{D}} [\nabla\cdot\mathbf{F}]\, dxdydt = \int\int_S [\mathbf{F}]\cdot\mathbf{n}\, dS \tag{7.641}$$

where

$$S = S_1 \cup S_2 \cup S_3 \tag{7.642}$$

and

$$\mathbf{n} = n_x\mathbf{i} + n_y\mathbf{j} + n_t\mathbf{k} \tag{7.643}$$

Referring to Figure 7.45, we have,

$$\text{on} \quad S_1 \;;\mathbf{n} = -\mathbf{k} \qquad ;\, dS_1 = dxdy \tag{7.644}$$

$$\text{on} \quad S_2 \;;\mathbf{n} = \mathbf{k} \qquad ;\, dS_2 = dxdy \tag{7.645}$$

$$\text{on} \quad S_3 \;;\mathbf{n} = \frac{dy}{ds}\mathbf{i} - \frac{dx}{ds}\mathbf{j} \;;\, dS_3 = dsdt \tag{7.646}$$

and $x = x(s)$ and $y = y(s)$ are parametric representations corresponding to the projection of $\mathcal{C}_s$ into the (x, y) plane. We can use the relationship (7.641) to express the domain integral in (7.640) as a surface integral, i.e.

$$I = \int\int\int_{\mathbb{D}}\left[\frac{1}{2}\frac{\partial}{\partial t}\left\{\widetilde{\rho}\left(\frac{\partial w^*}{\partial t}\right)^2 + T_m\left(\frac{\partial w^*}{\partial x}\right)^2 + T_m\left(\frac{\partial w^*}{\partial y}\right)^2\right\}\right.$$

$$+\frac{\partial}{\partial x}\left[-T_m\left(\frac{\partial w^*}{\partial t}\right)\left(\frac{\partial w^*}{\partial x}\right)\right]$$
$$+\frac{\partial}{\partial y}\left[-T_m\left(\frac{\partial w^*}{\partial t}\right)\left(\frac{\partial w^*}{\partial y}\right)\right]\bigg] dxdydt$$
$$=\int\int_S\left[-n_xT_m\left(\frac{\partial w^*}{\partial t}\right)\left(\frac{\partial w^*}{\partial x}\right)-n_yT_m\left(\frac{\partial w^*}{\partial t}\right)\left(\frac{\partial w^*}{\partial y}\right)\right.$$
$$\left.+\frac{n_t}{2}\left\{\widetilde{\rho}\left(\frac{\partial w^*}{\partial t}\right)^2+T_m\left(\frac{\partial w^*}{\partial x}\right)^2+T_m\left(\frac{\partial w^*}{\partial y}\right)^2\right\}\right] dS \tag{7.647}$$

The right hand side of (7.647) can be written as

$$I=-\frac{1}{2}\int\int_{S_1}\left[\widetilde{\rho}\left(\frac{\partial w^*}{\partial t}\right)^2+T_m\left(\frac{\partial w^*}{\partial x}\right)^2+T_m\left(\frac{\partial w^*}{\partial y}\right)^2\right]_{t=0} dxdy$$
$$+\frac{1}{2}\int\int_{S_2}\left[\widetilde{\rho}\left(\frac{\partial w^*}{\partial t}\right)^2+T_m\left(\frac{\partial w^*}{\partial x}\right)^2+T_m\left(\frac{\partial w^*}{\partial y}\right)^2\right]_{t=t_L} dxdy$$
$$+\int\int_{S_3}\left[-T_m\left(\frac{\partial w^*}{\partial t}\right)\left(\frac{\partial w^*}{\partial x}\right)\left(\frac{dy}{ds}\right)\right.$$
$$\left.-T_m\left(\frac{\partial w^*}{\partial t}\right)\left(\frac{\partial w^*}{\partial y}\right)\left(-\frac{dx}{ds}\right)\right]_{\substack{x=x(s)\\ y=y(s)}} dsdt \tag{7.648}$$

The first integral of (7.648) vanishes by virtue of the conditions (7.637) and (7.638) applicable to $w^*(x,y,0)$ and $(\partial w/\partial t)_{t=0}$. Considering the integrand of the third integral of (7.648) and (7.646) we have

$$-T_m\frac{\partial w^*}{\partial t}\left\{\frac{\partial w^*}{\partial x}\frac{dy}{ds}-\frac{\partial w^*}{\partial y}\frac{dx}{ds}\right\}=T_m\frac{\partial w^*}{\partial t}\left\{\frac{\partial w^*}{\partial n}\right\} \tag{7.649}$$

Hence, the third integral of (7.648) can be written as

$$I_3 = -T_m \int\int_{S_3} \left(\frac{\partial w^*}{\partial t}\right)\left(\frac{\partial w^*}{\partial n}\right) dsdt \tag{7.650}$$

We note that if the boundary condition (7.635) is satisfied we have $(\partial w^*/\partial t) = 0$ on $\mathcal{C}$ the boundary of S_3; and if the boundary condition (7.636) is satisfied on S_3, $\partial w^*/\partial n = 0$. The integral I_3 therefore vanishes identically. The remaining part of the integral I defined by (7.648) is

$$I = \frac{1}{2}\int\int_{S_2} \left[\widetilde{\rho}\left(\frac{\partial w^*}{\partial t}\right)^2 + T_m\left\{\left(\frac{\partial w^*}{\partial x}\right)^2 + \left(\frac{\partial w^*}{\partial y}\right)^2\right\}\right]_{t=t_L} \times dxdy \tag{7.651}$$

If we compare (7.651) with results such as (7.84) and (7.89) for, respectively, the *kinetic energy* and *potential energy* of the displaced stretched string during its motion, it is clear that the result (7.651) is an energy functional for the stretched membrane during its motion. Since $\widetilde{\rho}$ and T_m are non-zero and positive the integrand of (7.651) is positive definite. The integral can vanish only if

$$\frac{\partial w^*}{\partial t} = \frac{\partial w^*}{\partial x} = \frac{\partial w^*}{\partial y} = 0, \quad \text{for} \quad (x, y, t) \in \mathbb{D} \tag{7.652}$$

Hence

$$w^*(x, y, t) = \text{const.} \quad ; \quad (x, y, t) \in \mathbb{D} \tag{7.653}$$

Since the initial condition (7.637) is applicable, (7.653) implies that

$$w^*(x, y, t) \equiv 0 \tag{7.654}$$

Hence from (7.633) we have

$$w_1(x, y, t) = w_2(x, y, t) \tag{7.655}$$

and the solution to the initial boundary value problem posed by (7.628) to (7.632) is unique.

7.10 PROBLEM SET 7

7.1 Obtain the complete solution of the partial differential equation

$$\frac{1}{e^x}\frac{\partial^2 u}{\partial x^2} - \frac{4}{e^x}\frac{\partial^2 u}{\partial y^2} = 2(1+2x)\sqrt{e^y}$$

subject to the conditions

$$u(x,0) = x(x-4)e^x \quad ; \quad \left[\frac{\partial u}{\partial y}\right]_{y=0} = 0$$

7.2 A stretched string with tension T_0 occupying the region $-\infty < x < \infty$ is subjected to an initial displacement

$$w(x,0) = x^2$$

and an initial velocity

$$\left[\frac{\partial w}{\partial t}\right]_{t=0} = \cos x$$

Obtain an expression for $w(x,t)$.

7.3 Obtain a solution of the wave equation

$$c_0^2\frac{\partial^2 w}{\partial x^2} = \frac{\partial^2 w}{\partial t^2} \quad ; \quad -\infty < x < \infty$$

subject to the following initial conditions

$$\text{(i)}\ w(x,0) = x^3 \quad , \quad \frac{\partial w(x,0)}{\partial t} = x$$

(ii) $w(x,0) = 0 \quad , \quad \dfrac{\partial w(x,0)}{\partial t} = a = \text{const.}$

7.4 Using a Fourier transform technique show that the D'Alembert form of the solution of the wave equation for a taut string

$$c_0^2 \frac{\partial^2 w}{\partial x^2} = \frac{\partial^2 w}{\partial t^2} \quad ; \quad -\infty < x < \infty$$

subject to the following initial conditions

$$w(x,0) = f(x) \quad ; \quad \frac{\partial w(x,0)}{\partial t} = g(x)$$

takes the form

$$w(x,t) = \frac{1}{2}[f(x+c_0t) + f(x-c_0t)] + \frac{1}{2c_0}\int_{x-c_0t}^{x+c_0t} g(\tau)d\tau$$

7.5 A semi-infinite stretched string occupying the region $0 < x < \infty$ is subjected to an initial displacement

$$w(x,0) = xe^{-x^2}$$

If the string is released from this position with zero initial velocity, develop an expression for the wave motion $w(x,t)$ in the string.

7.6 A stretched string of infinite extent is subjected to an initial displacement as shown in Figure 7.46. If the velocity of the small amplitude waves in the stretched string is c_0, sketch the pattern of propagation of this initial disturbance at times $\ell/2c_0$, ℓ/c_0, $3\ell/c_0$,$2\ell/c_0$ and $3\ell/c_0$, if the initial displacement is released with zero velocity.

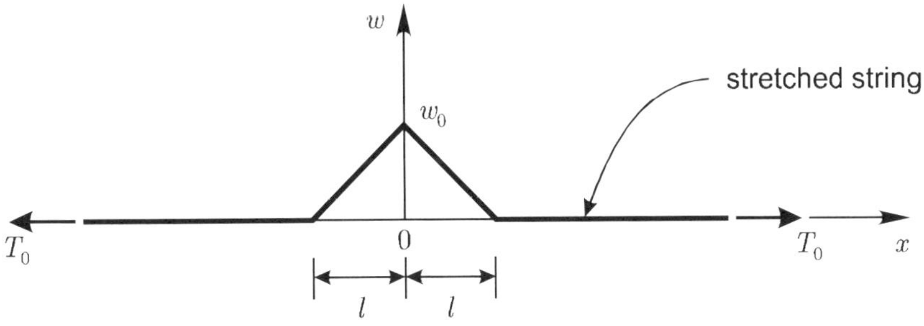

Figure 7.46: Initial displacement of a stretched string.

7.7 The Figure 7.47 shows a semi-infinite stretched string whose dynamic response is governed by the classical wave equation

$$c_0^2 \frac{\partial^2 w}{\partial x^2} = \frac{\partial^2 w}{\partial t^2} \quad ; \quad 0 \leq x < \infty$$

The location $x = 0$ is subjected to the time-dependent motion

$$w(0, t) = e^{-\lambda t} \sin(\omega_0 t)$$

where λ and ω_0 are constants. Derive, *from first principles*, an expression for the time-dependent deflections of the stretched string.

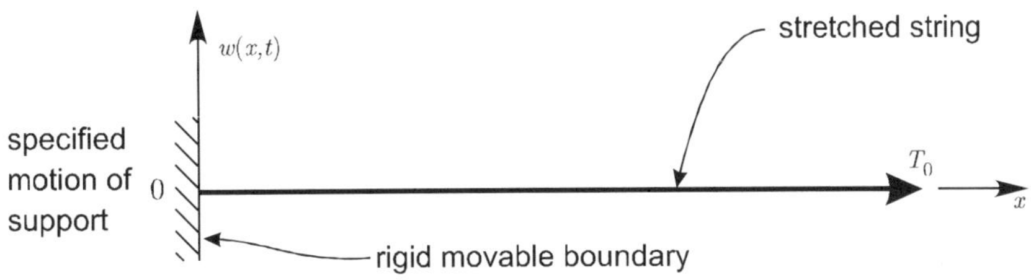

Figure 7.47: Boundary motion of a stretched semi-infinite string.

7.8 A semi-infinite string is attached to a rigid boundary and subjected to a tension T_0. The fixity of the string maintains a zero displacement at that end (Figure 7.48).

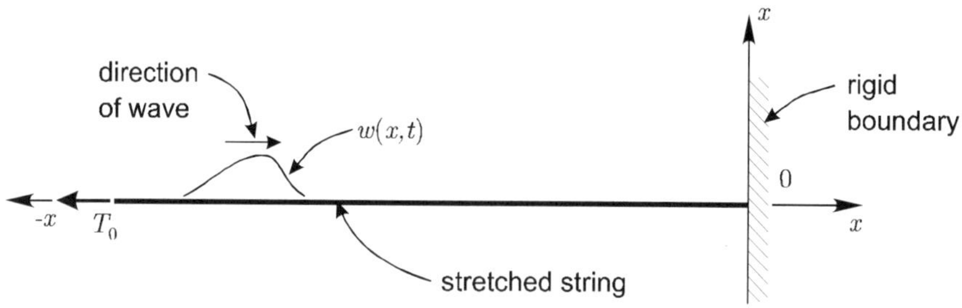

Figure 7.48: Wave in a semi-infinite string.

A wave $w(x,t) = F(x - c_0 t)$, propagates in the positive x-direction as shown in Figure 7.48. Derive an expression for the form of the reflected wave.

7.9 A *semi-infinite string* consists of two connected strings of differing mass density per unit length occupying the region $-l \leq x < \infty$. The string is fixed to rigid boundary at the location $x = -l$ and subjected to tension T_0. The mass density of the string varies as follows:

$$\rho^* = \rho_1 \quad ; \quad 0 < x < \infty \quad ; \quad \rho^* = \rho_2 \quad ; \quad -l < x < 0$$

Travelling waves of the type

$$w(x,t) = \Delta_0 \left\{ \cos \frac{\lambda}{c_1}(x - c_1 t) + \cos \frac{\lambda}{c_1}(x + c_1 t) \right\}$$

where λ and Δ_0 are constants, are generated in the region $0 < x < \infty$ by some imposed displacement. If there are no dissipative phenomena in the system and the deflections of the string are governed by the wave equation

$$\frac{\partial^2 w_i}{\partial x^2} = \frac{1}{c_i^2} \frac{\partial^2 w_i}{\partial t^2} \quad ; \quad i = 1, 2,$$

determine the form of the *first reflected wave* at the location $x = -l$.

7.10 The wave equation for the transverse motion of a stretched semi-infinite string ($x > 0$) is

$$c^2 \frac{\partial^2 w}{\partial x^2} = \frac{\partial^2 w}{\partial t^2} \quad ; \quad x \in (0, \infty) \quad ; \quad t > 0$$

Use a *Laplace transform technique* to determine the solution of $w(x,t)$ when the semi-infinite string is subjected to the following boundary condition and initial conditions:

$$w(x,0) = f(x) \quad ; \quad \left[\frac{\partial w}{\partial t}\right]_{t=0} = 0$$

$$w(0,t) = g(t)$$

7.11 The arbitrary motion of a string of length L is stretched between fixed supports is given by

$$w(x,t) = \sum_{n=1}^{\infty} \sin\left(\lambda_n \frac{x}{L}\right)\left\{A_n \cos(\frac{c_0 \lambda_n t}{L}) + B_n \sin(\frac{c_0 \lambda_n t}{L})\right\}$$

Derive an expression for the *total energy* present in the stretched string during its motion if the displacement field satisfies the wave equation.

7.12 A stretched string of finite length ℓ is subjected to an initial displacement as shown in Figure 7.49. If it is released from this position, derive an expression for the time-dependent displacement of the string.

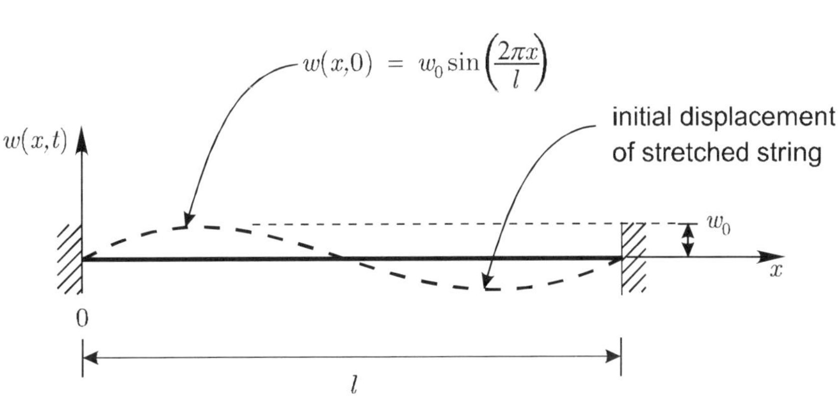

Figure 7.49: Initial diplacement profile of a stretched string.

7.13 A stretched guy-wire of a vertical axis wind turbine is under a tension T_0, and has a mass per unit length of ρ^*. (Figure 7.50). The wind action introduces small amplitude deflections of the guy wires. The wind-induced forced motion of the stretched guy wires is due to a force which varies linearly along the guy wire (from A to B) and acts *normal* to the stationary position of the guy wire (length of guy wire AB is ℓ).If the end locations of the guy wire are fixed, formulate the initial boundary value problem assuming that prior to the application of the wind forces the guy wire has *no initial displacement* and *no initial velocity*. Derive an expression for the time-dependent displacement of the guy wire normal to its equilibrium position. [An example of a vertical-axis wind turbine is the "*Darrieus-Egg Beater*" type wind turbine. In contrast to other wind turbines which need to be "oriented" to take advantage of the wind direction, the vertical axis turbines are *omnidirectional*. They can respond to wind in any direction. Since the axis of rotation is vertical, the drive train and the generator can be placed at the ground level which provides for easier maintenance.]

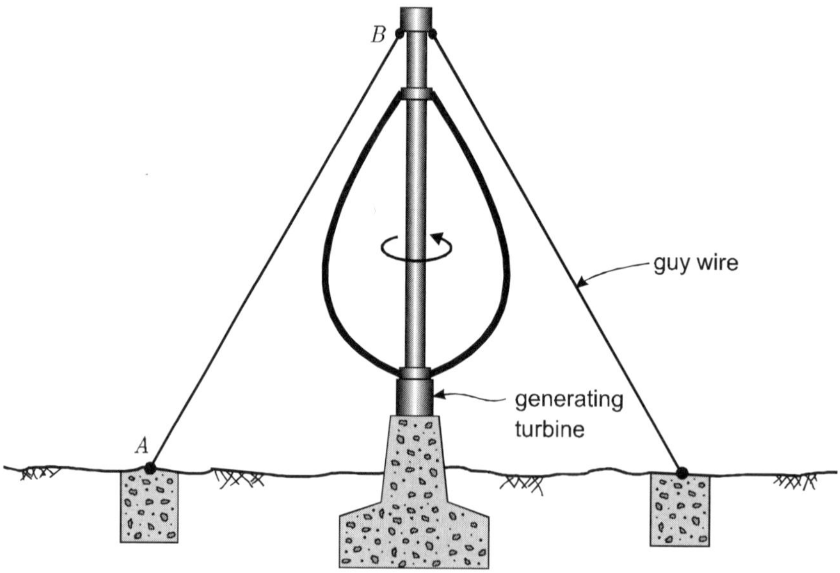

Figure 7.50: Vertical axis wind turbine.

7.14 A telecommunications cable of mass per unit length ρ^* is stretched over a length l with a tension T_0. The supports are rigid. Due to freezing rain, the cable has accumulated an *icicle* of weight W, at the centre of the cable (Figure 7.51). The *icicle* detaches from the cable without inducing any initial velocity in the cable. Derive an expression for the time-dependent displacements associated with the motion of the cable.[This particular problem was visualized prior to the great "Ice Storm" in Eastern Canada which ocurred in January 1998.]

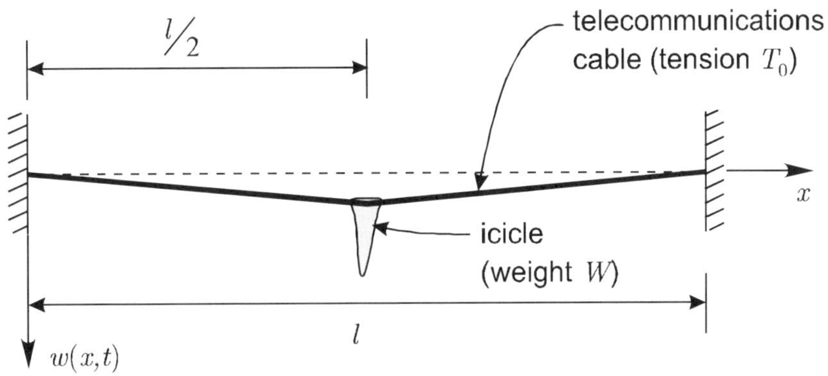

Figure 7.51: Deflection of stretched cable due to icicle.

7.15 Consider a stretched string of infinite length which possesses energy dissipation characteristics in the form of viscous damping. The equation of motion governing the dynamics of such a string is

$$T_0 \frac{\partial^2 w}{\partial x^2} - \eta_D^* \frac{\partial w}{\partial t} = \rho^* \frac{\partial^2 w}{\partial t^2}$$

where T_0 is the tension, ρ^* is the mass per unit length of the string and η_D^* is a *viscosity* term with dimensions (force)(time)/(length)^2. By considering harmonic wave forms of the type

$$w(x,t) = Ae^{i(\gamma x - \omega t)}$$

prove that the free propagation of harmonic waves is not possible when viscous damping is present.

7.16 The partial differential equation governing wave motion in a stretched string of finite length l is

$$c_0^2\frac{\partial^2 w}{\partial x^2}=\frac{\partial^2 w}{\partial t^2} \quad ; \quad x\in(0,l) \quad ; \quad t>0$$

where c is the wave speed which is related to the tension in the string T_0 and its mass per unit length ρ^* according to $c_0=[T_0/\rho^*]^{\frac{1}{2}}$.

The boundary conditions governing $w(x,t)$ are

$$w(0,t)=\zeta(t) \quad ; \quad w(l,t)=\chi(t) \quad ; \quad t\geq 0$$

and the initial conditions are

$$w(x,0)=\alpha(x) \quad ; \quad \left[\frac{\partial w}{\partial t}\right]_{t=0}=\beta(x) \quad ; \quad x\in(0,l)$$

where α, β, ζ and χ are assumed to be sufficiently smooth functions of their respective arguments (i.e. the functions are differentiable up to any required order). Prove that the solution to this initial boundary value problem is unique. [Hint: Examine the property of the integral

$$I=\int_0^l \frac{\partial w^*}{\partial t}\frac{\partial^2 w^*}{\partial x^2}dx$$

where $w^*(x,t)$ is the difference of two admissible solutions]

7.17 Show that

$$w(x,y,t)=f(x+i\gamma y-Vt)+g(x-i\gamma y-Vt)$$

is a solution of

$$\nabla^2 w(x,y,t)=\frac{1}{c_m^2}\frac{\partial^2 w}{\partial t^2}$$

where f and g are arbitrary functions which are twice continuously differential and

$$\gamma^2=1-\frac{V^2}{c_m^2}$$

and γ, V, c_m are constants.

7.18 A stretched membrane of infinite extent is subjected to an initial displacement of the form

$$w(r,0) = w_0 e^{-\lambda r}$$

where w_0 and λ are constants. If the membrane is released from this position which zero initial velocity, formulate the resulting initial boundary value problem and derive, from first principles, an expression for the time-dependent deflections of the stretched membrane.

7.19 A stretched membrane of infinite extent is subjected to a instantaneous air blast which develops an initial velocity

$$\left[\frac{\partial w}{\partial t}(x,y,t)\right]_{t=0} = v_0 \cos\left(\frac{\pi x}{a}\right)\cos\left(\frac{\pi y}{a}\right)$$

over a square patch (Figure 7.52). Formulate the initial boundary value problem and develop an expression for the time-dependent deflections of the stretched membrane.

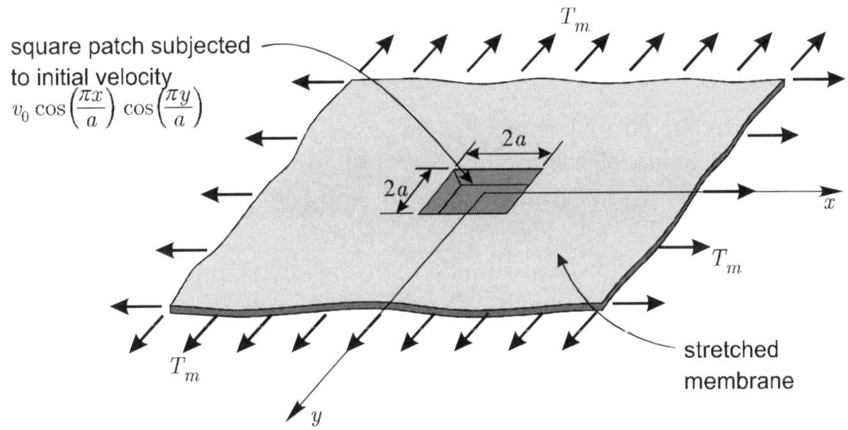

Figure 7.52: Dynamic initial indentation of a square patch of a stretched membrane.

7.20 Prove mathematically that the striking of a circular drum results in *noise* rather than *music*.

7.21 A stretched square membrane of side length $2a$ is subjected to an asymmetric initial displacement of the form

$$w(x,y,0) = w_0 \sin\left(\frac{\pi x}{a}\right) \sin\left(\frac{\pi y}{a}\right)$$

as shown in Figure 7.53.

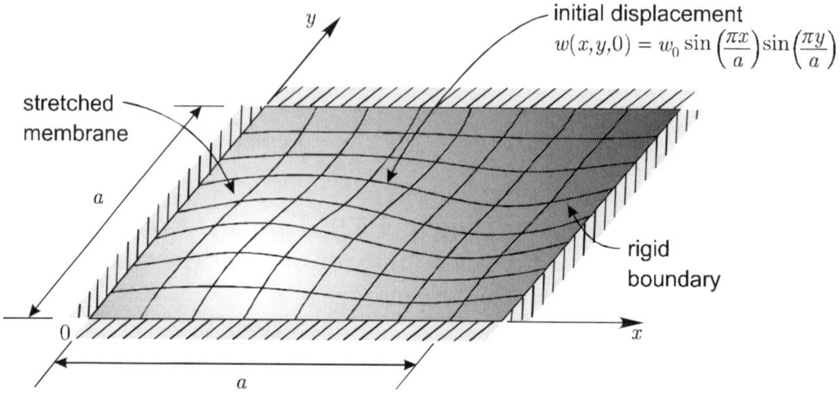

Figure 7.53: Initial displacement of stretched membrane.

Derive from first principles an expression for the time-dependent displacements of the membrane, if the membrane is released from this initially displaced position with zero velocity.

7.22 Show that the slowest mode of vibration of a stretched annular membrane stretched along the circular boundaries $r = a$ (outer radius) and $r = b$ (inner radius) is given by the roots of the transcendental equation

$$J_0(kb)Y_0(ka) - J_0(ka)Y_0(kb) = 0$$

7.23 Stretched circular membrane of radius a is subjected to a polar symmetric initial displacement of the form

$$w(r,0) = w_0 \left\{\frac{a-r}{a}\right\}^2$$

Derive an expression for the resulting dynamic displacement of the membrane if it is released from this displacement position with zero velocity.

7.24 (i) Derive from first principles the partial differential equation governing longitudinal vibrations of a bar element of cross sectional area A and Young's modulus E assuming that the *state of stress* in the bar is one dimensional. How long will this equation change if the *state of deformation* is one dimensional?

(ii) An elastic circular bar of cross sectional radius a and length L, where $L >> a$, is fixed at both ends. The bar is subjected to a longitudinal concentrated force P which acts at the mid-section of the bar (Figure 7.54). This force is suddenly removed at $t = 0$.

Formulate the *initial boundary value problem* associated with the longitudinal wave motion in the circular bar and derive a *general solution* for the governing partial differential equation.

Derive the expression for the resulting longitudinal motion of the circular rod.

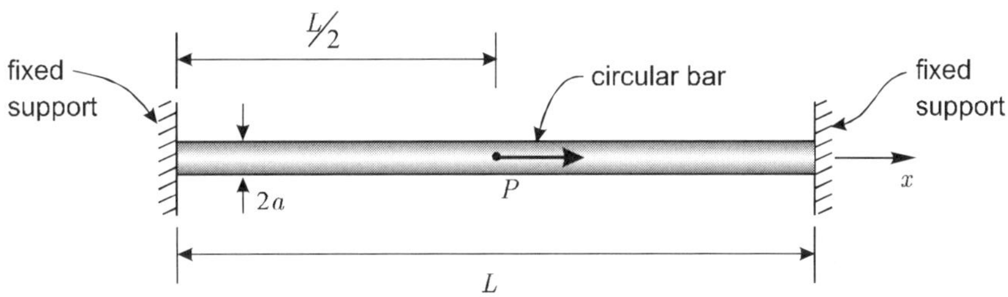

Figure 7.54: Longitudinal vibrations of a circular elastic bar.

7.25 A circular elastic bar of uniform cross section is *fixed* at the location $x = 0$ and *free* at the location $x = l$ (Figure 7.55). It is subjected to a static torque T_0 at the location $x = a$, which induces a static rotation of Ω_0 at the location $x = a$. If the torque T_0 is suddenly removed, determine the resulting torsional motion of the bar which is governed by the following partial differential equation for the rotation $\varphi(x, t)$, i.e.

$$c_S^2 \frac{\partial^2 \varphi}{\partial x^2} = \frac{\partial^2 \varphi}{\partial t^2}$$

where $c_S \left(= \sqrt{\mu/\rho_0}\right)$ is the shear wave velocity of the elastic material.

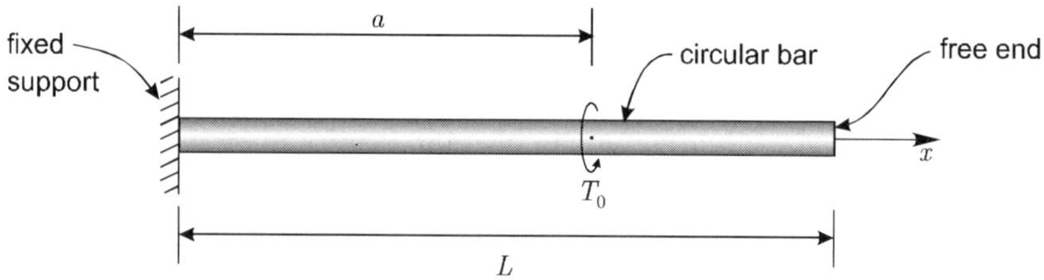

Figure 7.55: Torsional vibrations of a circular elastic bar.

7.26 A slender elastic bar is made of a material with unit weight γ. It is suspended from the end $x = 0$ (Figure 7.56) and the end $x = l$ is maintained free. The elastic bar is subjected to an initial displacement $f(x)$ and an initial velocity $g(x)$. Derive an expression for the axial motion of the bar.

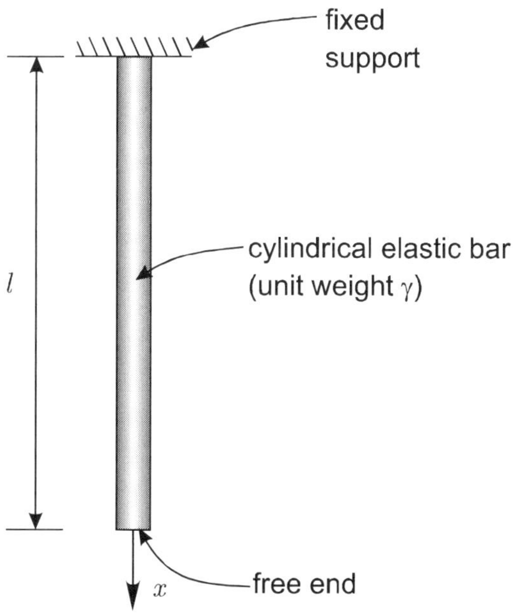

Figure 7.56: Axial motion of a circular bar in the presence of self weight.

7.27 A cylindrical bar used in a control device consists of a porous elastic solid which is impregnated with a viscous gel (Figure 7.57). The constitutive behaviour of the impregnated porous solid exhibits a Maxwell-type (named after James Clerk Maxwell (1831-1879)) one-dimensional stress-strain-time relationship of the form

$$E\sigma_{xx} + \eta \frac{\partial \sigma_{xx}}{\partial t} = \eta E \frac{\partial \epsilon_{xx}}{\partial t}$$

where E is the elastic modulus and η is a viscosity (E and η are non-zero).

(i) Derive from first principles, the partial differential equation governing equation of motion of the bar.

(ii) Use a variables separation technique to develop a general solution for the longitudinal motion of the bar.

(iii) Develop the expression for the longitudinal wave motion in the bar, if the activation of the device establishes contact, inducing only an initial displacement of the bar.

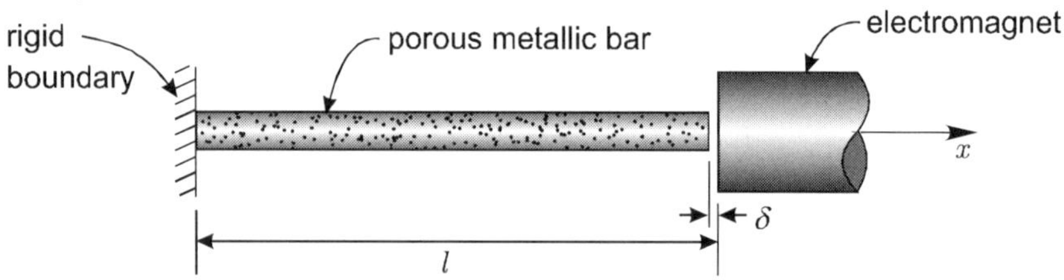

Figure 7.57: Wave motion in a cylindrical bar with internal damping.

7.28 The Figure 7.58 shows the cross section through a shallow water region in the vicinity of a geological fault. Due to seismic activity, a section of the fault experiences uplift. The upheaval region has a width $2l$ and extends a significant distance in the y-direction.

(i) Formulate the initial boundary value problem associated with the propagation of shallow water waves in the region due to the geological discontinuity.

(ii) Use an integral transform technique to develop a solution to the propagation of shallow water waves in the fluid region.

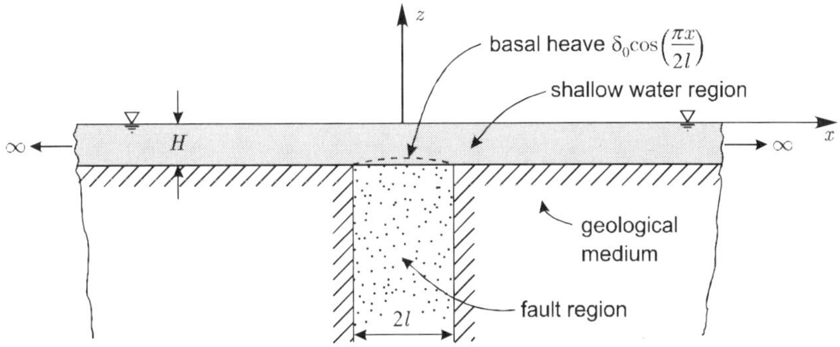

Figure 7.58: Shallow water waves due to basal heave
$[H \sim 10\text{m};\ 2l \sim 100\text{m};\ \delta_0 \sim 1\text{m}]$.

7.29 In the case of one-dimensional propagation of shallow water waves (i.e. wave length much greater than the depth of water H), the vertical displacement of the water surface in relation to its static position, $h(x,t)$, is governed by the one-dimensional wave equation

$$gH\frac{\partial^2 h}{\partial x^2} = \frac{\partial^2 h}{\partial t^2}$$

If the horizontal displacement of the water surface is denoted by $f(x,t)$ such that the horizontal velocity

$$v_x = \frac{Df}{Dt} = \frac{\partial f}{\partial t}$$

and

$$h(x,t) = -H\frac{\partial f}{\partial x}$$

Show that $f(x,t)$ is also governed by the partial differential equation

$$gH\frac{\partial^2 f}{\partial x^2} = \frac{\partial^2 f}{\partial t^2}$$

Bibliography

Advanced Topics in Engineering Mathematics

Bender, C.M. and Orzag, S.A. (1978) *Advanced Mathematical Methods for Scientists and Engineers*, McGraw-Hill, New York.

Boas, M.L. (1966) *Mathematical Methods in the Physical Sciences*, 2nd edition, John Wiley, New York.

Courant, R. and Hilbert, D. (1962) *Methods of Mathematical Physics*,Vols.1 and 2, Interscience, New York.

Dennery, P. and Krzywicki, A. (1967) *Mathematics for Physicists*, Harper and Row, New York.

Dettman, J.W. (1988) *Mathematical Methods in Physics and Engineering*, Dover Publications, New York.

Greenberg, M.D. (1978) *Foundations of Applied Mathematics*, Prentice-Hall, Englewood Cliffs, N.J.

Greenberg, M.D. (1998) *Advanced Engineering Mathematics*, Prentice-Hall, Englewood Cliffs, N.J.

Haug, E. and Choi, K.K. (1993) *Methods of Engineering Mathematics*, Prentice-Hall, Englewood Cliffs, N.J.

Keener, J.P. (1988) *Principles of Applied Mathematics*, Addison Wesley, New York.

Kraut, E.A. (1967) *Fundamentals of Mathematical Physics*, McGraw-Hill, New York.

Kreyszig, E. (1992) *Advanced Engineering Mathematics*, John Wiley, New York.

O'Neil, P.V. (1995) *Advanced Engineering Mathematics*, Brooks/Cole Publishing Company, Boston.

Pipes, L.A. and Harvill, L.R. (1970) *Applied Mathematics for Engineers and Physicists*, McGraw-Hill, New York.

Potter, M.C. (1978) *Mathematical Methods in the Physical Sciences*, Prentice-Hall, Englewood Cliffs, N.J.

Riley, K.F., Hobson, M.P. and Bence, S.J. (1998) *Mathematical Methods for Physics and Engineering*, Cambridge University Press, Cambridge.

Spencer, A.J.M., et al. (1977) *Engineering Mathematics*, Vols. 1 and 2, Van Nostrand Reinhold, New York.

Spiegel, M.R. (1971) *Advanced Mathematics for Engineers and Scientists*, McGraw-Hill, New York.

Strang, G. (1986) *Introduction to Applied Mathematics*, Wellesley-Cambridge Press, Cambridge, Mass.

von Karman, T. and Biot, M.A. (1940) *Mathematical Methods in Engineering*, McGraw-Hill, New York.

Wylie, C. (1958) *Advanced Engineering Mathematics*, McGraw-Hill, New York.

Fourier Series, Integral Transforms, Special Functions and Complex Variables

Andrews, L.C. (1985) *Special Functions for Engineers and Applied Mathematicians*, Macmillan, New York.

Boyce, W.E. and DiPrima, R.C. (1992) *Elementary Differential Equations and Boundary Value Problems*, 5th edition, John Wiley, New York.

Brown, J.W. and Churchill, R.V. (1993) *Fourier Series and Boundary Value Problems*, McGraw-Hill, New York.

Byerly, W.E. (1959) *Fourier Series*, Dover Publications, New York.

Carslaw, H.S. (1980) *An Introduction to the Theory of Fourier Series and Integrals*, Dover Publications, New York.

Churchill, R.V. (1960) *Complex Variables and Applications*, McGraw-Hill, New York.

Davis, H.F. (1963) *Fourier Series and Orthogonal Functions*, Allyn and Bacon, Boston.

Doetsch, G. (1963) *Guide to the Applications of the Laplace Transform*, Van Nostrand, New York.

Edwards, R.E. (1967) *Fourier Series: A Modern Introduction*, Holt, Rinehart and Winston, New York.

Franklin, P. (1949) *An Introduction to Fourier Methods and the Laplace Transformation*, Dover Publications, New York.

Hochstadt, H. (1961) *Special Functions of Mathematical Physics*, Holt Rinehart, New York.

Lebedev, N.N. (1972) *Special Functions and Their Applications* (Translated and Edited by R.A. Silverman), Dover Publications, New York.

Lighthill, M.J. (1958) *An Introduction to Fourier Analysis and Generalized Functions*, Cambridge University Press, Cambridge.

MacRobert, T.M. (1948) *Spherical Harmonics*, Dover Publications, New York.

McLachlan, N.W. (1955) *Bessel Functions for Engineers*, Clarendon Press, Oxford.

Miles, J.W. (1971) *Integral Transforms in Applied Mathematics*, Cambridge University Press, Cambridge.

Rainville, E.D. (1963) *The Laplace Transform: An Introduction*, McMillan, New York.

Seeley, R. (1966) *Introduction to Fourier Series and Integrals*, Benjamin, New York.

Sneddon, I.N. (1951) *Fourier Transforms*, McGraw-Hill, New York.

Sneddon, I.N. (1972) *The Use of Integral Transforms*, McGraw-Hill, New York.

Titchmarsh, E.C. (1967) *Introduction to the Theory of Fourier Integrals*, Clarendon Press, Oxford.

Tolstov, G.P. (1962) *Fourier Series*, Prentice-Hall, Englewood Cliffs, N. J.

Tranter, C.J. (1968) *Bessel Functions with Some Physical Applications*, English Universities Press, London.

Tranter, C.J. (1971) *Integral Transforms in Mathematical Physics*, Chapman and Hall, London.

Watson, E.J. (1981) *Laplace Transforms and Applications*, Van Nostrand Reinhold, New York.

Watson, G.N. (1944) *A Treatise on the Theory of Bessel Functions*, Cambridge University Press, Cambridge.

Whittaker, E.T. and Watson, G.N. (1950) *Modern Analysis*, Cambridge University Press, Cambridge.

Widder, D.V. (1941) *The Laplace Transform*, Princeton University Press, Princeton, N.J.

Ordinary Differential Equations

Agnew, R.P. (1960) *Differential Equations*, McGraw-Hill, New York.

Arnold, V.I. (1982) *Ordinary Differential Equations*, MIT Press, Boston.

Ayres, Jr. F. (1952) *Theory and Problems of Differential Equations*, Schaum Publishing Co., New York.

Bellman, R.E. and Cooke, K.L. (1971) *Modern Elementary Differential Equations*, Addison-Wesley, Reading, Mass.

Birkhoff, G. and Rota, G.-C. (1989) *Introduction to Ordinary Partial Differential Equations*, 4th edition, John Wiley, New York.

Boyce, W.E. and DiPrima, R.C. (1992) *Elementary Differential Equations and Boundary Value Problems*, 5th edition, John Wiley, New York.

Brand, L. (1966) *Differential and Difference Equations*, John Wiley, New York.

Brauer, F. and Nobel, J. (1973) *Ordinary Differential Equations*, 2nd edition, Benjamin, New York.

Braun, M. (1983) *Differential Equations and the Applications: An Introduction to Applied Mathematics*, 3rd Edition, Springer-Verlag, New York.

Burkill, J.C. (1962) *The Theory of Ordinary Differential Equations*, 2nd edition, Interscience Publishers, New York.

Carrier, G.F. and Pearson, C.E. (1968) *Ordinary Differential Equations*, Ginn (Blaisdell), Boston, Mass.

Chorlton, F. (1965) *Ordinary Differential and Difference Equations*, Van Nostrand, London.

Coddington, E.A. (1989) *An Introduction to Ordinary Differential Equations*, Dover Publications, New York.

Coddington, E.A. and Levinson, N. (1984) *Theory of Ordinary Differential Equations*, Krieger Publ., Melbourne, Florida.

Cole, R.H. (1968) *Theory of Ordinary Differential Equations*, Irvington, New York.

Collatz, L. (1986) *Differential Equations: An Introduction with Applications*, John Wiley, New York.

Fattorini, H.O. (1985) *Second Order Linear Differential Equations in Banach Spaces*, North-Holland, New York.

Feineman, G., Garrett, S.J. and Karaus, A.D. (1965) *Applied Differential Equations*, Spartan Books, Washington.

Ford, L.R. (1955) *Differential Equations*, 2nd edition, McGraw-Hill, New York.

Forsyth, A.R. (1921) *Differential Equations*, Macmillan, London.

Golomb, M. and Shanks, M.E. (1965) *Elements of Ordinary Differential Equations*, 2nd edition, McGraw-Hill, New York.

Hagin, F. (1975) *A First Course in Differential Equations*, Prentice-Hall, Englewood Cliffs, N.J.

Hartman, P. (1982) *Ordinary Differential Equations*, 2nd edition, Birkhauser, Boston.

Henrici, P. (1962) *Variable Methods in Ordinary Differential Equations*, John Wiley, New York.

Hill, J.M. (1992) *Differential Equations and Group Methods for Scientists and Engineers*, CRC Press, Boca Raton, Florida.

Hochstadt, H. (1964) *Differential Equations: A Modern Approach*, Holt, New York. (reprinted Dover Publications, New York., 1975).

Hunt, R.W. (1973) *Differential Equations and Related Topics for Science and Engineering*, Brooks-Cole, Monterey, California.

Hurewicz, W. (1958) *Lectures on Ordinary Differential Equations*, M.I.T. Press, Cambridge, Mass.

Ince, E.L. (1956) *Ordinary Differential Equations*, Dover Publications, New York.

Ince, E.L. and Sneddon, I.N. (1987) *The Solution of Ordinary Differential Equations*, 2nd edition, John Wiley, New York.

Jordan, D.W. and Smith, P. (1987) *Nonlinear Ordinary Differential Equations*, 2nd edition, Oxford University Press, Oxford.

Kamke, E. (1948) *Differentialgleichungen Losungsmethoden und Losungen*, Chelsea, New York.

Lambert, J.D. (1973) *Computational Methods in Ordinary Differential Equations*, John Wiley, New York.

Leighton, W. (1966) *Ordinary Differential Equations*, 2nd edition, Wadsworth, Belmont, California.

Lomen, D. and Mark, J. (1986) *Ordinary Differential Equations with Linear Algebra*, Prentice-Hall, Englewood Cliffs, N.J.

Martin, W.T. and Reissner, E. (1961) *Elementary Differential Equations*, Addison-Wesley, Reading, Mass.

Michel, A.N. (1982) *Ordinary Differential Equations*, Academic Press, New York.

Miller, K.S. and Michel, A.N. (1982) *Ordinary Differential Equations*, Academic Press, New York.

Morris, M. and Brown, O.E. (1952) *Differential Equations*, Prentice-Hall, Englewood Cliffs, N.J.

Murphy, G.M. (1960) *Ordinary Differential Equations and Their Solution*, Van Nostrand, Princeton, N.J.

Petrovskii, I.G. (1966) *Ordinary Differential Equations*, Prentice-Hall, Englewood Cliffs, N.J.

Piaggio, H.T.H. (1982) *An Elementary Treatise on Differential Equations and Their Applications*, Bell and Hyman, London.

Pinney, E. (1958) *Ordinary Difference-Differential Equations*, University of California Press, Berkeley.

Rabenstein, A.L. (1972) *Introduction to Ordinary Differential Equations*, Academic Press, New York.

Rainville, E.D. (1964) *Intermediate Differential Equations*, 2nd edition, Macmillan, New York.

Rainville, E.D. and Bedient, P.E. (1989) *Elementary Differential Equations*, Macmillan, New York.

Reid, W. (1980) *Sturmian Theory for Ordinary Differential Equations*, Springer-Verlag, Berlin.

Reid, W.T. (1971) *Ordinary Differential Equations*, John Wiley, New York.

Ross, S. (1964) *Differential Equations*, Blaisdell, New York.

Rubinstein, Z. (1969) *A Course in Ordinary and Partial Differential Equations*, Academic Press, New York.

Spiegel, M.R. (1981) *Applied Differential Equations*, 3rd edition, Prentice-Hall, Englewood Cliffs, N.J

Van Iwaarden, J.L. (1985) *Ordinary Differential Equations with Numerical Techniques*, Harcourt Brace Jovanovitch, Orlando, Florida.

Wilcox, L.R. and Curtis, H.J. (1961) *Elementary Differential Equations*, International Textbook Co., Scranton, Pennsylvania.

Williamson, R.E. (1986) *Introduction to Differential Equations: ODE, PDE and Series*, Prentice-Hall, Englewood Cliffs, N.J.

Partial Differential Equations

Agnew, R.P. (1960) *Differential Equations*, McGraw-Hill, New York.

Andrews, L.C. (1986) *Elementary Partial Differential Equations*, Academic Press, New York.

Bateman, H. (1959) *Partial Differential Equations of Mathematical Physics*, Cambridge University Press, London.

Berg, P.W. and McGregor, J.L. (1966) *Elementary Partial Differential Equations*, Holden-Day, San Francisco.

Bers, L., John, F. and Schetcher, M. (1964) *Partial Differential Equations*, John Wiley, New York.

Bleecker, D. and Csordas, G. (1996) *Basic Partial Differential Equations*, International Press, Cambridge, Massachusetts.

Bluman, G.W. and Kumei, S. (1989) *Symmetries and Differential Equations*, Springer-Verlag, New York.

Boyce, W.E. and DiPrima, R.C. (1992) *Elementary Differential Equations and Boundary Value Problems*, 5th edition, John Wiley, New York.

Braun, M. (1978) *Differential Equations and their Applications*, Springer-Verlag, New York.

Broman, A. (1989) *Introduction to Partial Differential Equations*, Dover Publications, New York.

Carrier, G.F. and Pearson, C.E. (1988) *Partial Differential Equations*, 2nd edition, Academic Press, New York.

Carrol, R.W. (1969) *Abstract Methods in Partial Differential Equations*, Harper and Row, New York.

Chester, C.R. (1971) *Techniques in Partial Differential Equations*, McGraw-Hill, New York.

Colton, D.L. (1976) *Partial Differential Equations in the Complex Domain*, Pitman, San Francisco.

Copson, E.T. (1975) *Partial Differential Equations*, Cambridge University Press, New York.

de Rham, G. (1960) *Varietes Differentiables*, Hermann, Paris.

Dennemeyer, R. (1968) *Introduction to Partial Differential Equations and Boundary Value Problems*, McGraw-Hill, New York.

Dezin, A.A. (1987) *Partial Differential Equations: An Introduction to a General Theory of Linear Boundary Value Problems*, Springer-Verlag, New York

DiBenedetto, E. (1995) *Partial Differential Equations*, Birkhauser, Boston.

Driver, R.D. (1977) *Ordinary and Delay Differential Equations*, Springer-Verlag, Berlin.

Driver, R.D. (1978) *Introduction to Ordinary Differential Equations*, Harper and Row, New York.

DuChateau, P. and Zachmann, D. (1986) *Theory and Problems of Partial Differential Equations*, McGraw-Hill, New York.

DuChateau, P. and Zachmann, D. (1989) *Applied Partial Differential Equations*, Harper and Row, New York.

Duff, G.F.D. (1956) *Partial Differential Equations*, Toronto Univ. Press, Toronto.

El'sgol'ts, L.E. and Norkin, S.B. (1973) *Introduction to the Theory and Application of Differential Equations with Derivating Arguments*, Academic Press, New York.

Epstein, B. (1962) *Partial Differential Equations*, McGraw-Hill, New York.

Evans, L.C. (1993) *Partial Differential Equations*, Berkeley Mathematics Lecture Notes, Vols. 3A 3B, Berkeley, Calif.

Farlow, S.J. (1993) *Partial Differential Equations for Scientists and Engineers*, Dover Publications, New York.

Folland, G.B. (1995) *Introduction to Partial Differential Equations*, 2nd edition, Princeton University Press, N.J.

Friedman, A. (1969) *Partial Differential Equations*, Holt, Rinehart and Winston, New York.

Garabedian, P.R. (1964) *Partial Differential Equations*, John Wiley, New York.

Gilbert, R.P. (1969) *Function Theoretic Methods in Partial Differential Equations*, Academic Press, New York.

Gribben, R.J. (1975) *Elementary Partial Differential Equations*, Van Nostrand Reinhold, New York

Gustafson, K.E. (1987) *Introduction to Partial Differential Equations and Hilbert Space Methods*, 2nd edition, John Wiley, New York.

Haberman, R. (1998) *Elementary Applied Partial Differential Equations*, 3rd edition, Prentice-Hall, Englewood Cliffs, N.J.

Hellwig, G. (1977) *Partial Differential Equations: An Introduction*, B.G. Teubner, Stuttgart.

Hill, J.M. (1992) *Differential Equations and Group Methods for Scientists and Engineers*, CRC Press, Boca Raton, Florida.

Hormander, L. (1964) *Linear Partial Differential Operators*, Springer, Berlin.

Hormander, L. (1985) *The Analysis of Linear Partial Differential Operators*, Vols. 1-4, Springer-Verlag, Berlin.

John, F. (1982) *Partial Differential Equations*, 4th edition, Springer-Verlag, New York.

Kervorkian, J. (1990) *Partial Differential Equations, Analytical Solution Techniques*, Wadsworth, Belmont, California.

Krein, M.G. and Gobberg, L. (1983) *Topics in Differential and Integral Equations and Operator Theory*, Birkhauser Verlag, Boston.

Krzyzanski, M. (1971) *Partial Differential Equations of Second Order* (trans. H. Zorski), Polish Scientific Publishers, Warszawa.

Lamb, Jr. G.L. (1995) *Introductory Applications of Partial Differential Equations*, John Wiley, New York.

Langer, R. (Ed) (1960) *Boundary-Value Problems in Differential Equations*, Univ. of Wisconsin Press, Madison.

Langer, R. (Ed) (1961) *Symp. Partial Differential Equations and Continuum Mechanics*, Univ. of Wisconsin Press, Madison.

Lieberstein, H.M. (1972) *Theory of Partial Differential Equations*, Academic Press, New York.

Lions, J.L. (1957) *Lectures on Elliptic Partial Differential Equations*, Tata, Bombay.

McOwen, R. (1995) *Partial Differential Equations: Methods and Applications*, Prentice-Hall, Englewood Cliffs, N.J.

Miller, F.H. (1941) *Partial Differential Equations*, John Wiley, New York.

Miller, K.S. (1953) *Partial Differential Equations in Engineering Problems*, Prentice-Hall, Englewood Cliffs, N.J.

Moon, P.H. and Spencer, D.E. (1969) *Partial Differential Equations*, D.C. Heath, Lexington, Mass.

Myint-U, T. and Debnath, L. (1987) *Partial Differential Equations for Scientists and Engineers*, 3rd edition, North-Holland, New York.

Noble, B. (1958) *Methods Based on the Wiener-Hopf Technique for the Solution of Partial Differential Equations*, Pergamon Press, New York.

Olver, P.J. (1986) *Applications of Lie Groups to Differential Equations*, Springer-Verlag, New York.

Ovsiannikov, L.V. (1982) *Group Analysis of Differential Equations*, Academic Press, New York.

Paris, R.B. and Wood, A.D. (1986) *Asymptotics of High Order Differential Equations*, John Wiley, New York.

Petrovskii, I.G. (1954) *Lectures on Partial Differential Equations*, Dover Publications, New York.

Pinsky, M.A. (1991) *Partial Differential Equations and Boundary-Value Problems with Applications*, 2nd edition, McGraw-Hill, New York.

Rahman, M. (1991) *Applied Differential Equations for Scientists and Engineers*, Computational Mechanics Publications, Boston.

Renardy, M. and Rogers, R.C. (1993) *An Introduction to Partial Differential Equations*, Springer-Verlag, New York.

Rubinstein, Z. (1969) *A Course in Ordinary and Partial Differential Equations*, Academic Press, New York.

Showalter, R.E. (1977) *Hilbert Space Methods for Partial Differential Equations*, Pitman, London.

Smirnov, M.M. (1966) *Second-Order Partial Differential Equations*, Edited by S. Chomet, Noordhoff, Groningen.

Smith, M.G. (1967) *Introduction to the Theory of Partial Differential Equations*, Van Nostrand, London.

Sneddon, I.N. (1957) *Elements of Partial Differential Equations*, McGraw-Hill, New York.

Sobolev, S.L. (1964) *Partial Differential Equations of Mathematical Physics*, Pergamon Press, Oxford.

Spiegel, M.R. (1958) *Applied Differential Equations*, Prentice-Hall, Englewood Cliffs, N.J.

Stakgold, I. (1968) *Boundary Value Problems of Mathematical Physics*, Vols. I and II, Macmillan, New York.

Stephenson, G. (1996) *Partial Differential Equations for Scientists and Engineers*, Imperial College Press, London.

Strauss, W.A. (1992) *Partial Differential Equations, An Introduction*, John Wiley, New York.

Street, R.L. (1973) *Analysis and Solution of Partial Differential Equations*, Wadsworth, Belmont, California.

Taylor, M. (1996) *Partial Differential Equations - Basic Theory*, Springer-Verlag, Berlin.

Titchmarsh, E.C. (1958) *Eigenfunction Expansions Associated with Second-Order Differential Equations*, Oxford University Press, London.

Treves, F. (1975) *Basic Linear Partial Differential Equations*, Academic Press, New York.

Tricomi, F.G. (1961) *Differential Equations*, Blackie, London.

Trim, D.W. (1990) *Applied Partial Differential Equations*, PWS-KENT Publishing, Boston.

Troutman, J.L. (1994) *Boundary Value Problems of Applied Mathematics*, PWS Publishing, Boston.

Webster, A.G. (1966) *Partial Differential Equations of Mathematical Physics*, 2nd edition, Dover Publications, New York.

Weinberger, H.F. (1965) *A First Course in Partial Differential Equations*, Dover Publications, New York.

Williams, W.E. (1980) *Partial Differential Equations*, Clarendon Press, Oxford.

Williamson, R.E. (1986) *Introduction to Differential Equations: ODE, PDE and Series*, Prentice-Hall, Englewood Cliffs, N.J.

Wloka, J. (1987) *Partial Differential Equations*, Cambridge University Press, Cambridge.

Young, E.C. (1972) *Partial Differential Equations: An Introduction*, Allyn and Bacon, Boston.

Zachmanoglou, E.C. and Thoe, D.W. (1987) *Introduction to Partial Differential Equations with Applications*, Dover Publications, New York.

Zauderer, E. (1998) *Partial Differential Equations of Applied Mathematics*, 2nd edition, John Wiley, New York.

Non-Linear Partial Differential Equations

Bellman, R.E. (1953) *Stability Theory of Differential Equations*, McGraw-Hill, New York.

Cesari, L. (1963) *Asymptotic Behaviour and Stability Problems in Ordinary Differential Equations*, 2nd edition, Springer-Verlag, Berlin.

Davis, H.T. (1961) *Introduction to Nonlinear Differential and Integral Equations*, Dover Publications, New York.

Diaz, J.L. (1985) *Nonlinear Partial Differential Equations and Free Boundaries*, Vol. 1, Pitman, Boston.

Dresner, L. (1983) *Similarity Solutions of Nonlinear Partial Differential Equations*, Pitman, Boston.

Evans, L.C. (1990) *Weak Convergence Methods for Nonlinear Partial Differential Equations*, American Mathematical Society, R.I.

Fuchik, S. and Kufner, A. (1980) *Nonlinear Differential Equations*, Elsevier Scientific Publ., Amsterdam.

Hirsch, W.H. and Smale, S. (1974) *Differential Equations: Dynamical Systems and Linear Algebra*, Academic Press, New York.

Huntley, I. and Johnson, R.M. (1983) *Linear and Nonlinear Differential Equations*, Halsted Press, New York.

Leung, A.W. (1989) *Systems of Nonlinear Partial Differential Equations: Applications to Biology and Engineering*, Kluwer Academic, Boston.

Logan, J.D. (1997) *Introduction to Nonlinear Partial Differential Equations*, John Wiley, New York.

Perko, L. (1991) *Differential Equations and Dynamical Systems*, Springer-Verlag, New York.

Reissig, R., Sansone, G. and Conti, R. (1974) *Non-Linear Differential Equations of Higher Order*, Noordhoff International, Leyden.

Smoller, J.A. (Ed) (1983) *Nonlinear Partial Differential Equations, Contemporary Mathematics*, vol. 17, American Mathematical Society, Providence, R.I.

Struble, R.A. (1962) *Nonlinear Differential Equations*, McGraw-Hill, New York.

Verhulst, F. (1990) *Differential Equations and Dynamical Systems*, Springer-Verlag, New York.

Mathematical Methods and Boundary Value Problems

Ames, W.F., Harrell Jr., E.M. and Herod, J.V. (Eds.) (1993) *Differential Equations with Applications to Mathematical Physics*, Academic Press, New York.

Basmadjian, D. (1999) *The Art of Modelling in Science and Engineering*, Chapman and Hall/CRC Press, Boca Raton, FL.

Bateman, H. (1959) *Partial Differential Equations of Mathematical Physics*, Cambridge University Press, London.

Bick, T.A. (1993) *Elementary Boundary Value Problems*, Marcel Dekker, New York.

Chorlton, F. (1969) *Boundary Value Problems in Physics and Engineering*, Van Nostrand Reinhold, London.

Dennemeyer, R. (1968) *Introduction to Partial Differential Equations and Boundary Value Problems*, McGraw-Hill, New York.

Dixon, C. (1971) *Applied Mathematics of Science and Engineering*, John Wiley and Sons, London.

Duff, G.F.D. and Naylor, D. (1966) *Differential Equations of Applied Mathematics*, John Wiley, New York.

Edelstein-Keshet, L. (1988) *Mathematical Models in Biology*, Random House, New York.

Fowler, A.C. (1997) *Mathematical Models in the Applied Sciences*, Cambridge University Press, Cambridge.

Frank, F. and von Mises, R. (1937) *Differential and Integral Equations of Mathematical Physics*, United Scientific and Technical Press.

Friedman, A. (1990) *Mathematics in Industrial Problems*, Parts 1-5, Springer-Verlag, Berlin.

Genin, J. and Maybee, J.S. (1970) *Introduction to Applied Mathematics*, Vol. 1, Holt, Rinehart and Winston, New York.

Guenther, R.B. and Lee, J.W. (1996) *Partial Differential Equations of Mathematical Physics and Integral Equations*, Dover Publications, New York

Hildebrand, F.B. (1965) *Methods of Applied Mathematics*, Prentice-Hall, Englewood Cliffs, N.J.

Irving, J. and Mullineux, N. (1959) *Mathematics in Physics and Engineering*, Academic Press, New York.

Jeffreys, H.J. and Jeffreys, B.S. (1946) *Methods of Mathematical Physics*, Cambridge University Press, Cambridge.

Kovach, L.D. (1984) *Boundary Value Problems*, Addison Wesley, Reading, Mass.

Leis, R. (1983) *Partial Differential Equations of Applied Mathematics*, John Wiley, New York.

Levin, V.L. and Grosberg, Yu.N. (1951) *Differential Equations of Mathematical Physics*, Gostekhizdat, Moscow.

Lin, C.C. and Segel, L.A. (1974) *Mathematics Applied to Deterministic Problems in the Applied Sciences*, Macmillan, New York.

Mackie, A.G. (1965) *Boundary Value Problems*, Oliver and Boyd, London.

Margenau, H. and Murphy, G. (1964) *The Mathematics of Physics and Chemistry*, Vols. I and II, Van Nostrand, New York.

Mathews, J. and Walker, R. (1965) *Mathematical Methods of Physics*, W.A. Benjamin, New York.

Mikhlin, S.G. (1967) *Linear Equations of Mathematical Physics*, Holt, Rinehart and Winston, New York.

Mikhlin, S.G. (1970) *Mathematical Physics, An Advanced Course*, North-Holland Publishing Co., Amsterdam, The Netherlands.

Morse, P.M. and Feshbach, H. (1953) *Methods of Theoretical Physics*, Parts I and II, McGraw-Hill, New York.

Peskin, C.S. (1976) *Partial Differential Equations in Biology*, Courant Institute of Mathematical Sciences, New York.

Powers, D.L. (1979) *Boundary Value Problems*, Academic Press, New York.

Reed, M. and Simon B. (1972) *Methods of Mathematical Physics*, Vols. I and II, Academic Press, New York.

Ritger, P.D. and Rose, N.J. (1968) *Differential Equations and Applications*, McGraw-Hill, New York.

Rubinow, S.I. (1976) *Introduction to Mathematical Biology*, John Wiley, New York.

Rubinstein, I. and Rubinstein, L. (1993) *Partial Differential Equations in Classical Mathematical Physics*, Cambridge University Press, Cambridge.

Sagan, H. (1989) *Boundary Value Problems in Mathematical Physics*, Dover Publications, New York.

Schechter, M. (1977) *Modern Methods in Partial Differential Equations: An Introduction*, 2nd edition, McGraw-Hill, New York.

Segel, L.A. (1987) *Mathematics Applied to Continuum Mechanics*, Dover Publications, New York.

Sobolev, S.L. (1964) *Partial Differential Equations of Mathematical Physics*, Pergamon Press, Oxford.

Sokolnikoff, I.S. and Redheffer, R.M. (1966) *Mathematics of Physics and Modern Engineering*, McGraw-Hill, New York.

Sommerfeld, A. (1964) *Partial Differential Equations in Physics*, Academic Press, New York.

Stakgold, I. (1968) *Boundary Value Problems of Mathematical Physics*, Vols. I and II, Macmillan, New York.

Svobodny, T. (1998) *Mathematical Modelling for Industry and Engineering*, Prentice-Hall, Englewood Cliffs, N.J.

Taylor, A.B. (1986) *Mathematical Models in Applied Mechanics*, Clarendon Press, Oxford.

Tikhonov, A.N. and Samarskii, A.A. (1990) *Equations of Mathematical Physics*, Dover Publications, New York.

Troutman, J.L. (1994) *Boundary Value Problems of Applied Mathematics*, PWS Publishing, Boston.

Walter, W. (1964) *Differential- und Integral-Ungleichungen*, Springer-Verlag, Berlin.

Whitham, G.B. (1976) *Linear and Nonlinear Waves*, John Wiley, New York.

Wong, C.W. (1991) *Introduction to Mathematical Physics: Methods and Concepts*, Oxford University Press, Oxford.

Yosida, K. (1960) *Lectures on Differential and Integral Equations*, Wiley-Interscience, New York.

First-Order Partial Differential Equations: Advective Flows, Traffic Flow, Heat Exchangers

Aris, R. and Amundson, N. (1973) *Mathematical Methods in Chemical Engineering, Vol. II, First Order Partial Differential Equations*, Prentice-Hall, Englewood Cliffs, N.J.

Bird, R.B., Stewart, H.E. and Lightfoot, E.N. (1960) *Heat, Mass and Momentum Transfer*, Prentice-Hall, Englewood Cliffs, N.J.

Botha, J.F. and Pinder, G.F. (1983) *Fundamental Concepts in the Numerical Solution of Differential Equations*, John Wiley, New York.

Caratheodory, C. (1968) *Calculus of Variations and Partial Differential Equations of the First Order*, Vol. I and II, Holden-Day, San Francisco.

Haight, F.A. (1963) *Mathematical Theory of Traffic Flow*, Academic Press, New York.

Lapidus, L. and Pinder, G.F. (1982) *Numerical Solution of Partial Differential Equations in Engineering*, John Wiley, New York.

Moon, P.H. and Spencer, D.E. (1969) *Partial Differential Equations*, D.C. Heath, Lexington, Mass.

Rhee, H., Aris, R. and Amundson, N.R. (1986) *First Order Partial Differential Equations*, Prentice-Hall, Englewood Cliffs, N.J.

Whitham, G.B. (1976) *Linear and Nonlinear Waves*, John Wiley, New York.

Laplace's Equation: General Concepts, Mechanics of Fluid Flow, Flow in Porous Media

Basset, A.B. (1888) *A Treatise on Hydrodynamics*, Vols. 1 and 2, Dover Publications, New York.

Batchelor, G.K. (1967) *Introduction to Fluid Dynamics*, Cambridge University Press, Cambridge.

Bear, J. (1972) *Dynamics of Fluids in Porous Media*, Dover Publications, New York.

Bear, J. (1979) *Hydraulics of Groundwater*, McGraw-Hill, New York.

Bergman, S. and Schiffer, M. (1953) *Kernel Functions and Elliptic Differential Equations in Mathematical Physics*, Academic Press, New York.

Chorlton, F. (1967) *Textbook of Fluid Dynamics*, Van Nostrand, London.

Collins, R.E. (1961) *Flow of Fluids Through Porous Media*, Reinhold, New York.

Dullien, F.A.L. (1992) *Porous Media: Fluid Transport and Pore Structure*, Academic Press, New York.

Fichera, G. (1965) *Linear Elliptic Differential Systems and Eigen-value Problems*, Lecture Notes in Mathematics, 8, Springer-Verlag, Berlin.

Gilbarg, D. and Trudinger, N. (1977) *Elliptic Partial Differential Equations of the Second Order*, Springer-Verlag, Berlin.

Greenkorn, R.A. (1983) *Flow Phenomena in Porous Media*, Marcel-Dekker, New York.

Harr, M.E. (1962) *Groundwater and Seepage*, McGraw-Hill, New York.

Hermance, J.F. (1991) *A Mathematical Primer on Groundwater Flow*, Prentice-Hall, Englewood Cliffs, N.J.

Kellogg, O.D. (1953) *Foundations of Potential Theory*, Dover Publications, New York.

Kirkham, D. and Powers, W.L. (1972) *Advanced Soil Physics*, John Wiley, New York.

Kleinstreuer, C. (1997) *Engineering Fluid Dynamics*, Cambridge University Press, Cambridge.

Lamb, Sir Horace (1993) *Hydrodynamics*, Cambridge University Press, Cambridge.

Lin, C.C. and Segel, L.A. (1974) *Mathematics Applied to Deterministic Problems in the Applied Sciences*, Macmillan, New York.

Lions, J.L. (1957) *Lectures on Elliptic Partial Differential Equations*, Tata, Bombay.

Miranda, C. (1970) *Partial Differential Equations of Elliptic Type*, 2nd edition, Springer-Verlag, New York.

Muskat, M. (1937) *The Flow of Homogeneous Fluids Through Porous Media*, McGraw-Hill, New York.

Nikolaevskii, V.N. (1990) *Mechanics of Porous and Fractured Media*, World Scientific, Singapore.

Ozisik, M.N. (1993) *Heat Conduction*, John Wiley, New York.

Philips, O.M. (1991) *Flow and Reactions in Permeable Rocks*, Cambridge University Press, Cambridge.

Polubarinova-Kochina, P.Ya. (1962) *Theory of Ground Water Movement*, Princeton University Press, Princeton, N.J.

Sahimi, M. (1995) *Flow and Transport in Porous Media and Fractured Rock*, VCH Verlagsgesselschaft, Weinheim.

Scheidegger, A.E. (1974) *The Physics of Flow Through Porous Media*, University of Toronto Press, Toronto.

Strack, O.D.L. (1989) *Groundwater Mechanics*, Prentice-Hall, Englewood Cliffs, N.J.

Verruijt, A. (1970) *Theory of Groundwater Flow*, Gordon and Breach, New York.

Vukovich, M. (1997) *Groundwater Dynamics: Steady Flow*, Water Resources Publications, Littleton, CO.

White, F.M. (1984) *Heat Transfer*, Addison-Wesley, Reading, Mass.

Xiao, S. (Ed) (1992) *Flow and Transport in Porous Media*, World Scientific Publishing, Singapore.

Yih, C.-S. (1969) *Fluid Mechanics*, McGraw-Hill, New York.

Diffusion Equation: General Concepts, Heat Conduction, Pressure Transients, Chemical Diffusion

Banks, R.B. (1994) *Growth and Diffusion Phenomena*, Springer-Verlag, Berlin.

Barenblatt, G.I., Entov, V.M. and Ryzhik, V.M. (1990) *Theory of Fluid Flow Through Natural Rocks*, Kluwer Academic, Dordrecht, The Netherlands.

Carslaw, H.S. and Jaeger, J.C. (1959) *Heat Conduction in Solids*, Clarendon Press, Oxford.

Crank, J. (1975) *The Mathematics of Diffusion*, Clarendon Press, Oxford.

Cussler, E.L. (1984) *Diffusion: Mass Transfer in Fluid Systems*, Cambridge University Press, Cambridge.

Friedman, A. (1964) *Partial Differential Equations of Parabolic Type*, Prentice- Hall, Englewood Cliffs, N.J.

Ghez, R. (1988) *A Primer of Diffusion Problems*, John Wiley, New York.

Hill, J.M. and Dewynne, J. (1987) *Heat Conduction*, Blackwell Scientific Publications, Oxford.

Kirkham, D. and Powers, W.L. (1972) *Advanced Soil Physics*, John Wiley, New York.

Luikov, A.V. (1968) *Analytical Heat Diffusion Theory*, Academic Press, New York.

Mikhailov, M.D. and Ozisik, M.N. (1984) *Unified Analysis of Solutions of Heat and Mass Diffusion*, John Wiley, New York.

Nield, D.A. and Bejan, A. (1999) *Convection in Porous Media*, Springer-Verlag, Berlin.

Nikolaevskii, V.N. (1990) *Mechanics of Porous and Fractured Media*, World Scientific, Singapore.

Okubo, A. (1980) *Diffusion and Ecological Problems. Mathematical Models*, Springer-Verlag, Berlin.

Ozisik, M.N. (1993) *Heat Conduction*, John Wiley, New York.

Philips, O.M. (1991) *Flow and Reactions in Permeable Rocks*, Cambridge University Press, Cambridge.

Sahimi, M. (1995) *Flow and Transport in Porous Media and Fractured Rock*, VCH Verlagsgesselschaft, Weinheim.

Sun, N.-Z. (1996) *Mathematical Modelling in Groundwater Pollution*, Springer-Verlag, Berlin.

Taira, K. (1988) *Diffusion Processes and Partial Differential Equations*, Academic Press, Orlando.

Watson, N.A. (1989) *Parabolic Equations on an Infinite Strip*, Marcel Dekker, New York.

White, F.M. (1984) *Heat Transfer*, Addison-Wesley, Reading, Mass.

Widder, D. (1975) *The Heat Equation*, Academic Press, New York.

Wave Equation: General Concepts, Waves in Strings, Membranes and Elastic Solids

Achenbach, J.D.(1973) *Wave Propagation in Elastic Solids*, North-Holland Publishing Co., Amsterdam, The Netherlands.

Bland, D.R. (1988) *Wave Theory and Applications*, Clarendon Press, Oxford.

Brekhovskikh, L.M and Goncharov, V. (1994) *Mechanics of Continua and Wave Dynamics*, Springer-Verlag, Berlin.

Coulson, C.A. and Jeffrey, A. (1977) *Waves. A Mathematical Approach to the Common Types of Wave Motion*, Longman, London.

Elmore, W.C. and Heald, M.A. (1969) *Physics of Waves*, Dover Publications, New York.

Garabedian, P.R. (1964) *Partial Differential Equations*, John Wiley, New York.

Graff, K.F. (1975) *Wave Motion in Elastic Solids*, Ohio State University Press, OH.

John, F. (1981) *Plane Waves and Spherical Means Applied to Partial Differential Equations*, Springer-Verlag, Berlin.

Lax, P.D. (1963) *Lectures on Hyperbolic Partial Differential Equations*, Stanford University Press, California.

Lighthill, M.J. (1978) *Waves in Fluids*, Cambridge University Press, Cambridge.

Morse, P.M. (1948) *Vibration and Sound*, McGraw-Hill, New York.

Nowacki, W. (1963) *Dynamics of Elastic Systems*, Chapman and Hall, London.

Rayleigh, Lord (1945) *Theory of Sound*, Vols. 1 and 2, Dover Publications, New York.

Sakamoto, R. (1982) *Hyperbolic Boundary Value Problems*, Cambridge University Press, Cambridge.

Whitham, G.B. (1976) *Linear and Nonlinear Waves*, John Wiley, New York.

Witten, M. (1983) *Hyperbolic Partial Differential Equations*, Pergamon Press, New York.

Biharmonic Equation: Mechanics of Elastic Solids, Contact Problems, Theory of Plates, Slow Viscous Flow

Ambartsumyan, S.A. (1991) *Theory of Anisotropic Plates: Strength, Stability and Vibrations*, Hemisphere Publishing Corporation, New York.

Atkin, R.J. and Fox, N. (1980) *An Introduction to the Theory of Elasticity*, Longmans, London.

Basset, A.B. (1888) *A Treatise on Hydrodynamics*, Vols. 1 and 2, Dover Publications, New York.

Batchelor, G.K. (1967) *An Introduction to Fluid Dynamics*, Cambridge University Press, Cambridge.

Bird, R.B., Stewart, H.E. and Lightfoot, E.N. (1960) *Heat, Mass and Momentum Transfer*, Prentice-Hall, Englewood Cliffs, N.J.

Birkhoff, G. (1960) *Hydrodynamics: A Study in Logic, Fact and Similitude*, Princeton University Press, Princeton, N.J.

Boresi, A.P. and Chong, K.P. (1987) *Elasticity in Engineering Mechanics*, Elsevier Science Publ., New York.

Boresi, A.P. and Lynn, P.P. (1974) *Elasticity in Engineering Mechanics*, Prentice-Hall, Englewood Cliffs, N.J.

Chou, P.C. and Pagano, N.J. (1967) *Elasticity, Tensor, Dyadic and Engineering Approaches*, Dover Publications, New York.

Churchill, S.W. (1988) *Viscous Flows*, Butterworths, Boston.

Clements, D.L. (1981) *Boundary Value Problems Governed by Second-Order Elliptic Systems*, Pitman Publ. Ltd., London

Constantinescu, V.N. (1995) *Laminar Viscous Flow*, Springer-Verlag, Berlin.

Davis, R.O. and Selvadurai, A.P.S. (1996) *Elasticity and Geomechanics*, Cambridge University Press, Cambridge,

Drew, D.A. and Passman, S.L. (1999) *Theory of Multicomponent Fluids*, Applied Mathematical Sciences Vol. 135, Springer-Verlag, Berlin.

England, A.H. (1971) *Complex Variable Methods in Elasticity*, Wiley-Interscience, London.

Eringen, A.C. (1967) *Mechanics of Continua*, John Wiley, New York.

Flugge, S. (Ed.) (1959) *Encyclopedia of Physics*, Vol. VIII/1(C.Truesdell, Co-Editor), Springer-Verlag, Berlin.

Freeze, R.A. and Cherry, J.A. (1979) *Groundwater*, Prentice-Hall, Englewood Cliffs, N.J.

Fung, Y.C. (1965) *Foundations of Solid Mechanics*, Prentice-Hall, Englewood Cliffs, N.J.

Gladwell, G.M.L. (1980) *Contact Problems in the Classical Theory of Elasticity*, Sijthoff and Noordhoff, Alphen aan den Rijn, The Netherlands.

Godfrey, D.E.R. (1959) *Theoretical Elasticity and Plasticity for Engineers*, Thames and Hudson, London.

Gould, P.L. (1983) *Introduction to Linear Elasticity*, Springer-Verlag, Berlin.

Green, A.E. and Zerna, W. (1968) *Theoretical Elasticity*, Clarendon Press, Oxford.

Gurtin, M.E.(1972) Linear Theory of Elasticity, in *Handbook of Physics Vol. VIa/2, Mechanics of Solids II*(Ed. C. Truesdell), Springer-Verlag, Berlin.

Gurtin, M.E. (1981) *An Introduction to Continuum Mechanics*, Academic Press, New York.

Hahn, H.G. (1984) *Elastizitatstheorie*, B.G.Teubner, Stuttgart.

Happel, J. and Brenner, H. (1973) *Low Reynolds Number Hydrodynamics*, Martinus Nijhoff, The Hague, The Netherlands.

Herrmann, G. (Ed) (1974) *R.D.Mindlin and Applied Mechanics*, Pergamon Press, Oxford.

Hunter, S.C. (1983) *Mechanics of Continuous Media*, Ellis Horwood-Wiley, New York.

Jaeger, L.G. (1964) *Elementary Theory of Elastic Plates*, Pergamon Press, Oxford.

Kalandiya, A.I. (1975) *Mathematical Methods of Two-Dimensional Elasticity*, Mir publishers, Moscow.

Kleinstreuer, C. (1997) *Engineering Fluid Dynamics*, Cambridge University Press, Cambridge.

Karasudhi, P. (1991) *Foundations of Solid Mechanics*, Kluwer Academic Publishers, Dordrecht, The Netherlands.

Knops, R.J. and Payne, L.E. (1971) *Uniqueness Theorems in Linear Elasticity*, Springer-Verlag, Berlin.

Lamb, Sir Horace (1993) *Hydrodynamics*, Cambridge University Press, Cambridge.

Landau, H.G. and Lifshitz, E.M. (1970) *Theory of Elasticity, Course in Theoretical Physics*, Vol. 7, Pergamon Press, Oxford.

Landau, L.D. and Lifshitz, E.M. (1987) *Fluid Mechanics*, 2nd edition, Pergamon Press, Oxford.

Langlois, W.E. (1964) *Slow Viscous Flow*, Macmillan, New York.

Leipholz, H. (1974) *Theory of Elasticity*, Noordhoff, Leyden, The Netherlands.

Lin, C.C. and Segel, L.A. (1974) *Mathematics Applied to Deterministic Problems in the Applied Sciences*, Macmillan, New York.

Little, R.W. (1973) *Elasticity*, Prentice-Hall, Englewood Cliffs, N.J.

Love, A.E.H. (1944) *An Introduction to the Mathematical Theory of Elasticity*, Dover Publicatons, New York.

Lur'e, A.I., (1964) *Three-Dimensional Problems of the Theory of Elasticity*, Wiley-Interscience, New York.

Mal, A.K. and Singh, S.J. (1991) *Deformation of Elastic Solids*, Prentice-Hall, Englewood Cliffs, N.J.

Malvern, L.E. (1969) *Introduction to the Mechanics of a Continuous Medium*, Prentice-Hall, Englewood Cliffs, N.J.

Mansfield, E.H. (1989) *The Theory of Bending and Stretching of Plates*, Cambridge University Press, Cambridge.

Mase, G.E. and Mase, G.T. (1991) *Continuum Mechanics for Engineers*, CRC Press, Boca Raton, Florida.

Milne-Thomson, L.M. (1960) *Plane Elastic Systems*, Springer-Verlag, Berlin.

Muskhelishvili, N.I. (1963) *Some Basic Problems in the Mathematical Theory of Elasticity*, Noordhoff, Groningen, The Netherlands.

Nadeau, G. (1964) *Introduction to Elasticity*, Holt, Rinehart and Winston, New York.

Nield, D.A. and Bejan, A. (1999) *Convection in Porous Media*, Springer-Verlag, Berlin.

Ockendon, H. and Ockendon, J.R. (1995) *Viscous Flow*, Cambridge University Press, New York.

Pai, S.-I. (1959) *Slow Viscous Flow Theory, Vol. 1, Laminar Flow*, Van Nostrand, Princeton, N.J.

Panc, V. (1975) *Theories of Elastic Plates*, Noordhoff, Leyden, The Netherlands.

Panton, R.L. (1984) *Incompressible Flow*, John Wiley, New York.

Parton, V.Z. and Perlin, P.I. (1982) *Integral Equations in Elasticity*, Mir Publishers, Moscow.

Paterson, A.R. (1983) *A First Course in Fluid Dynamics*, Cambridge University Press, Cambridge.

Pedley, T. (1980) *The Fluid Mechanics of Large Blood Vessels*, Cambridge University Press, Cambridge.

Philips, O.M. (1991) *Flow and Reactions in Permeable Rocks*, Cambridge University Press, Cambridge.

Reddy, J.N. (1999) *Theory and Analysis of Elastic Plates*, Taylor and Francis, Philadelphia, Pennsylvania.

Reismann, H. (1988) *Elastic Plates: Theory and Application*, John Wiley, New York.

Reismann, H. and Pawlik, P.S. (1980) *Elasticity, Theory and Applications*, John Wiley, New York.

Scheidegger, A.E. (1974) *The Physics of Flow Through Porous Media*, University of Toronto Press, Toronto.

Sedov, L.I. (1972) *A Course in Continuum Mechanics*, Vols.I-IV, Wolters-Noordhoff, Groningen, The Netherlands.

Segel, L.A. (1987) *Mathematics Applied to Continuum Mechanics*, Dover Publications, New York.

Selvadurai, A.P.S. (1979) *Elastic Analysis of Soil-Foundation Interaction*, Elsevier Scientific Publishing, Amsterdam, The Netherlands.

Sherman, F.S. (1990) *Viscous Flow*, McGraw-Hill, New York.

Sokolnikoff, I.S. (1956) *Mathematical Theory of Elasticity*, McGraw-Hill, New York.

Southwell, R.V. (1969) *An Introduction to the Theory of Elasticity*, Dover Publications, New York.

Spencer, A.J.M. (1980) *Continuum Mechanics*, Longmans, London.

Sun, N.-Z. (1996) *Mathematical Modelling in Groundwater Pollution*, Springer-Verlag, Berlin.

Szilard, R. (1974) *Theory and Analysis of Plates, Classical and Numerical Methods*, Prentice-Hall, Englewood Cliffs, N.J.

Timoshenko, S.P. and Goodier, J.N. (1970) *Theory of Elasticity*, McGraw-Hill, New York.

Timoshenko, S.P. and Woinowsky-Krieger, S. (1959) *Theory of Plates and Shells*, McGraw-Hill, New York.

Volterra, E. and Gaines, J.H. (1971) *Advanced Strength of Materials*, Prentice-Hall, Englewood Cliffs, N.J.

Wang, C.T. (1953) *Applied Elasticity*, McGraw-Hill, New York.

Westergaard, H.M. (1952) *Theory of Elasticity and Plasticity*, Harvard University Press. Cambridge, Mass.

Yih, C.-S. (1969) *Fluid Mechanics*, McGraw-Hill, New York.

Poisson's Equation: Flow in Porous Media, Axial Viscous Flows, Torsion of Prismatic Elastic Solids

Boresi, A.P. and Lynn, P.P. (1974) *Elasticity in Engineering Mechanics*, Prentice-Hall, Englewood Cliffs, N.J.

Chou, P.C. and Pagano, N.J. (1967) *Elasticity, Tensor, Dyadic and Engineering Approaches*, Dover Publications, New York.

Kellogg, O.D. (1953) *Foundations of Potential Theory*, Dover Publications, New York.

Little, R.W. (1973) *Elasticity*, Prentice-Hall, Englewood Cliffs, N.J.

Muskat, M. (1937) *The Flow of Homogeneous Fluids Through Porous Media*, McGraw-Hill, New York.

Nield, D.A. and Bejan, A. (1999) *Convection in Porous Media*, Springer-Verlag, Berlin.

Polya, G. and Szego, G. (1951) *Isoperimetric Inequalities in Mathematical Physics*, Annals of Mathematical Studies, Princeton University Press, N.J.

Protter, M.H. and Weinberger, H.F. (1984) *Maximum Principles in Differential Equations*, Springer-Verlag, Berlin.

Reismann, H. and Pawlik, P.S. (1980) *Elasticity, Theory and Applications*, John Wiley, New York.

Sherman, F.S. (1990) *Viscous Flow*, McGraw-Hill, New York.

Sokolnikoff, I.S. (1956) *Mathematical Theory of Elasticity*, McGraw-Hill, New York.

Southwell, R.V. (1969) *An Introduction to the Theory of Elasticity*, Dover Publications, New York.

Stakgold, I. (1968) *Boundary Value Problems of Mathematical Physics*, Vols. I and II, Macmillan, New York.

Szarski, J. (1965) *Differential Inequalities*, Polish Scientific Publishers, Warsaw.

Timoshenko, S.P. and Goodier, J.N. (1970) *Theory of Elasticity*, McGraw-Hill, New York.

Green's Functions

Barton, G. (1989) *Elements of Green's Functions and Propagation: Potentials, Diffusion and Waves*, Clarendon Press, Oxford.

Greenberg, M.D. (1971) *Application of Green's Functions in Science and Engineering*, Prentice-Hall, Englewood Cliffs, N.J.

Roach, G.F. (1982) *Green's Functions*, Cambridge University Press, Cambridge.

Sneddon, I.N. (1972) *The Use of Integral Transforms*, McGraw-Hill, New York.

Stakgold, I. (1998) *Green's Functions and Boundary Value Problems*, 2nd edition, John Wiley, New York.

Williams, W.E. (1980) *Partial Differential Equations*, Clarendon Press, Oxford.

Zauderer, E. (1998) *Partial Differential Equations of Applied Mathematics*, 2nd edition, John Wiley, New York.

Collections of Problems

Budak, B.M., Samarskii, A.A. and Tikhonov, A.N. (1988) *A Collection of Problems in Mathematical Physics*, Dover Publications, New York.

Lebedev, N.N., Skalskaya, I.P. and Uflyand, Ya.S. (1965) *Problems in Mathematical Physics*, Pergamon Press, Oxford.

Smirnov, M.M. (1967) *Problems on the Equations of Mathematical Physics*, Noordhoff, The Netherlands.

Mathematical Tables and Handbooks

Abramowitz, M. and Stegun, I.A. (Eds.) (1972) *Handbook of Mathematical Functions with Formulas, Graphs and Mathematical Tables*, Dover Publications, New York.

CRC Standard Mathematical Tables, (1991) CRC Press, Boca Raton, Florida.

Dwight, H.B. (1961) *Tables of Integrals and Other Mathematical Data*, Macmillan, New York.

Erdelyi, A. et al. (1954) *Tables of Integral Transforms*, Vols. 1 and 2, McGraw-Hill, New York.

Gradshteyn, I.S. and Ryzhik, I.M. (1980) *Table of Integrals, Series and Products*, Academic Press, New York.

Hughes, W.F. and Gaylord, E.W. (1964) *Basic Equations of Engineering Science*, McGraw-Hill, New York.

Jolley, L.B.W. (1961) *Summation of Series*, Dover Publications, New York.

Korn, G. and Korn, T. (1968) *Mathematical Handbook for Scientists and Engineers*, McGraw-Hill, New York.

Petit Bois, G. (1961) *Tables of Indefinite Integrals*, Dover Publications, New York.

Roberts, G.E. and Kaufman, H. (1966) *Table of Laplace Transforms*, Saunders, Philadelphia, PA.

Spiegel, M.R. (1968) *Mathematical Handbook of Formulas and Tables*, McGraw-Hill, New York.

Wheelon, A.D. (1968) *Tables of Summable Series and Integrals Involving Bessel Functions*, Holden-Day, San Francisco, California.

Zwillinger, D. (1992) *Handbook of Differential Equations*, Academic Press, Boston.

Computer Based Symbolic Mathematics

Abell, M. and Braselton, J. (1994) *Differential Equations with Maple V*, Academic Press, New York.

Abell, M. and Braselton, J. (1994) *Mathematica by Example*, Academic Press, New York.

Articolo, G.A. (1998) *Partial Differential Equations and Boundary Value Problems with Maple V*, Academic Press, San Diego.

Betounes, D. (1998) *Partial Differential Equations for Computational Science: with Maple and Vector Analysis*, Springer-Verlag, New York.

Kreyszig, H.E. (1994) *Maple Computer Manual for Advanced Engineering Mathematics*, John Wiley, New York.

Kythe, P.M., Puri, P., and Schäferkotter, M.R. (1997) *Partial Differential Equations and Mathematica*, CRC Ress, Boca Raton, FL.

Redfern, D. (1994) *Maple Handbook: Maple V Release 3*, Springer-Verlag, Berlin.

Vvedensky, D. (1992) *Partial Differential Equations with Mathematica*, Addison-Wesley, New York.

Wilson, H.B. and Turcotte, L.H. (1994) *Advanced Mathematics and Mechanics Applications using MATLAB*, CRC Press, Boca Raton, Florida.

Numerical Methods for Partial Differential Equations

Ames, W.F. (1977) *Numerical Methods for Partial Differential Equations*, 2nd edition, Academic Press, New York.

Aziz, A.K. (Ed.) (1972) *The Mathematical Foundations of the Finite Element Method with Applications to Partial Differential Equations*, Academic Press, New York.

Bellman, R. and Cooke, K.L (1963) *Differential-Difference Equations*, Academic Press, New York.

Bellman, R.E. and Adomian, G. (1985) *Partial Differential Equations: New Methods for their Treatment and Solution*, Reidel Publishing, Hingham, Mass.

Botha, J.F. and Pinder, G.F. (1983) *Fundamental Concepts in the Numerical Solution of Differential Equations*, John Wiley, New York.

Brand, L. (1966) *Differential and Difference Equations*, John Wiley, New York.

Collatz, L. (1960) *The Numerical Treatment of Partial Differential Equations*, 3rd edition, Springer-Verlag, Berlin.

Colombini, F., Marino, A., Modica, L. and Spagnolo, S. (Eds) (1989) *Partial Differential Equations and the Calculus of Variations*, Vol. II, Birkhauser, Boston.

Crandall, S.H. (1956) *Engineering Analysis - A Survey of Numerical Procedures*, McGraw-Hill, New York.

Forsythe, G.E. (1958) *Numerical Analysis and Partial Differential Equations*, John Wiley, New York.

Forsythe, G.E. and Wasow, W.R. (1967) *Finite-Difference Methods for Partial Differential Equations*, John Wiley, New York.

Gear, C.W. (1971) *Numerical Initial Value Problems in Ordinary Differential Equations*, Prentice-Hall, Englewood Cliffs, N.J.

Gladwell, I. and Wait, R. (Eds) (1979) *A Survey of Numerical Methods for Partial Differential Equations*, Oxford University Press, Oxford.

Johnson, C. (1993) *Partial Differential Equations by the Finite Element Method*, Cambridge University Press, Cambridge.

Kocak, H. (1989) *Differential and Difference Equations through Computer Experiments*, 2nd edition, Springer-Verlag, New York.

Lapidus, L. and Pinder, G.F. (1982) *Numerical Solution of Partial Differential Equations in Engineering*, John Wiley, New York.

Meis, T.H. and Marcowitz, U. (1981) *Numerical Solutions of Partial Differential Equations*, Springer-Verlag, Berlin.

Mitchell, A.R. (1969) *Computational Methods in Partial Differential Equations*, John Wiley, New York.

Mitchell, A.R. and Griffiths, D.F. (1980) *The Finite Difference Method in Partial Differential Equations*, John Wiley, New York.

Mitchell, A.R. and Wait, R. (1977) *The Finite Element Method in Partial Differential Equations*, John Wiley, New York.

Noye, J. (1984) *Computational Techniques for Differential Equations*, North-Holland Mathematical Studies 83, North-Holland, New York.

Salvadori, M.G. and Schwarz, R.J. (1965) *Differential Equations in Engineering*, Prentice-Hall, Englewood Cliffs, N.J.

Sewell, G. (1988) *The Numerical Solution of Ordinary and Partial Differential Equations*, Academic Press, Boston

Shampine, L.F. and Gordon, M.K. (1975) *Computer Solution of Ordinary Differential Equations: The Initial Value Problem*, Freeman, San Francisco.

Smith, G.D. (1985) *Numerical Solution of Partial Differential Equations: Finite Difference Equations*, 3rd edition, Oxford University Press, Oxford.

Vichnevetsky, R. (1981) *Computer Methods for Partial Differential Equations*, Vol. I, Prentice-Hall, Englewood Cliffs, N.J.

von Rosenberg, D.U. (1969) *Methods for the Numerical Solution of Partial Differential Equations*, American Elsevier Publ. Co., New York.

Weinstein, A. (1966) *Numerical Solution of Partial Differential Equations*, Academic Press, New York.

Historical Notes

Abraham, R. and Marsden, J.E. (1978) *Foundations of Mechanics*, Benjamin/Cummings, Reading, Massachusetts.

Albers, D.J. and Alexanderson, G.L. (Eds.) (1985) *Mathematical People. Profiles and Interviews*, Birkhauser-Verlag, Basel.

Anglin, W.S. (1994) *Mathematics, a Concise History and Philosophy*, Springer-Verlag, Berlin.

Baron, M.E. (1987) *The Origins of the Infinitesimal Calculus*, Dover Publications, New York.

Boyer, C.B. (1968) *A History of Mathematics*, John Wiley, New York.

Bucciarelli, L.L. and Dworsky, N. (1980) *Sophie Germain: An Essay in the History of the Theory of Elasticity*, D.Reidel Publishing, Dordrecht, The Netherlands.

Cajori, F. (1985) *A History of Mathematics*, Chelsea, New York.

Cannell, D.M. (1993) *George Green: Mathematical Physicist, 1793-1841: The Background to His Life and Work*, Athlone Press, Atlantic Highlands, N.J.

Cannon, J.T. and Dostrovsky, S. (1981) *The Evolution of Dynamics: Vibration Theory from 1687 to 1742*, Springer-Verlag, Berlin.

Dugas, R. (1955) *A History of Mechanics*, Central Book Co., New York.

Duhem, P.-M.-M. (1980) *The Evolution of Mechanics*, Sijthoff and Noordhoff, Alphen aan den Rijn, The Netherlands.

Engelsman, S.B. (1984) *Families of Curves and the Origins of Partial Differentiation*, North Holland, Amsterdam, The Netherlands.

Eves, H.W. (1964) *An Introduction to the History of Mathematics*, Holt, Rinehart and Winston, New York.

Gellert, W., Küstner, H., Hellwich, M. and Kästner, H. (1975) *Mathematics at a Glance: A Compendium*, VEB Bibliographisches Institut, Leipzig.

Gellert, W. et al. (1975) *Mathematics at a Glance*, VEB Bibliographisches Institut, Leipzig, Germany.

Gillespie, C.C. (Ed.) (1972) *Dictionary of Scientific Bibliography*, Charles Scribner and Sons, New York.

Girvin, H.F. (1948) *A Historical Appraisal of Mechanics*, International Textbook Co., Scranton, PA.

Heath, Sir Thomas (1981) *A History of Greek Mathematics*, Dover Publications, New York.

Herival, J.W. (1975) *Joseph Fourier, The Man and the Physicist*, Clarendon Press, Oxford.

Kelvin, Lord and Tait, P.G. (1903) *A Treatise on Natural Philosophy*, Cambridge University Press.

Mach, E. (1907) *The Science of Mechanics: A Critical and Historical Account of its Development*, The Open Court Publishing Co., London.

O'Neil, P.V. (1995) *Advanced Engineering Mathematics*, Brooks/Cole, Boston, Mass.

Singer, C. (1959) *A Short History of Scientific Ideas to 1900*, Oxford University Press, Oxford.

Struik, D.J. (1969) *A Source Book in Mathematics 1200-1800*, Harvard University Press, Cambridge, Massachusetts.

Szabo, I. (1987) *Geschichte der Prinzipe der Mechanik*, Birkhauser-Verlag, Basel.

Timoshenko, S.P. (1953) *History of the Strength of Materials*, McGraw-Hill, New York.

Todhunter, I. and Pearson, K. (1886) *A History of the Theory of Elasticity*, Vol. I, Cambridge University Press, Cambridge.

Todhunter, I. and Pearson, K. (1893) *A History of the Theory of Elasticity*, Vol. II, Parts I and II, Cambridge University Press, Cambridge.

Truesdell, C. (1968) *Essays in the History of Mechanics*, Springer-Verlag, Berlin.

Truesdell, C. (1984) *An Idiot's Fugitive Essays on Science. Methods, Criticism, Training, Circumstances*, Springer-Verlag, Berlin.

Young, R.V. (Ed.) (1998) *Notable Mathematicians. From Ancient Times to the Present*, Gale Research, Detroit, Michigan.

Index

Production: Druckhaus Beltz, Hemsbach